Lecture Notes in Physics

339

F. Ehlotzky (Ed.)

Fundamentals of Laser Interactions II

Proceedings of the Fourth Meeting on Laser Phenomena
Held at the Bundessportheim in Obergurgl, Austria
26 February – 4 March 1989

Springer-Verlag
Berlin Heidelberg GmbH

Editor

Fritz Ehlotzky
Institute for Theoretical Physics, University of Innsbruck
Technikerstraße 25, A–6020 Innsbruck, Austria

ISBN 978-3-662-13723-9 ISBN 978-3-540-48123-2 (eBook)
DOI 10.1007/978-3-540-48123-2

Originally published by Springer-Verlag Berlin Heidelberg New York in 1989
Softcover reprint of the hardcover 1st edition 1989

2158/3140-543210 – Printed on acid-free paper

FOREWORD

The Seminar on Fundamentals of Laser Interactions II was the fourth meeting on Laser Phenomena held at the Bundessportheim in Obergurgl. It was attended by 52 physicists from Austria, Canada, Denmark, the Federal Republic of Germany, France, Hungary, Italy, The Netherlands, the Soviet Union, Sweden and the United States, who work actively in the rapidly developing field of laser interactions.

The present Seminar offered the opportunity to discuss at leisure problems of mutual interest to theoreticians and experimentalists who are working on various aspects of the field of laser interactions. The intention was to bring together people who are doing research on multiphoton ionization, atoms in intense laser fields, electron correlations in radiative processes, Rydberg atoms in external fields, and other laser interactions.

At the seminar 19 Invited Lectures were given by:

M. Aymar (Orsay)
J.E. Bayfield (Pittsburgh)
R. Blümel (Garching)
S.L. Chin (Laval)
W.E. Cooke (Los Angeles)
M. Crance (Orsay)
M.V. Fedorov (Moscow)
D. Feldmann (Bielefeld)
R.R. Freeman (Murray Hill)
M. Gavrila (Amsterdam)
H. Helm (Menlo Park)
H. Klar (Freiburg)
H. B. van Linden van den Heuvell (Amsterdam)
L.–A. Lompré (Saclay)
W. Sandner (Freiburg)
C.K. Rhodes (Chicago)
R. Shakeshaft (Los Angeles)
K.H. Welge (Bielefeld)
G. Wendin (Göteborg)

In addition, there were 18 contributed papers (posters) presented at the meeting.

The following pages present the full text of the invited lectures and the abstracts of the contributed papers. The invited lecture of P. Lambropoulos (Los Angeles) was not presented at the Seminar but has been accepted for publication in the Proceedings. The editor is grateful to the contributors for their collaboration in preparing their typescripts for rapid publication.

The active yet relaxed atmosphere of the Bundessportheim at Obergurgl, surrounded by the snow–capped peaks of the Ötztal Alps, provided a congenial setting for a very stimulating and rewarding meeting. It is a pleasure to thank all participants for their interest and enthusiasm. The most valuable secretarial assistance of Miss C. Hell is gratefully acknowledged.

Innsbruck, May 1989 F. Ehlotzky

For the convenience of the reader we quote below the proceedings of the previous meetings on laser phenomena held at Obergurgl:

Fundamentals of Quantum Optics
(February 26 – March 3, 1984)
Acta Physica Austriaca, Vol. 56, No. 1–2, 1984.

Fundamentals of Laser Interactions
(February 24 – March 2, 1985)
Lecture Notes in Physics, Vol. 229, 1985.

Fundamentals of Quantum Optics II
(February 22 – 28, 1987)
Lecture Notes in Physics, Vol. 282, 1987.

ACKNOWLEDGEMENTS

THE SEMINAR ON FUNDAMENTALS OF LASER INTERACTIONS II
HAS BEEN SUPPORTED BY:

Bundesministerium für Wissenschaft und Forschung
Bundesministerium für Unterricht, Kunst und Sport
Amt der Tiroler Landesregierung
Magistrat der Stadt Innsbruck
Rektor der Universität Innsbruck
Bundeskammer der Gewerblichen Wirtschaft
Kammer der Gewerblichen Wirtschaft für Tirol
Österreichische Forschungsgemeinschaft

CONTENTS

INVITED LECTURES

Part I: Multiphoton Ionization

Part II: Atoms in Intense Laser Fields

INVITED LECTURES

PART I: Multiphoton Ionization

RESONANCE MULTIPHOTON IONIZATION UNDER STRONG FIELDS

P. Lambropoulos*, A. Lyras* and X. Tang
Department of Physics, University of Southern California
Los Angeles, CA 90089-0484, USA

and

*University and Research Center of Crete
Box 1527, Heraklion, Crete, 71110, Greece

I. INTRODUCTION

Do atoms lose their identity under strong laser pulses? Perusal of the literature of the last five years or so leaves one confused. On the one hand, we have seen a tendency to assume that somewhere around 10^{13} W/cm^2 all atoms look alike, and simple, as if they were all made either by Keldysh[1] or by the more recent multinational firm KFR.[2] Such "atoms" have only one electron which can be either in the ground state or in the continuum, as there is nothing else in between. On the other hand, we have seen arguments[3] suggesting the simultaneous and collective response of many electrons (in fact, of a whole shell) at these intensities. Obviously, these two possibilities are diametrically opposite to each other. Both, however, would entail an atomic response which has little to do with the structure of the bare atom.

Viewed against this background, recent experiments[4-7] and interpretations thereof seem surprising as they reveal resonance behavior attributable to bound levels of the atom which, while undergoing rather large AC Stark shifts, participate in the overall process with their identity intact. Two different types of experiments are of interest from this point of view.

The first type[4,5] has appeared more recently and where it was least expected. Two experiments on Multiphoton Ionization (MPI) in connection with Above Threshold Ionization (ATI) have revealed new structure in the photoelectron energy spectra. It has appeared under short (subpicosecond) pulses of relatively high intensity (up to about 4×10^{14} W/cm^2) and of wavelengths in the range of ~ 620 nm. The first (closest to threshold) peak of the photoelectron energy (under higher resolution) was found to exhibit a number of spikes which become more pronounced as the duration of the pulse becomes shorter, with its total energy held constant; which implies increasing intensity. Freeman et al[4] have proposed a rather interesting interpretation which is based on the idea that bound states of the atom shift into resonance under the large (mostly ponderomotive) shift which can be of the order of 2 eV at these intensities and wavelengths. The reason this effect is seen when the pulse duration is short has to do with the fact, Freeman argues, that at longer pulses the ejected electron regains part or all of the shift energy, as it remains under the influence of the decreasing pulse much longer. The effect, according to this reasoning, is there even for longer pulses, but it is washed out. This has led Freeman to the exuberant declaration that "all high intensity multiphoton ionization is resonant." Well, he may turn out to be right, at least in some sense which time and further work will clarify. For the moment Freeman's interpretation appears to be compatible with the observations, especially after the distribution of the laser intensity in the interaction volume is taken into consideration.

The second type of experiment has been performed by Landen et al.[6] and Perry and Landen,[7] who have reported 3-photon-resonant 4-photon ionization of Kr at intensities ranging from $3x10^{12}$ to $7.8x10^{14}$ W/cm^2 and pulse duration 2ps. It is the more traditional resonance experiment in which frequency is scanned around the 3-photon resonance. Data have been obtained at different laser intensities. A clear resonance profile is observed and is found to shift and broaden as the laser intensity is increased. By fitting the data to a resonance MPI model, the experimentalists have been able to extract values for the AC Stark shifts and the photoionization cross sections of the resonant states as well as the 3-photon Rabi frequencies coupling the resonant excited states to the ground state.

Both of these elegant experiments pose a number of interesting questions in connection with Resonantly Enhanced Multiphoton Ionization (REMPI). The purpose of this article is to present independent theoretical studies of these phenomena in an effort to assess what REMPI theory can say about the way atomic structure enters in these strong-field interactions. In all cases, the temporal and spatial distribution of the laser pulse is as important as anything else in these problems and is taken fully into consideration in all calculations reported here.

II. RESONANCES SHIFTING RAPIDLY UNDER SHORT LASER PULSES.

1. Theoretical Formulation

We consider here a system (atom) sufficiently simple to allow quantitative calculations, and sufficiently complicated to contain the main features of the phenomenon under study. Our system has a ground state $|0\rangle$, two intermediate bound states $|1\rangle$ and $|2\rangle$ of the same parity, an ionization potential and infinitely many other states which contribute a non-resonant MPI background. The energy of $|j\rangle$ is denoted by $\hbar\omega_j$. We study the behavior of this system under conditions such that the two excited states shift in and out of resonance with a certain number of photons (to be specified in each case). We will use the abbreviated standard terminology (m+n)-REMPI, when m photons are necessary to reach near-resonance with the states $|1\rangle$ and $|2\rangle$, and n photons to ionize from there. Realistic numbers corresponding to real atoms (specified below) are employed in all of our calculations.

At each position in the interaction volume (which is determined by the light focusing system), the atoms see a pulse of a certain duration and temporal shape whose peak value is maximum at the center of the focal region and decreases outwards. Thus for a pulse of Fourier-limited bandwidth and frequency ω, the field amplitude can be written as $\mathcal{E}(\vec{R},t)$ where the indicated time variation of $\mathcal{E}$ is much slower than $2\pi/\omega$, being of the order of 200 fs or longer, in the present context.

A system of density matrix equations governing the dynamical behavior of the atom at a (macroscopic) space-time point $(\vec{R},t)$ can be derived, within the constraints of our model. The derivation parallels that of an earlier paper[8] in which a precursor to the present phenomenon was studied. The resulting differential equations (in a most compact form) for the slowly varying density matrix elements σ_{kj} are:

$$\frac{d}{dt}\sigma_{00} = \mathrm{Im}(\Omega_{01}\sigma_{10} + \Omega_{02}\sigma_{20}) - \Gamma_0\sigma_{00} \qquad (1)$$

$$\frac{d}{dt}\sigma_{jj} = -\Gamma_j\sigma_{jj} - \mathrm{Im}(\Omega_{0j}\sigma_{j0}) + \mathrm{Im}C_{jj}\ ;\ j=1,2 \qquad (2)$$

$$\left(\frac{d}{dt} - i\Delta_j + \frac{1}{2}\Gamma\right)\sigma_{j0} = \frac{1}{2}i\Omega_{j0}(\sigma_{jj}-\sigma_{00}) - \frac{1}{2}i\Omega_{jj'}\sigma_{j'0} + $$

$$+ \frac{1}{2}i\Omega_{j'0}\sigma_{jj'}\ ;\ j,j' = 1,2(j\neq j') \qquad (3)$$

$$\left(\frac{d}{dt} + i\bar{\omega}_{12} + \frac{1}{2}(\Gamma_1+\Gamma_2)\right)\sigma_{12} = -\frac{1}{2}i\Omega_{10}\sigma_{02} + \frac{1}{2}i\Omega_{02}\sigma_{10} + $$

$$\frac{1}{2}i(\Omega^*_{12}\sigma_{11} - \Omega_{12}\sigma_{22}) \qquad (4)$$

where $C_{11} = \Omega^*_{12}\sigma_{12}$ and $C_{22} = -\Omega_{12}\sigma_{12}$. Ω_{0j} is the m-photon Rabi frequency coupling the states $|0\rangle$ and $|j\rangle$, and is given by $\mathcal{E}^m\mu^{(m)}_{0j}$ with $\mu^{(m)}_{0j}$ being the corresponding m-photon dipole matrix element. Γ_j is the ionization width of state $|j\rangle$, given by the product $\hat{\sigma}_n F^n$ of the respective multiphoton ionization generalized cross section $\hat{\sigma}_n$ and the photon flux F. The detuning from resonance with the intermediate state is $\Delta_j = m\omega - \bar{\omega}_{j0}$ where all atomic energy differences (in units of rad/s) of the type $\bar{\omega}_{j0} = \bar{\omega}_j - \bar{\omega}_0$ and $\bar{\omega}_{12} = \bar{\omega}_1 - \bar{\omega}_2$ bear a horizontal bar to indicate that the AC Stark shift S_j has been added to the respective free-atom energy. These shifts S_j are proportional to the laser intensity $I(W/cm^2)$ and approach the ponderomotive shift $(e^2 \mathcal{E}^2 / 4m\omega^2)$ as the excited state approaches the ionization threshold. The complex quantity Ω_{12} is proportional to $\mathcal{E}^2$ and to a two-photon matrix element connecting the excited states $|1\rangle$ and $|2\rangle$. It is complex whenever the energy $\bar{\omega}_j + \omega$ falls above the ionization threshold (see ref. 8 for details). All of the above atomic parameters, Ω_{0j}, Γ_j, Δ_j and Ω_{12} are time-dependent on a time scale over which the amplitude (t) varies. As a consequence, the differential equations (1)-(4) have time-dependent coefficients.

The matrix elements $\sigma_{jk}(t)$ are slowly-varying in the sense that the transformations $\sigma_{0j}(t) = \rho_{0j}(t)\, e^{-im\omega t}$ have removed the rapid (resonant) variation of the primary off-diagonal density matrix elements $\rho_{0j}(t)$. Our equations are therefore valid as long as the envelope amplitude of the pulse does not vary significantly over a time of order $2\pi/\omega$.

The ponderomotive shift causes the ionization potential E_∞ to shift upward, becoming thus time-dependent as the pulse-intensity rises and falls. The kinetic energy E_k of the ejected photoelectrons will therefore vary during the pulse, obeying the equation $E_k(t) = E_o + S_o + (m+n)\hbar\omega - E_\infty(t)$.

The dynamical equations enable us to calculate the angle-integrated photoionization signal P_{ION} through the relation $P_{ION}(t) = 1 - \sigma_{00}(t) - \sigma_{11}(t) - \sigma_{22}(t)$. The number of photoelectrons generated between t and $t+\Delta t$ is proportional to $\Delta P_t = P(t+\Delta t) - P(t)$ which represents the number of photoelectrons of energy $E_k(t)$. Our calculation is implemented essentially as described above, with the additional integration over the interaction volume using a Gaussian form. We can thus calculate the total number of photoelectrons of energy E_k generated anywhere in the interaction volume at the end of the pulse. The temporal pulse shape has also been taken to be Gaussian.

Before embarking upon the solution of the dynamical Eqs. (1)-(4), we must calculate all relevant atomic parameters such as Ω_{ij}, Γ_j, S_j. In each case, we have performed such calculations quite accurately employing multiphoton techniques[9,10] based on quantum defect theroy. For excited states not very close to the ionization threshold, the shift often differs somewhat from the ponderomotive shift. In our calculations, we have always included the actual shift of the state obtained with the same wavefuncions and techniques that yield the other parameters.

Here are the questions we propose to explore: If this system with its known and calculable structure and dynamics is exposed to a short intense laser pulse, will its photoelectron spectrum exhibit peaks? Will they correlate with the bound states having shifted into resonance? What other features could be expected, and under what conditions would they be observable?

2. Results and discussion

We consider first 4+1 ionization in Na with $|0\rangle=|3s\rangle$, $|1\rangle=|7s\rangle$, $|2\rangle=|6d\rangle$, and radiation of frequency $\hbar\omega = 9625\ cm^{-1}$. The relevant parameters (all in rad/s) are: $\Omega_{01} = 2.4\times10^{-11}\ I^2$, $\Omega_{02} = 3.6\times10^{-11}\ I^2$, $\Gamma_1 = 0.65\ I$, $\Gamma_2 = 22.6\ I$, $S_1 = 200\ I$, $S_2 = 150\ I$, $S_0 = -58\ I$, with I in units of W/cm^2. The ponderomotive shift is $S_p = 8\ cm^{-1}$ per $10^{10}\ W/cm^2$. As in all cases considered here, we choose the frequency such that, at zero field, both excited states lie below $4\hbar\omega$, the idea being to examine the structure of the photoelectron spectra as these states shift into and away from resonance during the pulse. The result for a pulse of 1 ps and peak intensity $5\times10^{11}\ W/cm^2$ is shown in Fig. 1a where one quite pronounced peak appears very close to the photoelectron energy corresponding to the 6d having shifted into resonance. The peak does in fact appear 23 cm^{-1} (in electron energy) after 6d has gone through resonance. No peak, however, appears when the 7s goes through resonance. Inspection of the ionization widths shows that only the state with the larger ionization width produces a peak. This may at first seem contrary to conventional rules of thumb about resonance ionization where states with larger ionization rates appear with broader profiles, when frequency is scanned. The important difference here lies in that frequency is not scanned but the state goes through resonance rather quickly and it requires a certain rate of ionization in order to produce signal significantly above the non-resonant background.

Equally important in shaping this behavior is the magnitude of Γ_j relative to Ω_{0j}. From the photoelectron energy corresponding to the peak, we infer that most of the signal comes from a region of the interaction volume of peak intensity around $2\times10^{11}\ W/cm^2$. At this intensity $\Omega_{01}/\Gamma_1 \simeq 8$ while $\Omega_{02}/\Gamma_2 \lesssim 1$. Thus for the state 7s, the Rabi oscillations being much faster than Γ_1^{-1} prevent efficient ionization, while for the state 6d, Ω_{02} and Γ_2 are better matched.

To test the validity of this interpretation we have repeated the calculation leaving all parameters unchanged, except for the width of 7s which was set equal to that of 6d. Two essentially identical peaks, one for each state almost on resonance, have now appeared (dashed line in Fig. 1a). It is important to note here that a pulse of 1ps at $5\times10^{11}\ W/cm^2$ does not cause

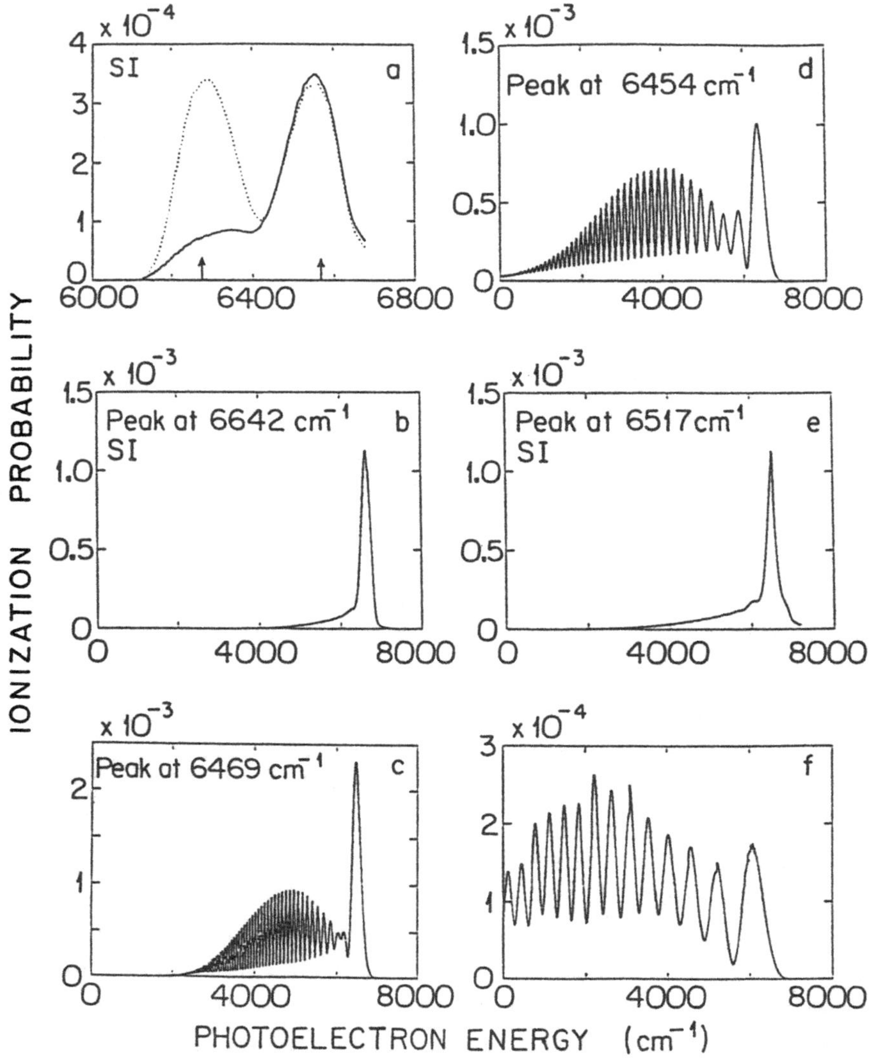

Fig. 1. Examples of photoelectron spectra at various pulse durations τ_L photon frequencies ω and intensities I, for a 4+1 process in Na.

Photon frequency: ω = 9625 cm^{-1} for (a) and 9725 cm^{-1} for (b) through (f).

Peak intensity: I = 5x10^{11} W/cm^2 for (a); I=5x10^{12} W/cm^2 for (b) and (c); I=10^{13} W/cm^2 for (d) and (e); I = 2.5x10^{13} W/cm^2 for (f).

Pulse duration: τ_L = 1 ps. for (a) through (c);

τ_L = 0.5 ps for (d) and (e); τ_L = 0.2 ps for (f).

State |1> is in resonance at 6656 cm^{-1} for (b) through (f) and state |2> is in resonance at 6454 cm^{-1} also for (b) through (f).

SI indicates signal integrated over space coordinates.

saturation in this case. The total amount of ionization at the end of the pulse integrated over the interaction volume is 9%. The saturation intensity in this case is ~ 10^{12} W/cm^2.

We change now the photon frequency to 9725 cm^{-1} leaving the pulse duration the same (1ps). The atomic parameters also remain practically unchanged. What is different now is that the states must shift by an extra 400 cm^{-1} for substantial ionization to occur, which means that the majority of ionization takes place at a slightly higher intensity region, but still below saturation, if the intensity were the same as before (5×10^{11} W/cm^2). A plot of the results (not shown here) looks much like Fig. 1a. In fact, the broad hump is hardly present because the second state does not have the chance to come near resonance.

Let us now increase the peak intensity to 5×10^{12} W/cm^2 (which is slightly above the saturation intensity) leaving everything else unchanged. The resulting photoelectron spectrum is shown in Fig. 1b. Only one peak has emerged, but both intermediate states have coalesced under the same peak; one having come into resonance 25 cm^{-1} before and the other 75 cm^{-1} after the photoelectron peak. When we reran the calculation with the ionization widths equal to each other, the plot looked practically the same; no particular sensitivity to the ionization rates in this case. The reason can be traced to the fact that, owing to the higher intensity, the states shift through resonance rather quickly and their contributions to the main ionization peak overlap. Moreover, most of the ionization is now produced at higher intensity than it was in Fig. 1a.

Further insight into the physics of this behavior can be gained by examining this photoelectron spectrum before the integration over space is performed. This of course can not be done experimentally, but the theoretical analysis has that flexibility. A sequence of results is shown in Figs. 1c-e. The first of these (Fig. 1c) corresponds to conditions identical to those of Fig. 1b. The oscillatory structure following the main peak represents Rabi oscillations with a rapidly changing generalized Rabi frequency $\sqrt{\Omega^2+\Delta^2}$ as the states shift away from resonance. If we decrease the pulse duration to 0.5 ps (with a corresponding increase in the intensity so as to keep the pulse energy constant), we can cause these oscillations to become more pronounced, as shown in Fig. 1d. In fact, the prominence of the main peak of Fig. 1c is now somewhat deemphasized, as more signal is contained under the oscillatory structure. Would any of this structure survive the integration over the interaction volume? The answer is given in Fig. 1e where only a slight hump can be discerned to the left of the main peak. The rest of the structure has been smoothed out by the integration becoming thus inaccessible to experiment.

To illustrate this behavior further, we have performed a calculation at an even higher intensity and correspondingly shorter duration, with the result shown in Fig. 1f (before integration over space). The spectrumm is now dominated by the structure of Rabi oscillations which would not be smoothed out completely by space-integration. Having made the point, we should caution the reader that this result is correct within the model, but unrealistic for the real atom (Na) and the states we have chosen. The reason is that other states, such as 5g, 5d, 6s, 4d, 5s, will shift into resonance just as the second peak begins developing, thus complicating but most probably not obliterating the structure. Only the signal for the initial part of the pulse is shown in Fig. 1f. Owing to the large intensity, a long spectrum of photoelectron energies results. At the point marked 0, the threshold has shifted by ~ 1.2 eV, the energy of one photon. From there on, it takes 6 photons to ionize the atom. Our calculation would be closer to reality if the two excited states were the two lowest ones.

We have examined a variety of other processes, with overall similar results. We quote here 6+1 ionization in H, with $\hbar\omega = 17367\ cm^{-1}$, $|1\rangle = |4s\rangle$ and $|2\rangle = |4d\rangle$. A new twist appears in this case. Although these two states are practically degenerate at zero field, they are about 1200 cm^{-1} apart at $10^{13}\ W/cm^2$; because their shifts are $S_1 = 72\ I$, $S_2 = 50.5\ I$ and $S_0 = -1.8\ I$, all in rad/s. In the same units, the rest of the parameters are: $\Omega_{01} = -6.2\times10^{-30}\ I^3$, $\Omega_{02} = 4.3\times10^{-28}\ I^3$, $\Gamma_1 = 15.8\ I$, $\Gamma_2 = 8.8\ I$ and $S_p = 2.5\ cm^{-1}$ per $10^{10}\ W/cm^2$. After integration over space (for 1 ps and peak intensity $4\times10^{13}\ W/cm^2$ which is well below the saturation intensity), we obtain the result of Fig. 2. The two states have come into resonance around the narrow peak. The broad peak reflects the fact that most of the ionization, in this case, comes from a region around the center of the interaction volume and at the peak intensity.

The results of our studies suggest that the conjecture of Freeman et al[4] is basically valid, in that states shifting through resonance are apt to produce structure. Integration over the interaction volume has a decisive effect on the structure, which looks significantly different below than it does above saturation. It appears that both experiments[4,5] of this type so far were performed below saturation. The Rabi oscillations structure will appear only above saturation but in almost all cases it will be washed out because of spatial effects. Our analysis has shown that the relation between experimental peaks and resonant states is rather subtle and, in general, not a one to one. Not every shifting state produces a peak and vice versa. From extensive calculations on a number of atoms, we can say that the shifts of many excited states differ from the ponderomotive shift significantly, so that we can not expect uniform and systematic shifting of all states. The shifts as well as the dynamical behavior will also exhibit some dependence on photon frequency.

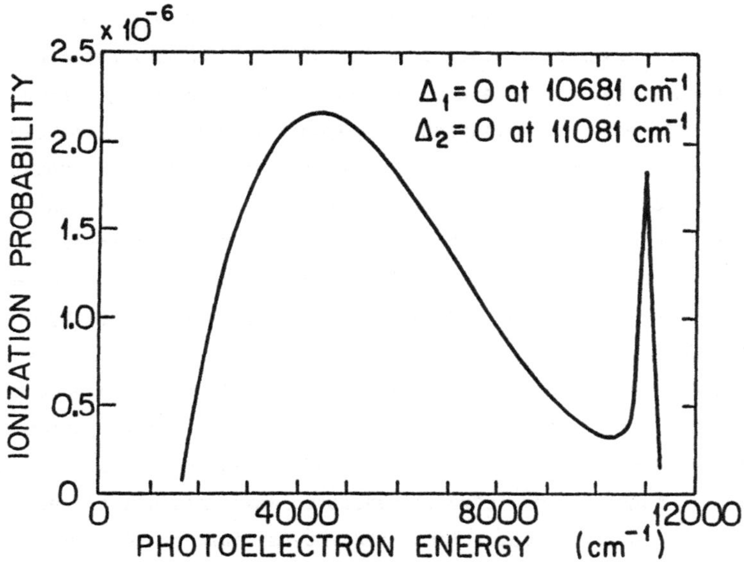

Fig. 2. Photoelectron spectrum for a 6+1 process in H. Photon frequency ω=17367 cm^{-1}; Pulse duration τ_L = 1 ps; Peak intensity I = 4x10^{13} W/cm^2 and the peaks at 10981 and 4565 cm^{-1}. The signal has been integrated over space coordinates.

Moreover, the relation between Rabi frequencies and ionization widths will also influence the observed structure. At larger intensities, a larger set of states must be included, but in any case realistic calculations for the atom and process under consideration are indispensable to a quantitative interpretation of experiments.

The structures we have seen in our calculations reveal an intriguing variety which can be investigated with presently available sources in a number of atoms. Depending on the atom and the order of the process they could be observed under rather modest intensities (10^{11}-10^{13} W/cm^2) and in processes of low order such as 4+1, 5+2, 6+1, etc.

III. REMPI IN Kr AT HIGH INTENSITY

We turn now to the second type of experiment in which light tunable in the wavelength range of 285 to 310 nm was employed in 3+1 resonant ionization of Kr. Around 285 to 290 nm, three photons can be swept into resonance with states of the type $4p^54d$, $4p^56s$ and $4p^55d$. States with total engular momentum J=1,3 can be reached with three such photons while a fourth photon causes ionization with the photoelectron energy being sufficient to lie above the second (fine-structure) ionization threshold $P_{1/2}$. The ion core can of course be left in the lower $P_{3/2}$ state as well, with a branching ratio depending on the resonant state. If no energy analysis of the photoelectrons is performed, the amount of measured ionization corresponds to the sum of the two. This is the case in the experiments of Landen et al.[6] and Perry and Landen[7] because ionization was detected through measurement of the number of ions.

To perform an independent theoretical calculation for these experiments, two important steps are required. First, one must calculate the necessary atomic parameters which in this case means the 3-photon coupling of the ground state $4p^6(J=0)$ to the appropriate excited state, the photoionization cross section of the excited state, and the AC Stark shift of both ground and excited states. It is the difference of the two that determines the apparent shift of the observed resonance. The calculation of these parameters is a task in itself as it requires knowledge of realistic wavefunctions as well as performing the multiple summations over intermediate states. Both of these pieces of the calculation are quite complex and far from routine because transitions in a rare gas like Kr can not be treated in terms of one-electron approximations. We were in the position to perform these calculations using multichannel wavefunctions obtained in the course of related work finished at about the time the results of Landen et al[6] appeared. We need not elaborate here on the details of the method and its application to similar problems as they can be found in our recent paper.[11]

The second step of the theoretical calculation of resonance ionization involves the solution of density matrix equations such as those in Eqs. (1-4) with a realistic spatiotemporal description of the pulse. Using the realistic atomic parameters in these differential equations, we obtain resonant line-shapes which can be compared with the experimental ones. Space does not allow the detailed description and results of this procedure here and it will be published spearately.[12]

The most interesting test of theory and experiment, however, is the comparison of the calculated atomic parameters with those obtained from the fitting of the experimental data. We consider first the state $4d[\frac{3}{2}]_1$ for which the experimentally determined photoionization cross section was σ_{4d}^{exp} = $8\pm1\times10^{-18}$ cm^2 while our theoretical value is 4.42×10^{-18} cm^2. For the net AC Stark shifts, we have S_{4d}^{exp} = 645 cm^{-1} per 10^{13} W/cm^2 as compared to our theoretical value S_{4d} = 639 cm^{-1} per 10^{13} W/cm^2. The experimental 3-photon Rabi frequency is $\Omega_{4d}^{exp} = 1.74\times10^{-7}\ I^{3/2}$ s^{-1} while our theoretical value is $1.21\times10^{-7}\ I^{3/2}$, where I is the intensity in W/cm^2. Given the complexity of both theory and experiment, the above agreement is remarkable and leaves no doubt that what we have here is the imprint of the structure of Kr. It should be noted that the above three parameters test different aspects of the atomic wavefunctions which makes it even more remarkable that all three agree as well as they do with the expperimental values.

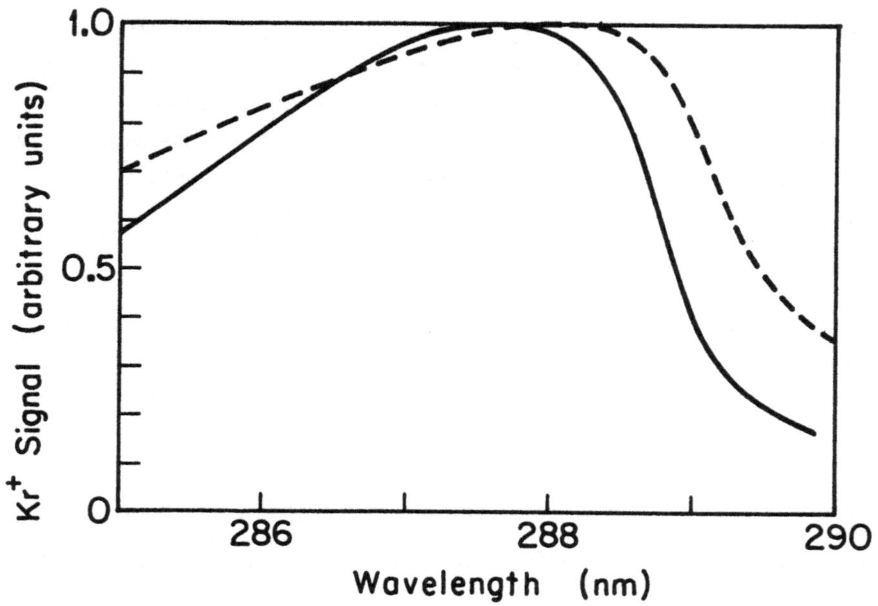

Fig. 3. Kr three-photon resonance four-photon ionization profile at peak intensity $I_0 = 4.5\mathrm{x}10^{13}$ W/cm^2. Solid line $5d[1/2]_1$ only; dashed line $4d'[5/2]_3$ and $5d[1/2]_1$ together.

We consider now the second resonance observed in the experiments which was initially identified as $5d[\frac{1}{2}]_1$ in ref. 6 but was reassigned as $4d'[\frac{5}{2}]_3$ in ref. 7. There is good reason for that ambivalence. These two states are very near each other and under the conditions of the experiment their contributions are not easily separated. We obtain the following parameters for each of these states: $\sigma_{4d'} = 10.65\mathrm{x}10^{-18}$ cm^2, $S_{4d'} = 556$ cm^{-1} per 10^{13} W/cm^2, $\Omega_{4d'} = 3.07\mathrm{x}10^{-8}\ I^{3/2}$ s^{-1}, $\sigma_{5d} = 5.56\mathrm{x}10^{-18}$ cm^2, $S_{5d} = 756$ cm^{-1} per 10^{13} W/cm^2 and $\Omega_{5d} = 1.89\mathrm{x}10^{-9}\ I^{3/2}$ s^{-1}. Note that the Rabi frequency for the 5d is about 16 times smaller than that for the 4d', but its ionization cross section is somewhat larger than that of the 4d. But since the states move during the pulse, it is not obvious that the 5d does not contribute significantly. One way to proceed theoretically is to calculate the resonance with one as well as with both states included and evaluate the difference between the two results.

One example of such a comparison is shown in Fig. 3 where we show the result of the calculation for an intensity 4.5×10^{13} W/cm^2. Obviously there are differences in the amount of ionization which suggests that if one of these curves is fitted to a set of rate equations with one resonant state, it will give different parameters from the other. The experimental values of the parameters given in ref. 7 are: $\sigma_{4d'} = 3\times10^{-18}$ cm^2, $S_{4d'} = 639$ cm^{-1} and $\Omega_{4d'} = 7.86\times10^{-8}\ I^{3/2}$ s^{-1}. We note that the shift is between the calculated shifts of the two states shown above. The other parameters are not too far either, but we would not expect better agreement since the presence of at least two overlapping resonances requires a somewhat different analysis of the experimental data. A more elaborate discussion of these effects will be published elsewhere.[12]

This work was supported by NSF Grant No. PHY-8609966 and by DOE Grant No. DE-FG03-87ER06504.

REFERENCES

1) L. V. Keldysh, Zh. Eksp. Teor. Fiz. 47, 1945 (1964) [Sov. Phys.-JETP 20, 1307 (1965)].
2) H. Reiss, Phys. Rev. 22, 1786 (1980); F.H.M. Faisal, J. Phys. B 6, L89 (1983).
3) K. Boyer and C. K. Rhodes, Phys. Rev. Lett. 54, 1490 (1965).
4) R. R. Freeman, P.. H. Bucksbaum, H. Milchberg, S. Darack, D. Schumacher and M. E. Geusic, Phys. Rev. Lett. 59, 1092 (1987).
5) H. G. Muller, H. B. van Linden van den Heuvell, P. Agostini, G. Petite, A. Antonetti, M. Franco and A. Migus, Phys. Rev. Lett. 60, 565 (1988).
6) O. L. Landen, M. D. Perry and E. M. Campbell, Phys. Rev. Lett. 59, 2558 (1987).
7) M. D. Perry and O. L. Landen, Phys. Rev. A 38, 2815 (1988).
8) S. N. Dixit and P. Lambropoulos, Phys. Rev. A 27, 861 (1983); experimental verification has been reported in W. Ohnesorge, F. Diedrich, G. Leuchs, D. S. Elliot, and H. Walter, Phys. Rev. A 29, 118 (1984).
9) P. Lambropoulos, Adv. At. Molec. Phys. 12, 87 (1976).
10) M. Edwards, X. Tang and R. Shakeshaft, Phys. Rev. A 35, 3758 (1987).
11) Anne L'Huillier, X. Tang and P. Lambropoulos, Phys. Rev. A 39, 1112 (1989).
12) X. Tang, Anne L'Huillier and P. Lambropoulos (in preparation).

On the Problem of the "New" Structure in Multiphoton Ionization Introduced by Short-Pulse High Intensity Lasers

R.R. Freeman, W.E. Cooke[a],
L.D. van Woerkom, and T.J. McIlrath[b]

AT&T Bell Laboratories, Murray Hill, NJ 07974

ABSTRACT

When an atom is exposed to high intensity radiation, its ground state will decay due to multiphoton ionization. In a classic measurement, Agostini *et.al* found that when the photoemitted electrons are energy analyzed they produce a spectrum containing a series of peaks, separated by the photon energy. This phenomena, dubbed "ATI", has been the subject of intense experimental and theoretical interest for over 10 years. Recently, Freeman *et al.* showed that the individual ATI peaks are not at all homogeneous, but are rather each a rich collection of sharp, well resolved peaks. In this paper we show how these distinct peaks arise and how they are intimately related to the energy structure of the target atom. We also predict that this structure in the energy spectrum is accompanied by a surprisingly complex *spatial* structure of the total ionization yield. We argue that this spatial structure, like the energy structure, is a general consequence of high intensity multiphoton ionization and is expected to appear in a wide variety of multiphoton experiments. Finally, we present a detailed procedure for the calculation of both the structure in the energy spectra and the spatial ionization yield, given a list atomic energy levels and transition strengths.

In a recent paper, Freeman *et al.*[1] reported the observation of an unexpected fine structure in the individual "ATI" peaks discovered by Agostini *et al.* [2] in the electron energy spectrum for short pulse, high intensity multiphoton ionization in xenon. This result, subsequently confirmed in xenon by other researchers[3] and also recently observed by us in krypton, argon and neon, is thought to be due to optical Stark shifts of atomic levels during the course of the intense laser pulse. Here we show that when such a structure appears in the energy spectrum, the resulting ionization yield is necessarily distributed spatially throughout the interaction volume in a complex, inhomogeneous manner. We demonstrate that the form of this spatial structure can be derived directly from that of the fine structure in the electron energy spectrum. Further, we show that the form of the structure in the energy spectrum as well as the structure in the spatial spectrum can be computed in detail from a knowledge of the energy levels of the target atom. This formalism also includes the direct computation of the dependence of the widths of the energy upon the pulse width of the ionizing budgets.

In an intense optical field with peak intensity I and frequency ω, the ionization threshold (I_{thres}) *increases* its energy relative to the ground state by the ponderomotive energy[4], *viz,*

$$U_p(I) = [2\pi e^2 I]/[m_e c\omega^2].$$

An electron ionized in a weak field after the absorption of n photons has kinetic energy of $KE_1 = (n\hbar\omega - I_{thres})$; if this same electron were photoemitted in an intense field of intensity I, it would possess at the moment of ionization a ponderomotive potential energy $U_p(I)$, but a reduced kinetic energy $KE_2 = KE_1 - U_p(I)$. The electron converts the potential energy into kinetic energy if it escapes the interaction volume before the optical pulse significantly changes intensity[5]; in so doing it exactly makes up the difference between KE_1 and KE_2. On the other hand, if the pulse duration is so short that the electron does not move substantially during the pulse, the potential energy $U_p(I)$ is given back to the optical field[1,6], and the electron arrives at the detector with a kinetic energy of KE_2.

The high lying Rydberg states of atoms or molecules increase in energy relative to the ground state by approximately[7] $U_p(I)$, opening up the possibility that they will shift into resonance with the mth harmonic of the laser frequency sometime during the pulse[8,9]. Such a resonance would produce a increase in the ionization rate and an increase in the observed number of photoemitted electrons having kinetic energy

$$E_R = q\hbar\omega - E_b$$

where E_b is the binding energy of the excited state which has shifted into resonance, and q is the number of photons necessary to yield a positive E_R, usually no more than two. States which can enhance the multiphoton ionization rate must have a zero field energy which is less than $m\hbar\omega$ above the ground state and have the appropriate parity. Figure 1 shows a portion of the record of electron kinetic energies obtained in high intensity ($\leq 7 \times 10^{13}$ W/cm^2) multiphoton photoemission of xenon for a laser at 616nm with a pulse width of $\approx$250 fsec. At 616 nm, xenon requires 7 photons to ionize the ground state, and the 6th harmonic ($6\hbar\omega$) lies just .05 eV below the (zero field) ionization threshold, I_{thres}. This relative positioning in energy of $6\hbar\omega$ and the xenon ionization threshold places the 6th harmonic just above the unshifted energy positions of a large family of xenon excited states. The position labeled $7\hbar\omega$ is the energy that electrons would have if photoemitted by 7 photon ionization without a ponderomotive shift.

Quite clearly, the more a state is bound in the atom, the less will be its measure kinetic energy upon short pulse photoemission, but the larger will be the ponderomotive shift necessary to bring it into resonance. Indeed, any resonance structure in the kinetic energy spectrum in Fig.(1) is uniquely related to the specific intensities required to shift the excited states into 6 photon resonance with the laser[1]. This intensity is the same for all atoms in the interaction volume; thus the abscissa on Fig.(1) may be relabeled according to the intensity I, with the conversion factor derived from the ponderomotive potential ($I(W/cm^2) = [(2.0 - E)/.37eV] \times 10^{13}$ at 616 nm). The states are labeled in Fig. (1) by assuming their optical Stark shifts to be approximately ponderomotive.

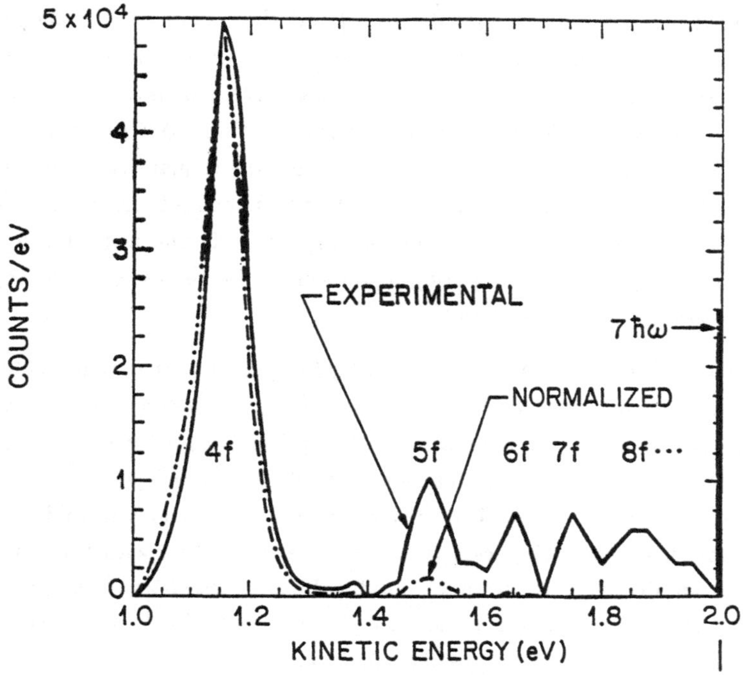

1

A portion of the experimental record of electron kinetic energies, N(E), obtained in high intensity ($\approx 1\times 10^{14}$W/cm^2) multiphoton photoemission of xenon for a laser at 616 nm with a pulse width of $\approx$ 250 fsec. The "normalized" spectrum is obtained by dividing the experimental spectrum at each energy by a value proportional to the effective spatial volume which can contribute photoelectrons at that energy (Eq.(2)).

The ponderomotive shift at any position (r,z) and time t within the ionization pulse is directly proportional to the local intensity[9], I(r,z,t); at the focus (z=0), $I(r,t) = I_o \exp[-(r/r_o)^2 - (t/\tau)^2]$ If the intensity required to shift one of the f states into 6 photon resonance is I_R, where $I_R < I_o$, electrons with energy E_R will be produced at the temporal peak of the pulse (i.e., t=0) from atoms in a narrow spatial band centered at $r_1 = r_o[\ln(I_o/I_R)]^{1/2}$. However, electrons with the *same* energy, E_R, from the *same* f state are also produced by different atoms whose radial position satisfy $r < r_1$. These atoms also produce their electrons at $I = I_R$, but at times other than t=0: once when the intensity passes through I_R on the leading edge of the pulse and again on the falling edge.

The ionization rate is a convoluted function of time and space[10]. However, the assumption that high intensity photoemission arises primarily from level crossings yields a significant simplification: Because each atom undergoes an m-photon resonance for a given state at precisely the same intensity, regardless of the spatial location of the atom, the ionization probability can vary from atom to atom only because the amount of time each atom experiences that intensity depends upon its location in the beam focus. This time is greatest for those atoms located where the resonance intensity, I_R, occurs at the peak in the local intensity; i.e., where $I(t) = I_R \exp[-(t/\tau)^2]$. At these points the resonance time interval is $t \approx \tau$, however, at spatial points where $I(t) = I_p \exp[-(t/\tau)^2]$ and $I_p > I_R$, resonance occurs only on the leading and falling edge of the pulse, and the time interval for resonance is much less than τ. *It is this spatial variation in "resonance time" that produces a non-monotonic variation of total ionization with radius.* Here we outline a quantitative model of the spatial dependence of the ionization probability based upon Landau-Zener curve crossing theory [11] extended[12] to account for curve crossings having not only the usual linear dependence on time (as on the steep edges of the pulse), but quadratic as well (as near the top of the pulse). The full derivation is given elsewhere[12].

Consider two states: state 1 is the excited Rydberg level of the atom and state 2, assumed to lie above state one, is the ground state dressed by m photons. These levels are initially separated in energy by $\hbar\omega_R$, but because of the optical Stark shift of state 1, the separation $\hbar\omega$ decreases with increasing laser intensity according to $\omega = \omega_R(1 - I/I_R)$, where I_R is the intensity for the levels to cross (i.e., "resonance"). The first and second derivatives of the energy separation are used to parameterize the level separation with time. For a laser beam with a gaussian distribution in time and for $I = I_R$, $\left|\partial\omega/\partial t\right| = (2\omega_R/\tau)[\ln(I_p/I_R)]^{1/2}$. If $I_R = I_p$, so that the resonance occurs at the peak of the local beam intensity where $\partial\omega/\partial t = 0$, then $\left|\partial^2\omega/\partial t^2\right| = 2\omega_R/\tau$ (ignoring terms of order $(t/\tau)^2$).

Conventional Landau-Zener theory for a curve crossing linear in time computes the probability for a transition from the ground (state 2) to the excited state (state 1) to be $P = 1 - \exp(-p)$, where $p = 2\pi V^2 \tau_c^2$ with $\hbar V$ equal to the interaction strength (proportional to $(I_R)^m$ in this instance) and $\tau_c^2 = \left|\partial\omega/\partial t\right|^{-1}$. (Assuming that the ionization cross section of state 1 is sufficiently large that no population is coherently returned to the ground state.) When $I_p = I_R$, $\partial\omega/\partial t \approx 0$, τ_c is dominated by the second order derivative and can be shown[12] by direct integration to be $\tau_c^2 = \alpha^2 \left|\partial^2\omega/\partial t^2\right|^{-2/3}$, where $\alpha^2 = (1/2\pi)(4/3)^{1/3}(\Gamma(1/3))^2$. Adding the linear and quadratic rates in quadrature, and assuming no coherence in the excitation, yields the total ionization probability from resonance R at any spatial location to be:

$$P = \frac{N'(r)}{N_o}[1 - \exp(-p)]$$

$$p = p(r,z,\omega_R) = 2\pi V_R^2 (\tau/\omega_R) \frac{\epsilon}{1 + \epsilon[\ln(I_p/I_R)]^{1/2}} \qquad (1)$$

where $\epsilon = \Gamma^2(1/3)(\omega_R\tau/3)^{1/3}/2\pi = 0.79(\omega_R\tau)^{1/3}$. Here $I_p = I_p(r)$ is the peak intensity at t=0 at position (r,z) in the beam; at z=0, I_p and r are related to I_o, the peak intensity at r=0, and r_o, the spot size at the focus by $r = r_o[\ln(I_o/I_p)]^{1/2}$. $N'(r)$, the ground state density remaining after the ionization due to all j atomic resonances requiring less intensity then I_R, is related to N_o, the initial ground state density, by $N'(r) = N_o \exp(-\sum_1^j p_i)$.

Under conditions far from saturation, $p << 1$, yielding $P \approx p$, but Eq.(1) correctly predicts the general case. We have elsewhere[13] considered explicitly how the spatial structure is distorted by saturation. For our purposes it is sufficient to note that short pulse multiphoton ionization measurements provide an empirical monitor of the role of saturation: the quality of the structure in the photoelectron energy spectrum. If there were saturation, the structure at the lower kinetic energies, corresponding to levels requiring more intensity to undergo resonance (and thus later in the pulse for all r), would be absent or seriously attenuated. However, none of the data we have taken in short pulse multiphoton ionization measurements show any sign of saturation for any resonance. Indeed, the ratio of signal strength of the 4f to 5f resonances in the volume normalized spectra in Fig.(1) is very nearly what one would expect from a ratio of resonance intensities raised to the 6th power. For these reasons our further discussion assumes the weak ionization limit, although we emphasize that Eq.(1) is quite general and applies to arbitrary ionization probabilities as well.

The total ionization contributed to the photoemission spectrum N(E) from resonance R is obtained by summing contributions from all points in the volume. That is, integrating $N_o P = N_o p$ over all space, we find (after the explicit radial integration):

$$N(\omega_R) = 8\pi^2 N_o V_R^2 \left(\frac{\tau}{\omega_R}\right) \int_0^{z_m} \left[(1+z^2)\left(u - \frac{1}{\epsilon}\log(1+\epsilon u)\right)\right] dz \qquad (2)$$

where $u = \left[\log \frac{I_o}{I_R(1+z^2)}\right]^{1/2}$ and $Z_m = \sqrt{I_o/I_R - 1}$

Equations (1) and (2) allow the calculation of both the spatial and energy distribution of a high intensity multiphoton experiment given a list of the atomic energy levels and the m-photon transition strengths: first calculate the spatial distribution from Eq.(1), then weight the contribution of each resonance in the energy distribution by computing its effective volume (Eq.(2)).

To calculate the spatial distribution from an experimental record of the energy spectrum N(E): 1) De-convolute the energy response function of the electron energy analyzer from the recorded data; 2) Determine the contribution of each E_R contained in N(E) to the spatial yield by first normalizing N(E) by $N(\omega_R)$. [We have shown the application of this normalization to the experimental data in Fig.(1). The effective volume of resonances with small values of I_R, e.g., 6f,7f..., is so large that they contribute significantly to the experimental photoemission spectrum, even though their intrinsic strength, shown in the normalized spectrum, is negligible.] 3) Apply Eq.(1) to determine the total ionization at (r,z).

That is

$$N(r,z) = \int N(E_R) N_o p(r,z,\omega_R)/N(\omega_R) dE_R$$

We have applied this procedure to the data in Fig.(1). Figure (2) shows the results: the peak intensity in the beam is taken to be $7\times 10^{13}W/cm^2$. The gaussian beam profile and the distribution of photoemitted electrons assuming truly nonresonant, 7 photon multiphoton ionization are also shown. Note that while the position of the peak in the nonresonant distribution is independent of I_o, that of the resonant ionization scales as $(\log I_o)^{1/2}$. The complex spatial structure introduced in Fig.(2) by the excited atomic states is dramatic[14].

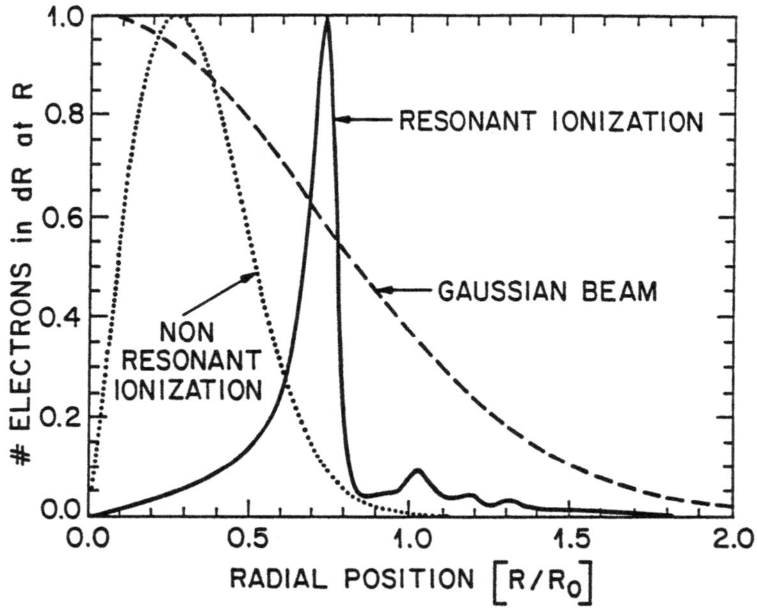

2

The resultant *spatial* distribution at z=0 of total ionization yield calculated from the *energy* spectrum shown in Fig. 1 for an intensity of $7\times 10^{13}W/cm^2$. Also shown is the ionization yield assuming conventional 7 order "nonresonant" multiphoton ionization. The position of the peak of the nonresonant distribution is independent of I_o, while that for resonant ionization scales as $(\log I_o)^{1/2}$.

Actually, the formalism presented here is even more powerful than suggested in the applications given above. We are now in a position to calculate not only the relative *yields* to the energy spectrum of a series of atomic states, but the energy *widths* of these states as well. This ability arises from our calculation of a time τ_c which is characteristic of the interaction time: τ_c^2 determines the transition strength, while $1/\tau_c$ determines the energy spread (uncertainty principle). Remember, τ_c is a function of position within the beam, and each resonance will contribute to the energy spectrum with many different values of $1/\tau_c$ (weighted by $2\pi V^2 \tau_c^2$).

We have taken the 4f,5f...10f states of xenon and have computed the expected energy spectrum assuming:

1. The coupling strengths vary as n^{-3}

2. The excited states couple to the laser on the 6 photon resonance; this assumption requires the laser to have at least 7 eV of ponderomotive potential ($\approx 3\times 10^{13}$W/cm^2). These are indeed the conditions for the data shown in Fig (1).

3. The resolution function of our electron energy analyzer is approximately 25 meV.

We then computed the distribution for each resonance to the electron energy distributions by evaluating $\hbar/\tau_c$ over the entire volume and weighting each point in the volume with $p(r,z,\omega_R)$ as given in Eq.(2). The results are shown in Fig.(3): We have plotted the

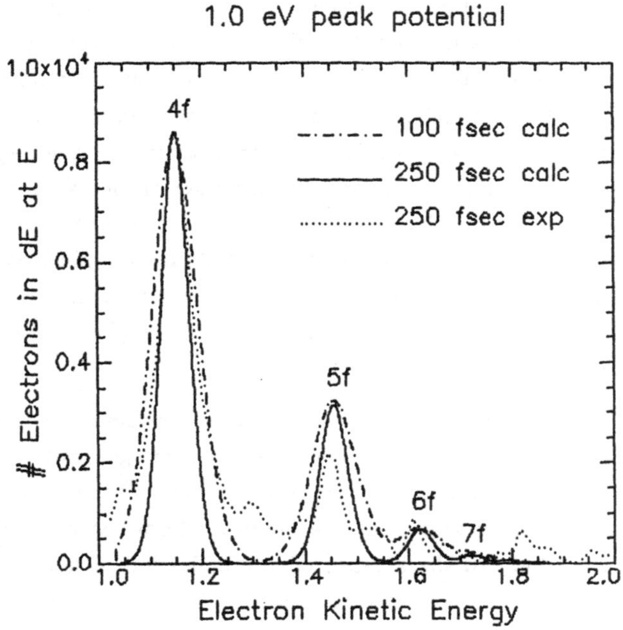

3

The results of a calculation of the energy spectrum arising from the 4f...10f states of xenon, including not only signal strength but energy widths as well. The experiment was performed on xenon with 1 eV max beam potential ($\approx 3\times 10^{13}$W/cm^2) and a 250 fsec laser pulse.

calculation on top of experimental results we have obtained under the same conditions. The calculation is normalized in amplitude to the experimental data on the 4f state. *There are no other adjustable parameters.* The experimental data contains Rydberg states other than the f states included in the calculation, the most prominent of which are p states. The correspondence between the calculation of the f states and the experimental results is excellent. As the laser pulse is increased out to a picosecond, the calculation shows that the resolution improves (not shown in Fig.(3)). There is one cautionary note concerning longer pulse lengths: one cannot simply increase the resolution arbitrarily by making the pulse length too long, for the requirement always holds that the electrons must not escape the interaction volume before the laser pulse leaves. For most practical experiments, the maximum length of pulse is less than 2-4 psec. It is particularly interesting to note what happens when the pulse is made shorter: We show the results of the calculation for even shorter pulses than we have used up to now in these experiments. For 100 fsec, the spectra broadens considerably and the resonances blur at the higher n values. Using too short a pulse can clearly be a detriment if the purpose is to identify clearly resonance positions.

The complex spatial distribution of ionization introduced by resonances is expected to influence several kinds of high intensity multiphoton experiments. In sequential ionization[15], for example, the rising laser pulse first ionizes the neutral atom and then, at higher intensities, produces higher stages of ionization a sequential manner. These experiments are usually analyzed by assuming that at saturation the resulting ions are produced primarily on axis, with a monotonically decreasing density with radius. These ions are then presumed to have been exposed to higher intensities where they nonresonantly ionize further. In contrast, this paper demonstrates that the ionization density for any stage will not necessarily fall off monotonically with radius, but may be instead concentrated in the regions of "spatial resonances". In general, the spatial resonances for one stage of ionization will not match up with the spatial regions of resonance for the succeeding ionization stage. Although saturation due to long laser pulse lengths or very large atomic transition strengths may limit the effects of this mismatch, we expect the complex ionization distributions to play a significant role in inhibiting multiple ionization; however, their effects can be explicitly calculated using the procedure outlined here.

Recent experiments have used higher order harmonic conversion as a probe of atomic nonlinearities at high optical intensities[16]. Since the harmonic generation reflects the same resonance structure as does absorption, the "spatial resonances" should produce an output mode which is far more complex that would be predicted from the usual gaussian focussed modes[17]. This may contribute to the absence of any indication of normal phase matching effects on the output[16]. Again, the approach presented here can be applied directly to the calculation of harmonic conversion.

The authors wish to thank A. Szoke for several important discussion at the inception of this work, P.H. Bucksbaum for a critical reading of the manuscript and S. Davey for technical help. W.E.C. was supported by NSF grant PHY85-0085; T.J.M. by NSF grant CPE-84-179333.

REFERENCES

[a] On leave from the Department of Physics, University of Southern California, Los Angeles, Ca. 90007
[b] Permanent address: Institute for Physical Science and Technology, U. of Maryland, College Park, Md 20742

1. R.R. Freeman, P.H. Bucksbaum, H. Milchberg, S. Darack, D. Schumacher and M.E. Geusic, Phys. Rev. Lett. *59*, 1092 (1987).

2. P. Agostini, F. Fabre, G. Mainfray, G. Petite, and N.K. Rahman, Phys. Rev. Lett. *42*, 1127 (1979).

3. H.G. Muller, H.B. van Linden van den Heuvell, P. Agostini, G. Petite, A. Antonetti, M. Franco, A. Migus, Phys. Rev. Lett., *60*, 565 (1988).

4. T.J. McIlrath, P.H. Bucksbaum, R.R. Freeman, and M. Bashkansky, Phys. Rev. A. *35*, 4611 (1987).

5. H.G. Muller, A. Tip and M.J. van der Wiel, J. Phys. B: At. Mol. Phys. *16*, L679 (1983); T.W.B. Kibble, Phys. Rev. *150*, 1060 (1966).

6. M. Crance, J. Phys. B: At. Mol. Opt. Phys. *21*, 2697 (1988); T.S. Luk, T. Graber, H. Jara, U. Johann, K. Boyer, and C.K. Rhodes, J. Opt. Soc. Am. B *4*, 847 (1987); P. Agostini, J Kupersztych, L.A. Lompre, G. Petite and F. Yergeau, Phys. Rev. A *36*, 4111 (1987); R.R. Freeman, P.H. Bucksbaum, T. J. McIlrath, IEEE JQE *24*, 1461 (1988).

7. L. Pan, L. Armstrong and J.H. Eberly, J. Opt. Soc. Am. B *3*, 1319 (1986); P. Avan, C. Cohen-Tannoudji, J. Dupont-Roc and C. Fabre, J. Phys. (Paris) *37*, 993 (1976); M.D. Perry, A. Szoke, O.L. Landen and E.M. Campbell, Phys. Rev. Lett. *60*, 1270 (1988).

8. M. Crance, J. Phys. B: At. Mol. Phys. *14*, 4301, (1981); O.L. Landen, M. Perry and E.M. Campbell, Phys. Rev. Lett. *59*, 2558 (1987).

9. Here we are interested only in deriving the spatial distribution at $z=0$. The method is quite general to all other points within the volume, only a more mathematically tedious expression for the gaussian beam is required for $z \neq 0$; see A. Yariv, *Quantum Electronics*, 2nd Ed., J.Wiley, N.Y.,1975 pg.110

10. M.D. Perry and O.L. Landen, Phys. Rev. A *38*, 2815 (1988).

11. A. Szoke, J. Phys. B: At. Mol. Opt. Phys. *21*, L125 (1988); M. Crance J. Phys. B: At. Mol. Opt. Phys. *21*, 1987 (1988).

12. W.E. Cooke, R.R. Freeman, T. J. McIlrath, and L.D. van Woerkom, to be published.

13. L.D. van Woerkom, R.R. Freeman, W.E. Cooke, and T.J. McIlrath, Journal of Optics, (to be published).

14. If the atom is exposed to even high intensities, there will be the possibility of additional level crossings, corresponding to m+2,m+4,...m+2n photons.

15. A. L'Huillier, L.A. Lompre, G. Mainfray, C. Manus, Phys. Rev. Lett. *48*, 1814 (1982), Phys. Rev. A *27*, 2503 (1983), J. Phys. B: At. Mol. Phys. *16*, 1363 (1983); T.S. Luk, U. Johann, H. Egger, H. Pummer, C.K. Rhodes, Phys. Rev. A *32*, 214 (1985); see also ref 6.

16. A. McPherson, G. Gibson, H. Jara, U. Johann, T.S. Luk, I.A. McIntyre, K. Boyer, C.K. Rhodes JOSA B *4*, 595 (1987); M. Ferray, A. L'Huillier, X.F. Li, L.A. Lompre, G. Mainfray, C. Manus J. Phys. B: At. Mol. Opt. Phys. *21*, L31 (1988).

17. Y.M. Yiu, T.J. McIlrath, R. Mahon Phys. Rev. A *20*, 2470 (1979).

RESONANT MULTIPHOTON IONIZATION OF XENON AND ATOMIC HYDROGEN IN INTENSE SUB-PS-LASER PULSES.

H. Rottke, B. Wolff, M. Tapernon, D. Feldmann and K.H. Welge
Fakultät für Physik, Universität Bielefeld
Postfach 8640, D-4800 Bielefeld 1

Introduction

Two kinds of multiphoton ionization (MPI) processes are often distinguished (1). Nonresonant ionization occurs, if the energies of excited states with proper parities are far detuned from multiples of the photon energy $N\hbar\omega$, and resonant ionization takes place, if one or more excited states serve as resonantly excited intermediate steps (or final states in the case of autoionizing states). This distinction is useful, if the laser intensities necessary for ionization are relatively low such that the states of the atom are almost unchanged in the radiation field. On the other hand it is well known and has been frequently studied that atomic states exhibit ac-Stark shifts and broadening in strong radiation fields. At moderate intensities these effects can be calculated in lowest order perturbation theory. It was recognized and calculated theoretically already some time ago (2) that this approach breakes down if the intensity is increased. Until recently there was no evidence that it would be possible to observe resonant ionization in very intense laser fields where the shifts become comparable with the photon energy. The influence of excited states on "nonresonant" ionization in intense fields could be deduced only indirectly from angular distributions of photoelectrons (3).

This situation has changed with the recent experiment of Freeman et al. (4) who observed that peaks in the photoelectron spectrum of xenon, belonging to a certain number of absorbed photons, develop a sharp substructure, if the pulse duration of the laser was reduced to below 1 ps. They proposed the following explanation for these substructures: High lying excited states of the atom are shifted into resonance with a virtual intermediate N-photon state in the intense laser field. By this intensity induced resonance the ionization probability is largely enhanced at a well defined intensity during the pulse in the laser focus. This results in a strong electron signal at a well defined electron energy; whereas at "nonresonant" intensities the ionization probability is low. This peak substructure can only be observed experimentally, if short laser pulses are used, such that the nascent photoelectrons cannot leave the high intensity region in the laser focus during the pulse; otherwise the electrons gain energy and change momentum when they leave the laser focus during the pulse due to the pondermotive potential. Therefore these sharp peaks cannot be observed in experiments with long laser pulses or at high kinetic energies of the photoelectrons.

Assuming that all high lying excited states of xenon shift by the same amount as the ionization limit, Freeman et al. (4) were able to identify excited states responsible for the main peaks of the substructure. The shift of the ionization limit is given by the quiver energy of the free electron in the laser field and a small shift of the initial ground state. Their experiment was performed at a wavelength of 616 nm with xenon atoms, and the identification of excited states was based on the energy of the peaks in the electron energy spectrum. With respect to the ac-Stark shifts, mentioned above, these results show that loosely bound electrons behave similar to free electrons in very intense laser fields. In a recent calculation by Kulander (5) the influence of a strongly shifted excited state on the MPI efficiency has also been identified.

We have performed experiments which give more detailed insight into this type of processes by measuring angular distributions of the photoelectrons in addition to their energy spectrum for Xenon near 600 nm. For atomic hydrogen, the ideal test case for comparison with theoretical calculations, we present the first electron energy spectra obtained in sub-ps MPI.

Experimental:

The experimental setup has been described in more detail elsewhere (3). One major modification is a sub-ps laser system, consisting of a synchronously pumped dye laser and an amplifier chain, which delivers pulses of about 0.6 ps duration. We have used wavelengths between 590 nm and 616 nm for the MPI experiments. The laser beam was focussed onto an effusive beam of xenon atoms or atomic hydrogen, which was produced in a microwave discharge. Laser intensities of up to some 10^{14} Wcm^{-2} were applied. Ions or electrons were detected after passage through a time-of-flight-(TOF) spectrometer perpendicular to the laser beam. Angular distributions of electrons for linearly polarized light were recorded by rotating the polarization with respect to the direction of detection. The energy scale of the electron TOF-spectrometer was checked by MPI-measurements of Xe with ns-pulses at 532 nm and 596 nm. In the case of hydrogen the electron signal contained a contribution from MPI of H_2-molecules, which was less than 10 % of the total signal as could be concluded from the respective ion signals. As we have used the TOF energy spectrometer without retardation at the entrance of the 50 cm long drift region, the experimental energy resolution is reduced with increasing electron energy. Large artificial signal fluctuations can occur at low electron energies because the signal within a small energy interval is distributed over many evenly spaced time channels of the transient digitizer. Therefore we have added an averaging full line in this energy region of the spectra shown below.

Results:

1. Xenon:

1.1. Electron energy spectra.

Figure 1) shows a typical energy spectrum of the photoelectrons from MPI of xenon at a wavelength of 596 nm with linearly polarized light parallel to the direction of detection ($\theta = 0^o$). Figure 1a) shows for comparison the spectrum obtained with pulses of 6 ns duration at the same wavelength and an intensity of the order of $5 \cdot 10^{12}$ Wcm^{-2}. On its upper scale the different ATI-peaks are identified by their S values, which give the number of photons absorbed in the continuum for the two final states of Xe^+: $^2P_{3/2}$ and $^2P_{1/2}$. Figure 1b) shows the spectrum for pulses of 0.7 ps duration and a laser pulse peak intensity of $4 \cdot 10^{13}$ Wcm^{-2}. Also here, the typical ATI-structure with sequences of features separated by the energy of one photon can be seen. At energies below 2 eV a well resolved substructure is observed, which is not well resolved experimentally at higher electron energies. The top scale shows the intervals corresponding to different numbers S of excess photons absorbed in the continuum. The S-labelling is different here as compared with that in figure 1a). As mentioned above, under long pulse conditions the photoelectrons leave the focal volume during the laser pulse and are detected outside with a kinetic energy $E_s \approx (N_o + S)\hbar\omega - IP_o$, where N_o is the minimum number of photons, which have to be absorbed to overcome the field free ionization limit IP_o. For xenon N_o is six at this wavelength. Under the short pulse conditions of figure 1b) the low energy electrons cannot leave the high intensity focal volume during the laser pulse. Therefore they are detected with their nascent excess energy $E_s' = (N_o' + S)\hbar\omega - W(I)$, where N_o' is the minimum number of photons to overcome the intensity shifted ionization limit W(I), which is the energy difference between the intensity shifted ionization threshold $IP_o(I)$ and the also shifted ground state energy. Therefore S = 0 now covers the energy interval between $E_s' = 0$ and $E_s' = \hbar\omega$, and other S-values cover the respective intervals shifted by $S \cdot \hbar\omega$.

In figure 1b) also the expected energies for electrons from ionization processes via nf-states, ac-Stark shifted into six-photon resonance, are indicated on the upper

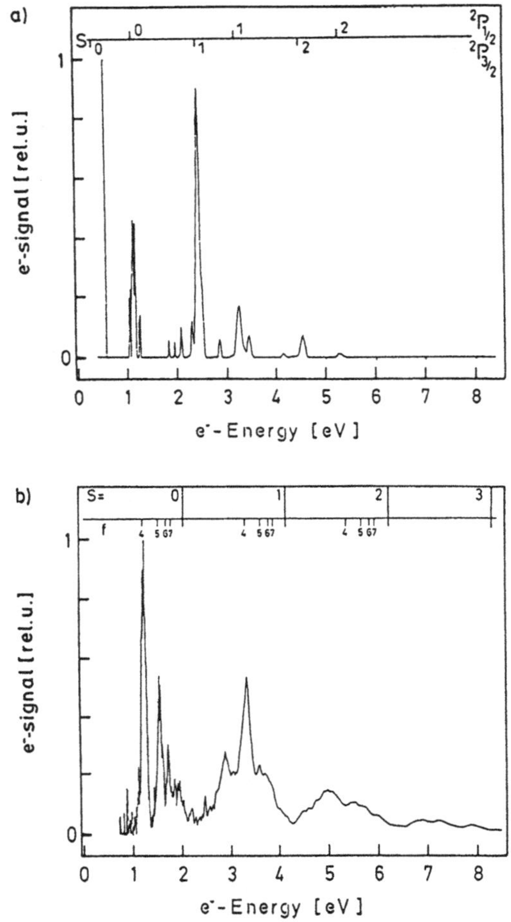

Figure 1: Energy spectrum of photoelectrons from MPI of xenon at λ = 596 nm. a) laser pulse duration 6 ns, intensity $I \approx 5 \cdot 10^{12}$ Wcm^{-2}; b) laser pulse duration 0.7 ps, laser pulse peak intensity $I \approx 4 \cdot 10^{13}$ Wcm^{-2}.

scale. These energy positions are calculated by assuming an equal shift for all nf-states and the ionization limit.

Figure 2) shows the low energy part of the electron energy spectrum for two different peak intensities of the laser pulse: $I = 0{,}5 \cdot 10^{14}$ Wcm^{-2} for the lower and $I = 10^{14}$ Wcm^{-2} for the upper curve. The peak positions do not shift with laser peak intensity, however, the relative signal strengths of the peaks vary. Especially the maximum below 0.9 eV increases more rapidly with intensity than the other ones. This

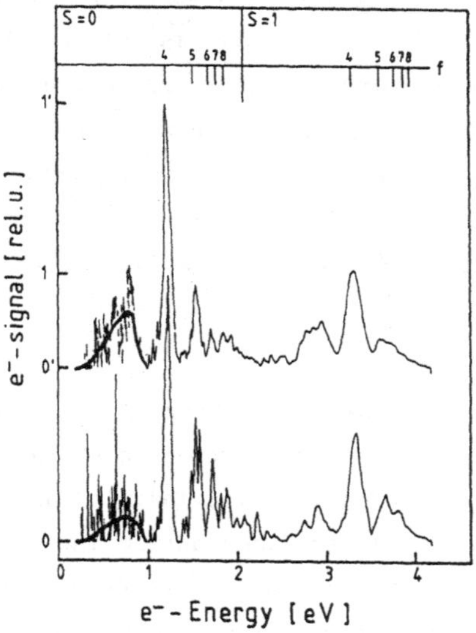

Figure 2: Energy spectrum of photoelectrons from MPI of xenon at λ = 596 nm for two different laser pulse peak intensities: lower curve $I \approx 5 \cdot 10^{13}$ Wcm^{-2}, upper curve $I \approx 10^{14}$ Wcm^{-2}. The relative scale is different for the two curves.

is in accordance with the model, that each excited state is shifted into resonance at its well defined resonance intensity which results in a well defined kinetic energy of the corresponding photoelectrons. States with higher resonance intensity like that correlated with the peak below 0.9 eV should be observable only at higher laser peak intensities. The restriction to a "well" defined resonance intensity should only be fulfilled within the intensity interval associated with the width of the states in the intense laser field. The assignment of the main peaks in the (S = 0)-interval above 1 eV to nf-states is straight forward and can be followed up to n $\approx$ 7 or 8. The relative signal height decreases with increasing principle quantum number n because the excitation cross section of the resonant state and its ionization cross section decrease with growing n. Additionally higher nf-states shift into 6-photon resonance at successively lower intensities, thus the excitation rate of these states and their ionization rate will also decrease due to the lower laser intensity.

At higher electron energies, in the (S = 1)-interval, only the 4f resonance is observed as an isolated peak at 3.3 eV, the 5f and 6f peaks are hardly resolvable. This is an experimental imperfection, as mentioned above. An additional broad maximum with some indication of substructure is observed at energies below 0.9 eV and the corresponding first ATI-maximum at about 2.9 eV. A unique identification of excited states responsible for these maxima is not possible, if one assumes equal ac Stark shifts for all excited states. Because one may expect that more thightly bound states can have different shifts, possible candidates might be the 7p and 5d states. They both have several J-sublevels with a considerable energy spacing among each other which might be responsible for the observed, large width of the maximum. Also resonant channels leading to Xe^+ in its low lying excited $^2P_{1/2}$-state can contribute.

An additional weak series of peaks was observed at higher electron energies which seems not to be correlated with the ATI structure of the peaks discussed before. One possible interpretation for this series is, that it consists of fast electrons which traveled some distance in the laser focus during the laser pulse and changed their energy in the ponderomotive potential. This reminds of long pulse behaviour. A different origin of this series could be double ionization. In the ion-spectrum Xe^{2+} ions were detected with a signal strength of 10 % of the Xe^+ ion signal at a laser peak intensity of about 10^{14} Wcm^{-2}. The corresponding electrons can be expected to have higher kinetic energies because a two step ionization process for double ionization requires a minimum of at least eleven additional photons to be absorbed by Xe^+. Therefore it should occur at correspondingly higher intensities where higher order ATI processes become more efficient.

Electron energy spectra were measured at various wavelengths between 592 nm and 616 nm. The spacing between peaks of different ATI-order changed with the photon energy as expected. For the 4f peak at S = 0 the shift was smaller than the observed peak-width. This is in accordance with the model of equal intensity shift for excited states $E_n(I)$ and the ionization limit $IP_o(I)$ which predicts a dependence of the kinetic energy $E_k(n, S = 0)$ on photon energy:

$$E_k(n, S = 0) = \hbar\omega - (IP_o(I = 0) - E_n(I = 0))$$

A variation in the wavelength from 616 nm to 592 nm corresponds to a change of $\hbar\omega$ by 80 meV, which is less than the half-width of the 4f peak in figure 2). If the electron peak energies were correlated with the energy of the total number of 7 absorbed photons, $E_k(n,S=0)$ should shift by up to about 0.55 eV. Therefore the lack of an observable shift with wavelength is another evidence for the interpretation of the peak substructure.

1.2. Angular distributions of photoelectrons from MPI of xenon

Figure 3) shows the experimental angular distributions for electrons within the series of nf-peaks. The laser wavelength was 592 nm and the pulse peak intensity

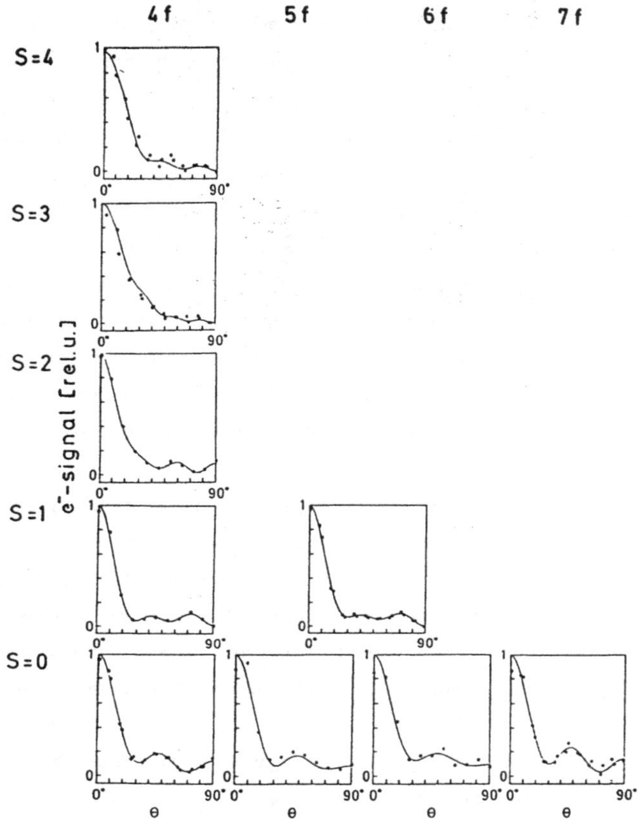

Figure 3: Angular distributions of photoelectrons from resonant ionization processes in xenon. S gives the number of excess photons absorbed in the continuum. The laser parameters are λ = 592 nm and $I = 8 \cdot 10^{13}$ Wcm^{-2}.

8×10^{13} Wcm^{-2}. Within the lowest ATI-order the angular distributions for resonant ionization via different nf states $4 \lesssim n \lesssim 7$ are very similar with a strong maximum in the polarization direction $\theta = 0^o$, minima at $\theta \approx 30^o$ and $\theta \approx 70^o$, and smaller maxima at $\theta \lesssim 50^o$ and $\theta = 90^o$. Visual inspection and the fit parameters from a fit of spherical harmonics to the data reveals a dominating g-wave (ℓ = 4) contribution. The maximum at $\theta = 90^o$ in the angular distributions for the S = 0 electron peaks shows clearly, that an odd number of 592 nm photons was necessary to produce the electrons in the S = 0 ATI group. The minimum odd photon number to ionize Xe is N_o = 7, if one takes into account the up-shift of the ionisation potential in the laser field. The dominating g-wave (ℓ = 4) contribution to the angular distributions for resonant ionization via bound states with principal quantum numbers n = 4,...,7 allows one to definitely fix the angular momentum of the resonant states. Starting at these states, which are shifted into 6-photon resonance, one photon suffices for ionization. One photon absorption in a linearly polarized laser beam changes the angular momentum from an initial value ℓ to ℓ-1 and ℓ+1. Thus the resonant states have angular momentum $\ell = 3$ or $\ell = 5$. For the states with principal quantum numbers n = 4,5 the angular momentum is then fixed to $\ell = 3$, while for n = 6,7 also $\ell = 5$ would be possible. Energetically nf (ℓ=3) and nh (ℓ=5) states should not be distinguishable in the electron spectra, because their quantum defects are very small ($\mu \sim 0.03$ for f-states with $n \gtrsim 6$). The propensity rule $\Delta\ell = +1$ for absorption in dipole approximation helps to specify the angular momentum for the resonant states n = 6,7. The dominating g-wave character of the corresponding electron groups together with the propensity rule identifies the resonant states as f-states (ℓ=3).

If we now consider the angular distributions of the electrons with increasing number S of excess photons (figure 3) one observes a change of the relative value at $\theta = 90^{\circ}$ from maximum to minimum and back with each additionally absorbed photon as expected. For the 4f resonance one can clearly observe the appearance of one additional maximum or minimum in the angle interval between $\theta = 0^{\circ}$ and $\theta = 90^{\circ}$ up to S = 2. This is typical for strong contributions of continuum states with successively higher angular momentum with increasing S. At S = 3 even the 4f contribution is no longer isolated in the experimental energy spectrum and the ondulations in the angular distributions are no longer resolvable which is partly also due to the finite experimental angular resolution of $\Delta\theta \approx \pm 3^{\circ}$.

Angular distributions of the other peak series observed in the energy spectrum as described above were also measured. The electrons in the maximum below 1 eV show a distribution with a maximum at $\theta = 0^{\circ}$ and no resolvable structure above $\theta \gtrsim 30^{\circ}$. The corresponding higher ATI-electrons have angular distributions similar to those of the corresponding 4f distributions. At the moment this does not allow a unique identification of the contributing intensity shifted excited states for this maximum in the energy spectrum.

Using a longer wavelength (λ = 616 nm) for the MPI process we have a strong indication that the main contribution to the resonant ionization peaks in the S = 0 electron group no longer comes from a 7-photon but from an 8-photon ionization process. The bound states with principal quantum numbers n = 4,5,6,7 are ac-Stark shifted into 7-photon resonance in regions of high pulse intensity in addition to 6-photon resonance at lower intensities. Between 593 nm and 616 nm there is a continuous change from 7- to 8-photon ionization contributing mainly to the S = 0 electron group. This change in the ionization process cannot be extracted from an electron energy spectrum alone, but also the angular distributions have to taken into account.

2. Atomic Hydrogen:

Figure 4 shows the electron energy spectrum of atomic hydrogen after multiphoton ionization at λ = 608 nm and a laser pulse peak intensity of 1.2 $\cdot$ 10^{14} Wcm^{-2}. Its overall structure is similar to that measured with xenon. ATI-structure with sequences of peaks at intervals of the photon energy $\hbar\omega$ can be distinguished and at low energies a substructure of narrow peaks is resolved. Below the peaks an unresolved background signal is observed which may partly originate from MPI of residual H_2

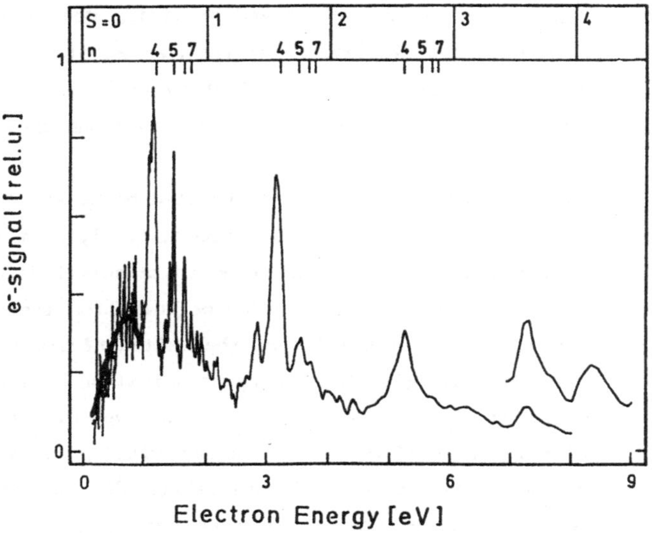

Figure 4: Energy spectrum of photoelectrons from MPI of atomic hydrogen at λ = 608 nm and I = 1.2 $\cdot$ 10^{14} Wcm^{-2}.

molecules. As for xenon the sharp resonance structure in the S = 0 electron group above 1 eV can be interpreted as resonant MPI with bound excited states ac-Stark shifted into resonance. Using the unperturbed hydrogen ionization threshold IP_o = 109678,77 cm^{-1} seven photons of the 608 nm radiation are sufficient to ionize the atom and produce photoelectrons with a kinetic energy of 0.68 eV. The assignment of principal quantum numbers to the resonances in S = 0 is possible, if one assumes not 7- but 8-photon ionization the dominant contribution to S = 0 with the hydrogen Rydberg states shifted into 7-photon resonance. The resonance positions for principal quantum numbers n = 4,5,6,7, indicated above the spectrum, were calculated assuming the shift for the bound states and the ionization threshold to be the same. For n = 5,6,7 the measured and calculated positions for the peaks in the spectrum are in agreement to within experimental accuracy.

The strong (n = 4)-peak shows two significant differences if it is compared with the 4f-peak for xenon: First, its energy position is slightly shifted from the "expected" value towards lower energies. This indicates that a higher resonance intensity is necessary than calculated in the simplest approximation described above. This is a departure from the first order assumption of equal intensity dependent ac-Starkshift of all excited states. Departures can be expected for excited states which are bound more tightly in the Coulomb field such that the quiver energy in the ac-field becomes comparable to the binding energy. Second, the (n = 4)-peak is significantly broader than the experimental energy resolution. Two effects can contribute to a larger peak width: a) An unresolved splitting due to two resonant ionization processes via the 4p and 4f sublevels. b) The (n = 4)-state can exhibit a larger broadening probably because of a higher ionization rate, and (or) a larger width of the "crossing" with the ground state 1s dressed with 7 photons.

In the case of atomic hydrogen the ac-Stark shift of excited bound states depends on the angular momentum of the state and it can deviate substantially from the shift of the ionization threshold. This behaviour has been found in calculations by Trahin at 608 nm in the low intensity limit where lowest order perturbation theory is applicable (6). This calculation shows that the shift of the exited $n\ell$-states approaches the shift of the ionization threshold with increasing principal quantum number n and for fixed $n \geq 4$ with increasing angular momentum ℓ (6). At the intensities, where the excited 4p and 4f states can be shifted into resonance with the 1s state dressed with 7 photons in our experiment, perturbation theory is probably not applicable to calculate the ac-Stark shift of these states. Thus we cannot decide whether the width of the n = 4 resonance in the electron spectrum is due to an unresolved resonance with both the 4p and 4f states ac-Stark shifted by different amounts. If case b) applies, here also the broadening should be largest for low principal quantum numbers. With increasing n the interaction between the $n\ell$ level and the 1s state dressed with 7 photons and the ionization broadening of the excited state should decrease because of a smaller coupling to the continuum and a lower intensity necessary to shift the excited state into 7 photon resonance.

Similar to the xenon spectrum an additional sequence of weak ATI-peaks can be seen at high electron energies which may again be attributed to electrons with "long" pulse behaviour which have travelled some distance in the focal volume during the laser pulse. Below 0.9 eV electron energy the spectrum shows a broad maximum. (The strong signal fluctuations in this energy region are of no significance, as mentioned above.) The relative signal height of this maximum increases with laser pulse peak intensity, indicating that it is caused by one or more excited states, which have a high resonance intensity. From the energy position of this maximum no resonantly excited state can be assigned to it, if one uses ac-Stark shifts of the excited states similar to the shift of the ionization threshold. First angular distributions of the photoelectrons also could not yet help to unambiguously identify the bound state leading to the resonance in the electron energy spectrum below 0.9 eV. We hope that current experimental improvements will give us cleaner and more detailed experimental results to interpret all the yet unidentified resonances in the spectrum.

Conclusions:

1. Angular distributions allow the identification of the angular momentum of excited states, which are shifted into resonance at well defined resonance intensities of the pulsed radiation field.
2. The electron energy spectra measured at different laser wavelengths and laser pulse peak-intensities confirm the model of intensity induced resonances, and exclude a connection of the peak energies with non-resonant processes in which the electron energy would be determined by the total energy of $N \cdot \hbar\omega$.
3. Within experimental resolution the states with quantum numbers $n \geq 5$ in xenon and hydrogen shift by the same amount given by the mean quiver energy of the free electron in the radiation field.

Our results indicate that for MPI of atoms at wavelengths, where intensities of the order of 10^{13-14} Wcm^{-2} are necessary, resonant ionization processes dominate, irrespective of the detuning of virtuell intermediate states from excited states of the atom. For high order MPI-processes with circularly polarized light the situation may be different because possible resonances can usually be found only in a small energy range below the ionization limit.

Acknowledgement:

We thank Dr. M. Trahin (6) for providing us with calculated values of ac-Stark shifts at the wavelengths we have used.
This work is supported by the DFG, Sonderforschungsbereich 216.

References:

1. A.T. Georges and P. Lambropoulos in "Advances in Electronics and Electron Physics", Vol. 54, p. 191 (1980) Academic Press
   J. Morellec, D. Normand and G. Petite in "Advances in Atomic and Molecular Physics", Vol. 18, p. 97 (1982)
   N.B. Delone and Krainov, " Atoms in Strong Light Fields", Springer Series in Chemical Physics (1985)
2. Y. Gontier and M. Trahin, Phys. Rev. A7, 1899 (1973)
3. B. Wolff, H. Rottke, D. Feldmann and K.H. Welge, Z. Phys. D -Atoms, Molecules and Clusters 10, 35 (1988)
   D. Feldmann, D. Petring, D. Otto and K.H. Welge, ibid. Vol. 6, 35 (1987)
4. R.R. Freeman, P.H. Buchsbaum, H. Milchberg, S. Darack, D.Schumacher and M.E. Geusic, Phys. Rev. Lett. 59, 1092 (1987)
5. K.C. Kulander, Phys. Rev. A 38, 778 (1988)
6. M. Trahin, private communication

MULTIPHOTON DETACHMENT FROM NEGATIVE IONS

Michèle Crance
Laboratoire Aimé Cotton, C.N.R.S II, Bât.505
91405 Orsay Cedex France

INTRODUCTION

Multiphoton ionisation of atoms has been widely studied both experimentally and theoretically [1-4] for a large variety of atoms, wavelengths and interaction time. The similar process of multiphoton detachment from negative ions has comparatively received very little attention. Negative ions have specific properties [5] which suggest some noticeable differences between multiphoton detachment and multiphoton ionisation. Negative ions have no bound excited states. Structures in the continuum appear for high energies, near excited states of the neutral. The ejected electron is submitted to a short range potential and can be considered, in many respects, as a free electron. This justifies the use of simplifying assumptions.

The first experiment in the field was two-photon detachment of negative iodine ion with a Ruby laser [6]. In order to interpret the latter experiment, an extensive calculation of two-photon detachment from negative halogen ions was carried out by Robinson and Geltman [7]. Then the subject of multiphoton detachment has been put aside for almost twenty years.

A new generation of experiments has begun recently involving two groups working on negative halogen ions, using either a beam [8,9] or trapped ions [10,11]. Another experiment is in progress at Los Alamos to study H^- with beams of relativistic velocity [12,13]. In relation with electron correlation effects in multiphoton ionisation, several theoretical groups became interested in two-photon detachment from H^- [14-18], and negative halogen ions [19,20].

Accurate methods which are tractable for two-photon detachment cannot be extended easily to a larger number of absorbed photons. On the contrary, an extensive use of the plane wave approximation allows one to obtain useful estimates of detachment cross sections for a large number of absorbed photons. It is then possible to compute quantities accessible to experiment such as angular distributions and probabilities for excess photon absorption. We shall first present the comparison between results obtained with various levels of approximation. Then extensive results obtained in the plane wave approximation will be given in some cases of experimental interest.

FROZEN CORE HARTREE FOCK APPROXIMATION

Calculations of multiphoton processes are rather difficult since they require the knowledge of the entire spectrum of the element considered. Actually, one often has to make drastic simplifications to be able to calculate transition probabilities when the number of absorbed photons exceeds two or three. The accurate methods developped for two-photon detachment from Cl^- [19] or I^- [20] would be difficult to extend to a large number of absorbed photons. We propose two types of symplifying assumptions, one for the ground state, the other one for excited states.

Ground states of many negative ions have been computed accurately. A good test for the quality of the wavefunctions is the value of the electron affinity [21]. It appears that a correct description must involve a large number of configurations and yields a rather complicated wavefunction. Such accurate representations would lead to untractable computations if taken as the basis for calculation of multiphoton transition probabilities. It is thus necessary to start with a reasonably simple representation of the ground state. For negative halogen ions, we shall use Hartree Fock wavefunctions. The part of the wavefunction which is left out, in such an approximation, plays a crucial role in the determination of the electron affinity but may be less important for the calculation of multiphoton detachment cross sections.

For clarity, we shall consider the specific case of negative fluorine ion. Hartree Fock wavefunction corresponds to the configuration $2p^6$ for outershell electrons. The wavefunction is greatly improved in a multiconfigurational Hartree Fock calculation (MCHF) by introduction of configurations such as $2p^4l^2$ (l=s, p, d) [22]. However, the weight of configuration $2p^6$ remains close to one. When we represent negative fluorine ion ground state by configuration $2p^6$ in the calculation of multiphoton detachment, we neglect quantum paths involving doubly excited states such as $2p^4ll'$ which can be described as continuum states above the detachment threshold leading to the formation of neutral fluorine in excited state $2p^4l$. Such intermediate states are rather high above the first detachment threshold and correspond to energy denominators, in multiphoton transition probabilities, several times larger than the electron affinity. It is likely that the quantum paths neglected when the ground state is represented by an Hartree Fock wavefunction do not contribute noticeably to a process leading to the ejection of an electron with an energy of the order of the electron affinity or smaller.

Hartree Fock wavefunction provides a good representation of the neutral ground state. For fluorine, it corresponds to the configuration $2p^5$ for outershell electrons. Obviously the "2p" monoelectronic wavefunctions appearing in neutral ground state and negative ion ground state are different, but rather close to each other. This can be measured by the overlap integral, A, between the neutral fluorine and the negative ion core: $A = \langle 1s^2\ 2s^2\ 2p^5 \mid 1s^2\ 2s^2\ 2p^5 \rangle$ (the bra contains monoelectronic wavefunctions of neutral fluorine and the ket contains monoelectronic wavefunctions of negative fluorine ion). By using the analytic Hartree Fock wavefunctions computed by Clementi and Roetti [23], we obtain $A^2 = 0.916$. This suggests to keep a same representation for neutral ground state and ionic core in all excited states taken into account in the calculation.

Starting from the above considerations, we have carried out a first set of calculations for 2-, 3-, 4-, 5-photon detachment from negative fluorine ion [24] and for 2-, 3-photon detachment from negative chlorine ion [25]. The light field is assumed to be monochromatic, linearly polarised. We use the complex dilatation method with a finite basis of square integrable functions. Excited states are described in the frozen core approximation and written as $|2p^5\ nl\ ^1L\rangle$. The radial part of the monoelectronic wavefunction nl is expanded on a basis of Slater type functions with complex exponent. In the dressed atom picture, the basis states are $|i\rangle|N\rangle$, $|i\rangle$ is an ionic state and $|N\rangle$ is a field state in number representation. One basis state, $|g\rangle|M\rangle$, is the ion ground state, g, dressed by M photons, the other states are excited states. We build an Hamiltonian matrix H_I+H_F+V, where H_I is the Hamiltonian of the bare ion, H_F is the Hamiltonian of the field, V is the light ion interaction restricted to dipole approximation in length form. Since basis states are complex, the Hamiltonian matrix is not hermitian, its eigenvalues are complex. The eigenvalues can be followed by continuity when the light intensity is varied. One of the eigenstates evolves from |g>|M> and can be interpreted as the ion ground state perturbed by the light field. The imaginary part of the corresponding energy is to be understood as one half of the detachment probability from the ground state. For weak intensity, this calculation is equivalent to a perturbative treatment applied at the lowest non vanishing order. This type of calculation will be referred to as Frozen Core Hartree Fock calculation (FCHF). The number of Slater integrals required increases when the number of absorbed photons increases and when the size of the ion increases. As a result, the computation time becomes exceedingly large, this is the reason why the method has been applied only for a few selected cases, for light halogens.

FROZEN CORE FREE ELECTRON APPROXIMATION

For all excited states of negative ions, the outer electron evolves in a short range potential. This is the basis of the plane wave approximation used in the calculation of detachment cross sections. This approximation is rather good when excited states involved in the calculation are close to the detachment threshold [5]. This is the case when multiphoton detachment is considered in the framework of perturbation theory : final states have a positive energy above the detachment threshold less than the electron affinity and it is reasonable to assume that the intermediate states which contribute most are in the same range of energy. Considering the outer electron as a free electron means the neglecting of any electrostatic interaction with the ionic core. We have carried out a second set of calculations (referred to as FCFE for frozen core free electron) using again the complex dilatation method with finite basis a square integrable functions for 2-, 3-, 4-, 5-photon detachment of negative halogen ions.

Ground state of neutral halogens has a fine structure, the splitting is respectively 0.00164 a.u. for fluorine, 0.004014 a.u. for chlorine, 0.01679 a.u. for bromine and 0.03464 a.u. for iodine. Since we have performed a non relativistic calculation, we take into account the fine structure only phenomenologically : a calculation of detachment probability is carried out for each threshold ($^2P_{3/2}$ and $^2P_{1/2}$), the total detachment probability is obtained as the sum of the previous results pondered by the relevant

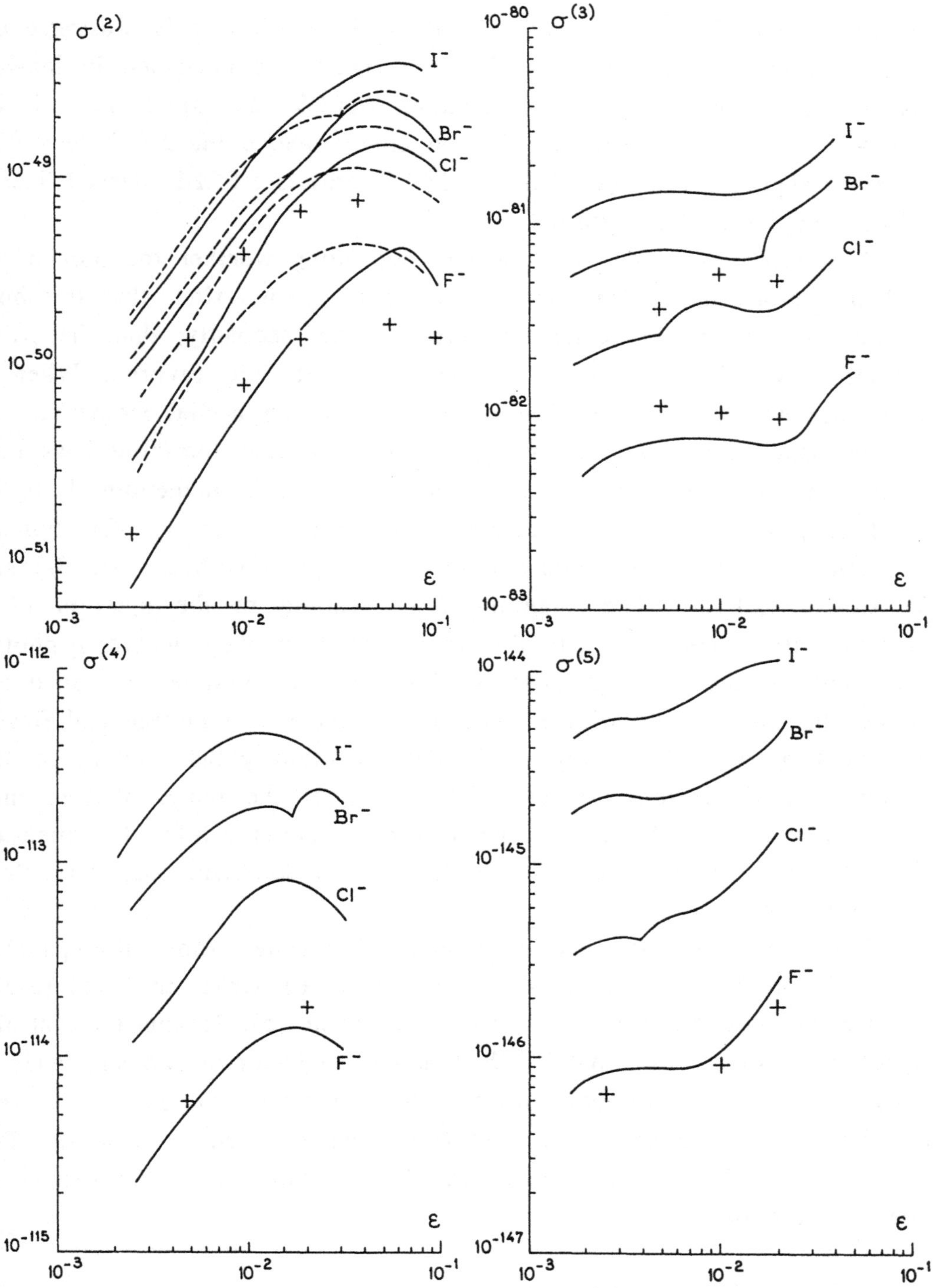

σ(2)
I⁻
Br⁻
Cl⁻
F⁻
10⁻⁴⁹
10⁻⁵⁰
10⁻⁵¹
ε
10⁻³
10⁻²
10⁻¹
σ(3)
10⁻⁸⁰
10⁻⁸¹
10⁻⁸²
10⁻⁸³
σ(4)
10⁻¹¹²
10⁻¹¹³
10⁻¹¹⁴
10⁻¹¹⁵
σ(5)
10⁻¹⁴⁴
10⁻¹⁴⁵
10⁻¹⁴⁶
10⁻¹⁴⁷

Fig. 1.

statistical weight. The result of FCHF and FCFE calculations is illustrated by Fig. 1. Because of threshold laws [26], the results are presented in log-log plots. q-photon detachment cross sections (q=2-5) are given in unit of $cm^{2q}s^{q-1}$ as a function of the electron energy above the J=3/2 threshold for various approximations (solid lines : FCFE; crosses : FCHF; dashed lines : results of Robinson and Geltman [7]).

Two types of curves are obtained depending whether the number of absorbed photons is odd or even. These results show, first, that threshold laws mainly govern the detachment process and, secondly, that the paths involving low orbital momenta in the final state are favored. When q photons are absorbed, an electron is ejected with an orbital momentum of parity opposite to q and at most equal to q+1. Wigner threshold laws [26] predict that the probability for an electron of orbital momentum k to be ejected is proportional to the electron energy at power (k+1/2). For an even number of absorbed photons q, the curves grow with a slope 3/2 and then slowly fall off. The quantum paths leading to the ejection of a p-electron are favored in all the range of frequency where q-photon detachment is the leading process. For an odd number of absorbed photons, the curves grow with a slope of 1/2, there is a plateau and finally the curves grow with a slope 5/2. The quantum paths leading to the ejection of a d-electron are favored in most of the range of frequency where q-photon detachment is the leading process while the quantum paths leading to the ejection of an s-electron are dominant only very close to the threshold.

The comparison between FCHF and FCFE shows some discrepancies, for two-photon detachment, the results of FCFE are larger than the results of FCHF as observed for one-photon detachment [5]. It appears that the discrepancy between FCFE and FCHF is at most a factor of two and tends to decrease when the number of absorbed photons increases. A few multiphoton detachment cross sections have been measured. For two-photon detachment, our results can be compared with the results of earlier calculations.

COMPARISON WITH PREVIOUS WORKS

For negative iodine ion, Hall et al [6] have obtained a cross section between 9 and 32 10^{-50} cm^4s. The error bar is large and contains all theoretical results obtained up to now. Experimental data obtained recently for 3-photon detachment are more accurate. For heavy halogens, experimental cross sections are weaker than FCFE predictions : for iodine, the cross section is measured to be between 2.2 and 4.9 10^{-82} cm^6s^2 for a prediction of 14 10^{-82} cm^6s^2; for bromine, the cross section is measured to be between 1. and 2.4 10^{-82} cm^6s^2 for a prediction of 5.9 10^{-82} cm^6s^2. For fluorine, FCFE prediction, 6.7 10^{-83} cm^6s^2, falls in the error bar of the two experiments : Blondel et al [9] measured a cross section between 4.3 and 8.7 10^{-83} cm^6s^2 while Kwon et al [11] measured a cross section between 4.3 and 14.5 10^{-83} cm^6s^2.

The first calculation of two-photon detachment was carried out by Robinson and Geltman [7]. They used a model potential involving adjustable parameters chosen to reproduce electron affinity and polarisability of the ground state. Near the threshold, they predict cross sections larger than the ones obtained in the FCFE approximation, the agreement is better for larger electron energy. The agreement is bad for fluorine but tends to improve for heavier halogens.

Using the Dalgarno method [27], Jiang and Starace [19] have calculated two-photon detachment from negative chlorine ion, in the frozen core approximation, starting with various representations of the ion ground state : Hartree Fock, multiconfigurational Hartree Fock, including or not interchannel interaction in excited states. The discrepancy between both Hartree Fock calculations is of the order of 20% at most. This can be explained as a lack of accuracy either in the use of finite basis or in the knowledge of asymptotic values of monoelectronic wavefunctions involved in the ground state. The use, in both calculations, of the length form for dipole operator, magnifies the possible small differences in monoelectronic wavefunctions for large radius while such differences have no effect on the determination of energy.

L'Huillier and Wendin have calculated two-photon detachment from iodine negative ion [20]. They start from Hartree Fock representation of the ion ground state but include screening effects in the framework of the random phase approximation. Their Hartree Fock calculation gives a cross section a little smaller than the one obtained in FCFE approximation. The result of screening effect is again a reduction of the cross section by a factor which varies from 0.7 near the threshold to 0.4 for an electron energy of 0.01 a.u..

Accuracy in the measurement of multiphoton detachment cross sections is determined by the same experimental problems as for multiphoton ionisation, since, in both types of experiments, the major source of error is the knowledge of the light beam characteristics (time and space distribution of intensity, statistical properties, absolute intensity measurements). In the present situation, it is not realistic to think that experimental data will allow one to discriminate between theoretical results. The comparison presented above shows that the plane wave approximation (FCFE) provides workable estimates of multiphoton detachment probabilities. It is thus interesting to use this method to make extensive calculations of measurable quantities in some cases of experimental interest. We present predictions for angular distributions of ejected electrons and probabilities for excess photon absorption.

ANGULAR DISTRIBUTIONS IN MULTIPHOTON DETACHMENT

The Dalgarno method consists of a direct determination of the perturbed wavefunction at any order and requires the resolution of a hierarchy of inhomogeneous differential equations. The initial state wavefunction provides the inhomogeous term in the first equation, each perturbed wavefunction provides the inhomogeneous term at the next step. In the plane wave approximation, the Dalgarno method becomes particularly simple if the final state is taken as the "source" wavefunction. Successive perturbed wavefunctions can be expressed analytically as a combination of free electron wavefunctions of various angular momentum involving an energy equal to the energy of the final state [28]. Typical results are presented in Figure 2.

The evolution of angular distributions is mainly governed by threshold laws as for total detachment cross sections. For frequencies just above detachment threshold, quantum paths leading to ejection of an electron with low angular momentum are dominant and angular distributions are typical of p- or s-electron depending on whether the number of absorbed photons is even or odd. For one-photon absorption, the anisotropy of angular distribution is characterised by a unique parameter which is positive for atoms and may be negative for negative ions. In other words, the interference between the partial waves in forward direction is constructive for atoms and may be destructive for negative ions. This is related to the l-dependent phase in asymptotic behavior for continuum wavefunctions. For electron energy much less than 1 a.u., this phase is close to $l\pi$ for an atom and $l\pi/2$ for a negative ion. For

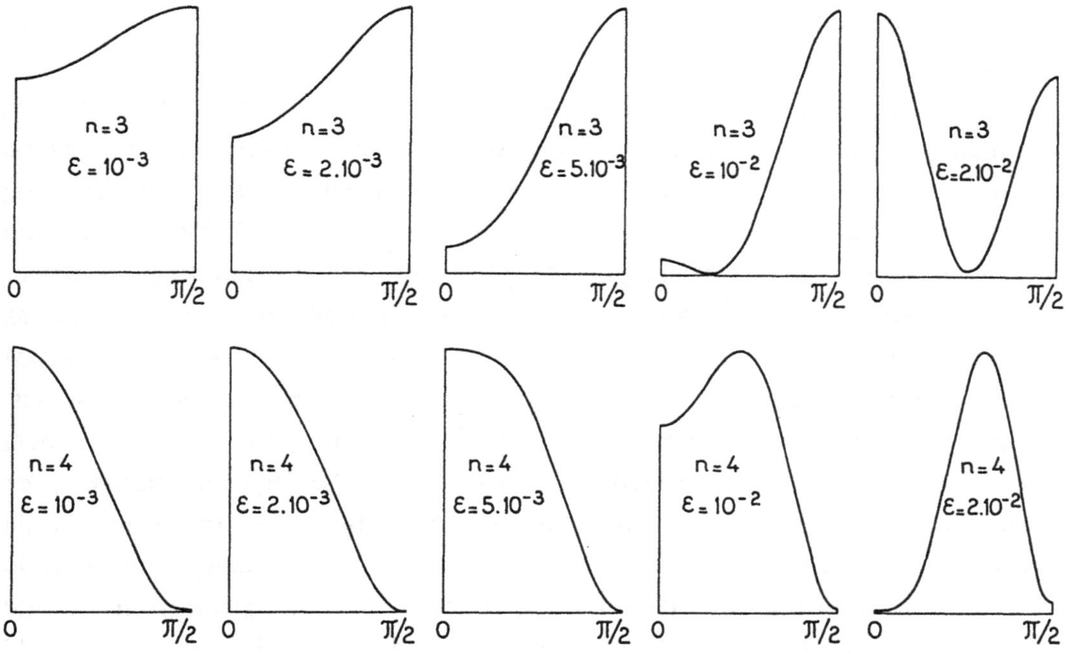

Fig. 2.

multiphoton detachment, the same features appear and destructive interference in forward direction is predicted between successive partial waves.

As an example, we give in Figure 2 the angular distributions predicted for negative iodine ion, in the FCFE approximation, for various frequencies. For each curve, the number of absorbed photons and the energy of the ejected electron are quoted. For 3-photon detachment, one observes the evolution from a typical s-wave, on the left, to a typical d-wave, on the right. For 4-photon detachment, one observes the evolution from a typical p-wave, on the left, to a completely destructive interference in forward direction between p-wave and f-wave, on the right. Interference features are extremely sensitive to the phase of continuum wavefunctions. Departure from free electron wavefunctions should be accurately detectable from angular distribution measurements and could provide valuable information on electron correlations for excited states of negative ions.

EXCESS PHOTON ABSORPTION

This process is well known in multiphoton ionisation and has been calculated for atoms quite early [29]. An atom may absorb one or several photons more than the minimum number required so that the electron energy spectrum consists of several peaks evenly spaced by one photon energy. In weak field, the magnitude of peaks varies according to power laws, the exponent being the net number of absorbed photons. This weak field regime is not easily observed for atoms and most experimental studies have concerned strong intensities delivered in short pulses. Extensive calculations of electron energy spectra for hydrogen have shown that no discontinuity is expected as a function of the light frequency when a threshold is passed (the ratio ionisation potential over frequency is an integer). The result is that the first peak is always dominant in weak field regime. For multiphoton detachment, excess photon absorption should occur. However, because of Wigner threshold laws, the process should present new features. The frequency being tuned across a threshold, the first peak will develop gradually and it should be possible to observe the first two peaks with comparable magnitudes in the weak field regime.

TRENDS IN MULTIPHOTON DETACHMENT

The use of plane wave approximation gives the possibility to obtain estimates of multiphoton detachment cross sections even when the number of absorbed photons reaches several tens. Negative halogens ions irradiated by a CO_2 laser would correspond to this situation. Varying the number of absorbed photons in a large range gives a way to investigate the mechanism of multiphoton detachment and allows a fruitful comparison with multiphoton ionisation.

Because of their difference in dimensionality, it is not possible to directly compare multiphoton absorption cross sections involving different numbers of absorbed photons. Once the interaction time is determined, the efficiency of a q-photon process can be characterised through the saturation intensity I(q). I(q) is the intensity for which the absorption probability is equal to one. In atoms, multiphoton ionisation cross sections present a resonance structure superimposed on a back ground. When a wide range of frequency is considered, it is possible to define an average cross section and thus an average saturation intensity. When the interaction time is taken as h/E (E is the ionisation potential), the saturation intensity I(q) hardly varies with q. It is of the order of $E^2/(h\pi\langle r\rangle^2)$ where $\langle r\rangle$ is the mean radius of the outershell electron orbitals [4]. This result suggests that the process of multiphoton ionisation is essentially governed by the absorbed energy : an electron is ejected from an atom if the light beam brings an amount of energy equal to the ionisation potential in a time short enough for the Heisenberg principle to be satisfied.

The same type of investigation can be carried out for negative ions. Because of the threshold laws, one has to keep constant the energy of the ejected electron. For negative fluorine ion, we have calculated saturation intensities for q-photon detachment, q varying from 4 to 30, for an ejected electron energy of 0.004 a.u. [28]. The trend seems rather different from the one obtained for atoms. I(q) is proportional to q^{-2}. This result suggests to interpret the saturation of multiphoton detachment in relation with the quiver motion that a light field imposes to a free electron. For a light intensity I, the quiver motion of a free electron is characterised by an oscillating velocity of amplitude v_Q proportional to the square root of the intensity and the inverse of the frequency. In the situation considered above, saturation of the detachment process occurs, whatever q is, when the quiver velocity amplitude v_Q becomes equal to the mean velocity of an outershell electron on its orbital.

CONCLUSION

We have carried out extensive calculations of multiphoton detachment from negative halogen ions. Comparison between various levels of approximation as well as confrontation with experimental data show that the plane wave approximation allows one to obtain workable estimates of cross sections and atomic quantities accessible to experiment. The results obtained can be used as a guide to choose interesting experimental situations. However, it is clear that the plane wave approximation is only a first step in the theoretical investigation of multiphoton detachment from negative ions and should be followed by more accurate studies including a treatment of electron correlations both in the ground state and in continuum states.

REFERENCES

1. Morellec J., Normand D., Petite G., Adv. At. Mol. Phys. 18 97 (1982)
2. L'Huillier A., Comments At. Mol. Phys. 18 289 (1986)
3. Lambropoulos P., Comments At. Mol. Phys. 20 199 (1987)
4. Crance M., Phys. Rep. 144 117 (1987)
5. Massey H., Negative Ions Cambridge University Press, Cambridge (1976)
6. Hall J. L., Robinson E. J. Branscomb L. M., Phys. Rev. Lett. 14 1013 (1965)
7. Robinson E. J., Geltman S., Phys. Rev. 153 4 (1967)
8. Blondel C., Champeau R. J., Crubellier A., Delsart C., Duong H. T., Marinescu D., Europhys. Lett. 4 1267 (1987)
9. Blondel C., Champeau R. J., Crance M., Crubellier A., Delsart C., Marinescu D., J. Phys. B to be published (1989)
10. Trainham R., Fletcher G. D., Larson D. J., J. Phys. B 20 L777 (1987)
11. Kwon N., Armstrong P. S., Olson T., Trainham R., Larson D. J. to be published
12. Bryant H. C., Dieterlé B. D., Donahue J., Sharifian H., Tootoonchi H., Wolfe D. M., Gram P. A. M., Yates-Williams M. A., Phys. Rev. Lett. 38 228 (1977)
13. Hamm M. E., Hamm R. W., Donahue J., Gram P. A. M., Pratt J. C., Yates M. A., Bolton R. D., Clark D. A., Bryant H. C., Frost C. A., Smith W. W., Phys. Rev. Lett. 43 1715 (1974)
14. Crance M., Aymar M., J. Phys. B 18 3529 (1985)
15. Fink M. G. J., Zoller P.,J. Phys. B 18 L373 (1985)
16. Arrighini G. P., Guidotti C., Durante N., Phys. Rev. A 35 1528 (1987)
17. Mercouris T., Nicolaides C. A., J. Phys. B 21 L285 (1988)
18. Shakeshaft R., Tang X., Phys. Rev. A 36 3193 (1988)
19. Jiang T. F., Starace A. F., Phys. Rev. A 36 2347 (1988)
20. L'Huillier A., Wendin G., J. Phys. B 21 L247 (1988)
21. Hotop H., Lineberger W. C., J. Phys. Chem. Ref. Data 14 731 (1985)
22. Froese C., The Hartree Fock Method for Atoms, Wiley, New York (1977)
23. Clementi E., Roetti C., At. Data Nucl. Data Tables 14 177 (1974)
24. Crance M., J. Phys. B 20 L411 (1987)
25. Crance M., J. Phys. B 20 6553 (1987)
26. Wigner E. P., Phys. Rev. 73 1002 (1948)
27. Dalgarno A., Lewis J. T., Proc. R. Soc. A 233 70 (1955)
28. Crance M., J. Phys. B 21 3559 (1988)
29. Zernik W., Klopfenstein R. W., J. Math. Phys. 6 262 (1965)

MULTIPHOTON PROCESSES IN INTENSE LASER FIELDS

R. M. Potvliege and Robin Shakeshaft
Physics Department, University of Southern California
Los Angeles, CA 90089-0484

I. INTRODUCTION

The subject of this paper is the theoretical description, within a time-independent framework, of multiphoton processes that occur when a one-particle atomic system is exposed to an intense radiation field. We pass to a time-independent framework by making the familiar Floquet ansatz [1,2]. We report below results of large scale nonperturbative calculations of partial and total ionization rates for high-order ionization of H(1s), based on solving the Floquet eigenvalue equations. We also present results for high-order harmonic generation.

In the next section we briefly address some of the formal and computational aspects of the theory. In section IIIA we show the quasienergy eigenvalue spectrum for hydrogen exposed to 1064 nm radiation, and we depict the behaviour of the eigenvalues with respect to the intensity of the light. The imaginary part of an eigenvalue is proportional to the total ionization rate from the corresponding dressed state. We find that at long wavelengths and moderate to high intensities, perturbation theory yields results for total ionization rates from the ground state that are orders of magnitude too large. Contrary to what perturbation theory implies, at long wavelengths an atom can experience the peak intensity of a fairly powerful pulse before undergoing ionization. We show the effect of intermediate resonances on total ionization rates, and we illustrate the resonance enhancement of the ionization yield for a realistic pulse. We present also a few results for differential ionization rates, again indicating the departure from perturbation theory. Finally, in section IIIB we report nonperturbative estimates of rates for harmonic generation by 1064 nm radiation of various intensities up to 2×10^{13} W/cm^2. These results exhibit a dependence on the order of the harmonic which is qualitatively different from that of perturbation theory results.

II. THEORY

We consider a monochromatic classical radiation field (frequency ω) which is spatially homogeneous over atomic dimensions. The interaction of the particle (mass μ, charge e)

with the field is

$$V(t) = -(e/\mu c)\mathbf{A}(t)\cdot\mathbf{p} \tag{1}$$

$$= V_+e^{-i\omega t} + V_-e^{i\omega t}, \tag{2}$$

where $\mathbf{p}$ is the canonical momentum operator and $\mathbf{A}(t)$ is the vector potential of the field :

$$\mathbf{A}(t) = \mathrm{Re}(\mathbf{A}_0e^{-i\omega t}). \tag{3}$$

The form (1) corresponds to a particular gauge which is reached from the Coulomb gauge by a transformation[3] with gauge function

$$\Lambda(t) = \frac{e}{2\mu c}\int^t \mathbf{A}^2(t')dt'. \tag{4}$$

We use this "velocity" gauge throughout. Note, however, that form (2) also holds in the "length" gauge (but not in the "acceleration" gauge).

The particle interacts with both the field and an atomic potential W, and its state vector $|\Psi(t)\rangle$ satisfies the time-dependent Schrödinger equation

$$i\hbar\frac{d}{dt}|\Psi(t)\rangle = H(t)|\Psi(t)\rangle, \tag{5}$$

where $H(t) = H_a + V(t)$ is the Hamiltonian, with $H_a = \mathbf{p}^2/2\mu + W$ the atomic hamiltonian. We make the Floquet ansatz

$$|\Psi(t)\rangle = e^{-iEt/\hbar}|F(t)\rangle, \tag{6}$$

where $F(t)$ is periodic with period $2\pi/\omega$. Substituting the Floquet ansatz into Eq. (5), and making the harmonic expansion

$$|F(t)\rangle = \sum_n e^{-in\omega t}|F_n\rangle, \tag{7}$$

we obtain the following coupled equations for the time-independent harmonic components $|F_n\rangle$:

$$(E + n\hbar\omega - H_a)|F_n\rangle = V_+|F_{n-1}\rangle + V_-|F_{n+1}\rangle. \tag{8}$$

We supplement these equations with the boundary conditions that in position space the $\langle\mathbf{x}|F_n\rangle$ are regular at $r \equiv |\mathbf{x}| \sim 0$ and have the following asymptotic form[4] for $r \sim \infty$:

$$\langle\mathbf{x}|F_n\rangle \sim C_ne^{ik_0\hat{\mathbf{u}}\cdot\mathbf{x}} + \sum_m f_{mn}(E,\hat{\mathbf{x}})e^{ik_mr}/r, \tag{9}$$

where $\hat{\mathbf{u}}$ is a unit vector and where

$$k_m = [(2\mu/\hbar^2)(E + m\hbar\omega)]^{1/2}. \tag{10}$$

The justification of (9), given more fully in Ref. 4, follows from noting that, if

$$\alpha(t) = -\frac{e}{\mu c}\int^t \mathbf{A}(t')dt', \tag{11}$$

$\exp[i\boldsymbol{\alpha}(t)\cdot\mathbf{p}/\hbar]|\Psi(t)\rangle$ would represent the state of a particle which at asymptotically large distances is free of both the atomic potential and the field ; thus, if the particle is incident in the presence of the field with mean momentum $\hbar k_0\hat{\mathbf{u}}$ we have, for $r\sim\infty$,

$$\langle\mathbf{x}|\exp[i\boldsymbol{\alpha}(t)\cdot\mathbf{p}/\hbar]|F(t)\rangle \sim Ce^{ik_0\hat{\mathbf{u}}\cdot\mathbf{x}} + \sum_m e^{-im\omega t} f_m(E,\hat{\mathbf{x}})e^{ik_m r}/r, \tag{12}$$

where C is the amplitude of the incident wave. It follows from Eqs. (7), (9) and (12) that

$$C_n = e^{in\chi_0(\hat{\mathbf{u}})}J_n[\rho_0(\hat{\mathbf{u}})]C, \tag{13}$$

$$f_{mn}(E,\hat{\mathbf{x}}) = e^{i(n-m)\chi_m(\hat{\mathbf{x}})}J_{n-m}[\rho_m(\hat{\mathbf{x}})]f_m(E,\hat{\mathbf{x}}), \tag{14}$$

where for any unit vector $\hat{\mathbf{n}}$, both $\chi_m(\hat{\mathbf{n}})$ and $\rho_m(\hat{\mathbf{n}})$ are real and are defined by

$$k_m\hat{\mathbf{n}}\cdot\boldsymbol{\alpha}(t) = \rho_m(\hat{\mathbf{n}})\sin[\omega t - \chi_m(\hat{\mathbf{n}})]. \tag{15}$$

If W has a Coulomb tail, the logarithmic phases must of course be included in (9) and (12). Note that $f_m(E,\hat{\mathbf{x}})$ is the amplitude for absorbing m real photons. There are infinitely many ways in which m photons may be absorbed, and $f_m(E,\hat{\mathbf{x}})$ is a coherent sum over the $f_{mn}(E,\hat{\mathbf{x}})$ with coefficients $\exp[i(m-n)\chi_m(\hat{\mathbf{x}})]J_{n-m}[\rho_m(\hat{\mathbf{x}})]$, as follows from Eq. (14) using the Graf addition formula. In solving Eqs. (8) we exploit the fact that the harmonic components $|F_n\rangle$ are coupled together only in tridiagonal form, a property which also holds in the length gauge[5,6], but not in the acceleration gauge[7–9].

Note that the boundary condition (9) holds only in the velocity gauge. In the length gauge, where the right-hand side of Eq. (1) is replaced by $-e\mathbf{F}(t)\cdot\mathbf{x}$ with $\mathbf{F}(t) = -(1/c)\partial\mathbf{A}(t)/\partial t$ the electric field, the boudary condition is quite different ; for example, if the field is circularly polarized, the right-hand side of Eq. (9) becomes[10]

$$\mathcal{U}_{0n}(\hat{\mathbf{u}},\mathbf{x})Ce^{ik_0\hat{\mathbf{u}}\cdot\mathbf{x}} + \sum_m \mathcal{U}_{mn}(\hat{\mathbf{x}},\mathbf{x})f_m(E,\hat{\mathbf{x}})e^{ik_m r}/r$$

where, if $a = (e/\hbar c)|\mathbf{A}_0|/\sqrt{2}$ and $\theta = \pm\arctan(y/x)$, with $\mathbf{x} = (x,y,z)$, and if the polarization plane is the $x-y$ plane, with the sign of θ fixed by the sign of the polarization, we have

$$\mathcal{U}_{mn}(\hat{\mathbf{n}},\mathbf{x}) = (2/\pi ar)^{1/2}e^{i(n-m)(\theta-\pi/2)} \\ \times\cos\{ar - \rho_m(\hat{\mathbf{n}})\sin[\chi_m(\hat{\mathbf{n}})-\theta] + (n-m)\pi/2 - \pi/4\}. \tag{16}$$

This boundary condition does not apply in the weak field limit, where $a\sim 0$, since we have assumed $ar\sim\infty$.

For $C\neq 0$, the asymptotic boundary condition (9) represents an electron which is incident from infinity with mean momentum $\hbar k_0\hat{\mathbf{u}}$ and which scatters from W, absorbing photons in the process. For $C=0$, this boundary condition represents an electron which is initially bound (or quasibound) to W, and which escapes by absorbing photons. Since the Schrödinger equation is linear, $f_m(E,\hat{\mathbf{x}})$ is proportional to C; therefore if at a particular value of the quasienergy E, say ϵ, there is a nontrivial solution with $C=0$, each amplitude $f_m(E,\hat{\mathbf{x}})$ must have a pole at $E=\epsilon$. In the absence of the inhomogeneous term in the asymptotic boundary condition (that is, $C=0$) the Floquet equations (8) become an eigenvalue system with complex eigenvalue E ; the eigensolution, at $E=\epsilon$, corresponds to a resonance, which may originate as a bound state or a nonradiative resonance in the

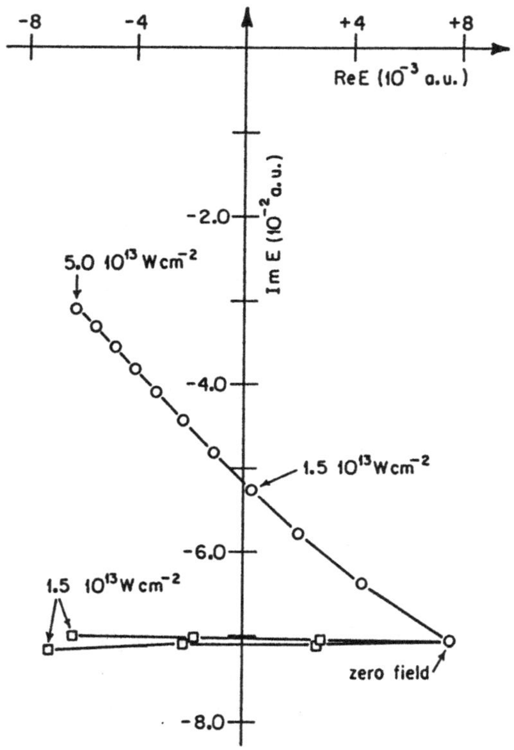

Fig. 1. Trajectories in the complex E plane of an s-wave dominant resonance pole (open circle) and two shadow poles (open squares) of $f_0(E, \hat{\mathbf{x}})$ vs laser intensity. The "atomic" potential is given by $W(r) = \beta_1 \exp(-\lambda_1 r) + \beta_2 \exp(-\lambda_2 r)$ with parameters (in a.u.) $\beta_1 = -1.85$, $\lambda_1 = 2.0$, $\beta_2 = 0.16$, $\lambda_1 = 0.30$. The results were calculated at intensity intervals of 0.5×10^{13} W/cm^2; we draw lines through the calculated points to guide the eye.

zero-field limit. Since k_m is a two-valued function of E, the amplitudes $f_m(E, \hat{\mathbf{x}})$ must be defined on a Riemann surface consisting of infinitely many sheets, with branch point singularities at $j\hbar\omega$, all integer j, and overlapping branch cuts extending along the real axis to infinity. We may define the physical sheet as the one on which $0 \leq \arg(k_j) < \pi$, all integer j ; the physical energy axis is the real axis on the physical sheet. The "dominant" poles of the amplitudes (that is, the poles that are physically most significant) are those that lie near to the physical energy axis. However, to each dominant pole there are associated infinitely many "shadow" poles lying farther from the physical energy axis, on other sheets. As the intensity varies, the poles move, and when a branch point threshold is passed, a shadow pole may change roles with a dominant pole[4]. As an example of the movement of poles (which does not, however, illustrate an interchange of roles) we show in Fig. 1 the trajectories, versus the intensity of 532 nm linearly polarized light, of a dominant pole and the shadow poles of $f_0(E, \hat{\mathbf{x}})$, for the case where these poles correspond to a resonance which begins as an s-wave resonance of a particle interacting with a superposition of two Yukawa potentials. In this example the dominant (resonance) pole shifts below the ionization threshold and, rather than becoming a "bound state" pole, moves further away from the physical energy axis.

The differential rate, denoted as $dR_N/d\Omega$, for a particle to absorb $N \geq N_0$ photons and emerge with mean speed $v = \hbar k_N/\mu$ into the solid angle $d\Omega$ along the direction $\hat{\mathbf{k}}_N$ is, with $|F(t)\rangle$ properly normalized,

$$dR_N/d\Omega = v_N |f_N(E, \hat{\mathbf{k}}_N)|^2. \tag{17}$$

Here N_0 is the smallest integer n for which $E + n\hbar\omega > 0$. If the particle is initially bound in a nondegenerate state we take E to be the real energy that is closest to the complex energy ϵ of the dominant pole which develops from the zero-field limit of the bound-state pole E_b of $f_0(E, \hat{\mathbf{x}})$: the particle starts out on the quasienergy eigenvalue curve corresponding to the dressed state that approaches the initial state in the limit of vanishing intensity, and remains adiabatically on this curve if the intensity is varied slowly. (However, the probability that the particle jumps to another curve may be large at "intermediate resonances", that is when the real parts of the quasienergies of two different dressed states differ by an integral multiple of $\hbar\omega$. This process should be described in a time-dependent framework[11].) In general $\epsilon = E_b + \Delta - i(\Gamma/2)$, where Δ is the shift and Γ is the (positive) width ; we put $E = E_b + \Delta$ in Eq. (17). Putting $E = \epsilon$ in Eq. (6) we see that $|\langle \mathbf{x}|\Psi(t)\rangle|^2$ decays with t as $\exp(-\Gamma t/\hbar)$; thus $\Gamma/\hbar$ is the total decay (ionization) rate of the initially bound atomic system, and conservation of current implies that

$$\Gamma/\hbar = \sum_N \int \frac{dR_N}{d\Omega} d\Omega. \tag{18}$$

Let $|\mathbf{k}\rangle$ denote an eigenstate of $\mathbf{p}$ with eigenvalue $\hbar\mathbf{k}$, normalized so that

$$\langle \mathbf{k}'|\mathbf{k}\rangle = a_k^2 \delta^3(\mathbf{k}' - \mathbf{k}).$$

Let $|\Phi_{\mathbf{k}}^-\rangle$ denote a scattering state eigenvector of H_a, such that $\langle \mathbf{x}|\Phi_{\mathbf{k}}^-\rangle$ approaches, for $r \sim \infty$, $\langle \mathbf{x}|\mathbf{k}\rangle$ plus an ingoing scattered wave. We showed previously[4] that, for $N \neq 0$, the N-photon absorption amplitude can be expressed in the two alternative forms :

$$-(2\pi)^{-3/2}(\mu/2\pi\hbar^2 a_{k_N})^{-1} f_N(E, \hat{\mathbf{k}}_N)$$

$$= [(\langle \Phi_{\mathbf{k}_N}^-|V_-|F_{N+1}\rangle + \langle \Phi_{\mathbf{k}_N}^-|V_+|F_{N-1}\rangle)]/J_0[\rho_N(\hat{\mathbf{k}}_N)] \tag{19}$$

$$= a_{k_N} \sum_n e^{i(N-n)\chi_N(\hat{\mathbf{k}}_N)} J_{n-N}[\rho_N(\hat{\mathbf{k}}_N)] \langle \mathbf{k}_N|W|F_n\rangle. \tag{20}$$

There remains the question of normalization of $|F(t)\rangle$. For a scattering process, the normalization is fixed by the amplitude C of the incident wave. For an ionization process, the normalization may be fixed by the current conservation condition (18), or independently by requiring either that [11,12]

$$\frac{\omega}{2\pi} \int_0^{2\pi/\omega} dt |\langle \Phi_b(t)|F(t)\rangle|^2 = 1, \tag{21}$$

or that[10]

$$\frac{\omega}{2\pi} \int_0^{2\pi/\omega} dt \{\langle F(-t+\pi)|T^\dagger\}|F(t)\rangle = 1, \tag{22}$$

where $|\Phi_b(t)\rangle$ represents the bound state, in the velocity gauge, immediately after the field has been turned on, and where T denotes the time-reversal operator. (If the initial

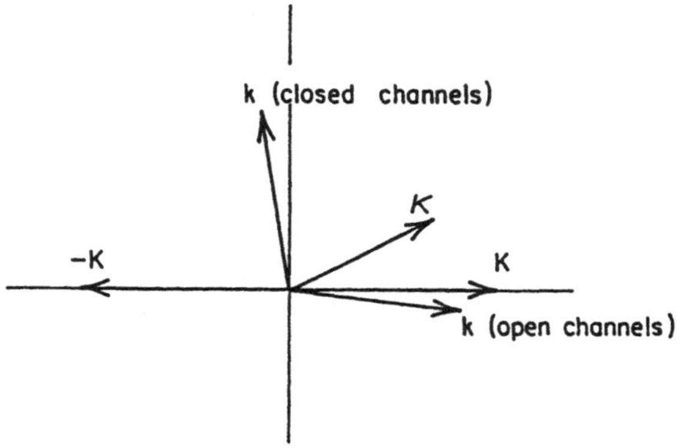

Fig. 2. Complex wave numbers k, K, and κ appearing in the integrand of the ionization matrix element.

bound state is not a s-state, or if the field is not linearly polarized, Eq. (22) must be modified slightly – see Ref. 10)

In the next section we report results for ionization of H(1s), and also harmonic generation. In this context, the electron is initally bound ; there is no inhomogeneous term in the asymptotic boundary condition. Thus $\langle \mathbf{x}|F(t)\rangle$ is a superposition of outgoing waves $\exp(ikr)$ where $k = k_m$ is complex ; k_m lies either in the upper left or the lower right quadrant of the k plane. The dominant poles are such that, for closed channels, $m < N_0$ and $\mathrm{Im}(k_m) > 0$, while for open channels $m \geq N_0$ and $\mathrm{Im}(k_m) < 0$, as illustrated in Fig. 2. To solve the Floquet eigenvalue equations we expand the harmonic components on a basis consisting of spherical harmonics $Y_{lm}(\hat{\mathbf{x}})$ and radial Sturmian functions $S^{\kappa}_{nl}(r)$:

$$S^{\kappa}_{nl}(r) = A_{nl}\,(-i\kappa r)^{l+1} e^{i\kappa r}\,{}_1F_1(l+1-n, 2l+2, -2i\kappa r), \tag{23}$$

where κ is complex and where A_{nl} is independent of κ and r. The expansion coefficients satisfy the set of homogeneous linear equations that results from the Floquet equations using the orthonormality of the $Y^*_{lm}(\hat{\mathbf{x}})$ and the orthonormality equation

$$\int_0^{\infty} dr\, S^{\kappa}_{ml}(r)(1/r)S^{\kappa}_{nl}(r) = \delta_{mn}. \tag{24}$$

These linear equations are solved using the method of inverse iteration[5]. We choose κ to lie in the upper right quadrant of the k-plane (see Fig. 2) ; this choice is fixed by noting that the expansion of a function behaving as $\exp(ikr)$ converges, in the sense that the expansion coefficients vanish as the quantum numbers of the basis functions increase, provided that[13,14] the argument of κ differs by less than $\pi/2$ from the argument of k. However, in evaluating $f_N(E, \hat{\mathbf{k}}_N)$ we project onto $|\Phi^-_{\mathbf{K}}\rangle$, where $\mathbf{K}$ is $\mathbf{k}_N$ with $E = E_b + \Delta$; but $\langle \mathbf{x}|\Phi^-_{\mathbf{K}}\rangle$ is a standing wave, comprised of both ingoing and outgoing waves, so that the resulting expansion of $f_N(E, \hat{\mathbf{k}}_N)$ does not converge. We obtain a finite result by using Padé summation[15] ; the convergence depends in part on how well the asymptotic boundary condition (9) is satisfied – this boundary condition cannot be exactly satisfied once the harmonic expansion of $|F(t)\rangle$ is truncated. The basis set (23) allows us to locate only those poles for which k_m lies in the lower right (upper left) quadrant of the k plane

Fig. 3 (next page). Real parts of various eigenvalues ϵ_j for hydrogen irradiated by linearly polarized light of wavelength 1064 nm. We show $\text{Re}(\epsilon_j) - N_j\hbar\omega$, where the integer index j labels the following atomic configurations (or superpositions of configurations), where known, in the weak intensity limit. (1-2): superpositions of 4s and 4d, $N_{1-2} = 10$; (3): 2s , $N_3 = 8$; (4): unknown ; (5): superposition of 8p, 8f, 8h, and 8j, $N_5 = 11$; (6-7): superpositions of 7p, 7f, and 7h, $N_{6-7} = 11$; (8-9): superpositions of 3s and 3d, $N_{8-9} = 10$; (10-12): superpositions of 6p, 6f, and 6h, $N_{10-12} = 11$; (13-14): superpositions of 5p and 5f, $N_{13-14} = 11$; (15): 1s, $N_{15} = 0$; (16-17): superpositions of 4p and 4f, $N_{16-17} = 11$; (18): 2p, $N_{18} = 9$; (19): unknown ; (20): superposition of 7s, 7d, 7g, and 7i, $N_{20} = 12$; (21): 3p, $N_{21} = 11$; (22-24): superpositions of 6s, 6d, and 6g, $N_{22-24} = 12$; (25-27): superpositions of 5s, 5d, and 5g, $N_{25-27} = 12$. We did not include states with orbital angular momentum quantum number greater than 7. The bold horizontal lines indicate multiphoton ionization thresholds.

when $m \geq m_0$ ($m < m_0$), where m_0 is an integer[4]. These are the poles which can be reached from the physical energy axis by a path which crosses the axis at most once, and therefore these are the poles of greatest physical interest ; unfortunately, it is usually very difficult, in practice, to observe shadow poles besides the dominant ones. Finally, we note that, for weak fields, when perturbation theory is valid, Eqs. (8) simplify enormously on the complex Sturmian basis set – this allows to calculate partial rates for high-order ionization[16], harmonic generation[17], and Raman scattering[17] from hydrogen at short and long wavelengths.

III RESULTS

IIIA Multiphoton Ionization

In Fig. 3 we show the real parts of twenty-seven different eigenvalue curves of the Floquet equations (8), versus intensity I, for linearly polarized light of wavelength 1064 nm. All of the eigenvalues shown correspond to states wich have parity $(-1)^{N_j}$ and which, for $I \sim 0$, are detuned from an N_j-photon resonance with the 1s state by no more than $\hbar\omega$ (the subscript j labels the different curves, in the order in which they appear in Fig. 3). We display $\text{Re}(\epsilon_j) - N_j\hbar\omega$ rather than ϵ_j. (Note that if ϵ_j is an eigenvalue, so is $\epsilon_j + m\hbar\omega$ for any integer m. We specify ϵ_j uniquely by requiring that $\epsilon_j \rightarrow E_j$, and hence $|F_{jm}\rangle \rightarrow |\phi_j\rangle\delta_{m0}$, as $I \rightarrow 0$.) However, when no confusion might arise we refer to $\epsilon_j - N_j\hbar\omega$ simply as ϵ_j, and we do not distinguish between ϵ_j and $\text{Re}(\epsilon_j)$. Thus, for example, ϵ_{18}, the eigenvalue corresponding to the 2p state when $I \sim 0$, is displayed as $\epsilon_{18} - 9\hbar\omega$ and appears below ϵ_{15}, the eigenvalue corresponding to the 1s state ($N_{15} = 0$). In the zero-field limit the eigenvectors for atomic hydrogen are, in general, superpositions of atomic states with the same principal quantum number and parity, but different orbital angular momentum quantum numbers. We show all of those eigenvalues corresponding to atomic states with principal quantum number ≤ 6 when $I \sim 0$, and a few more besides. We represent only the dominant poles in Fig. 3. They lie on three different sheets of the Riemann surface we introduced in Section II, depending whether $\text{Re}(\epsilon) > -11\hbar\omega$, $-11\hbar\omega > \text{Re}(\epsilon) > -12\hbar\omega$, or $\text{Re}(\epsilon) < -12\hbar\omega$. We comment on these results only briefly here ; they are discussed in more detail in Ref. 10.

First of all, we remark that the real parts of the eigenvalues plotted in Fig. 3 show multiple crossings, both real and avoided, as the intensity varies. These crossings, which

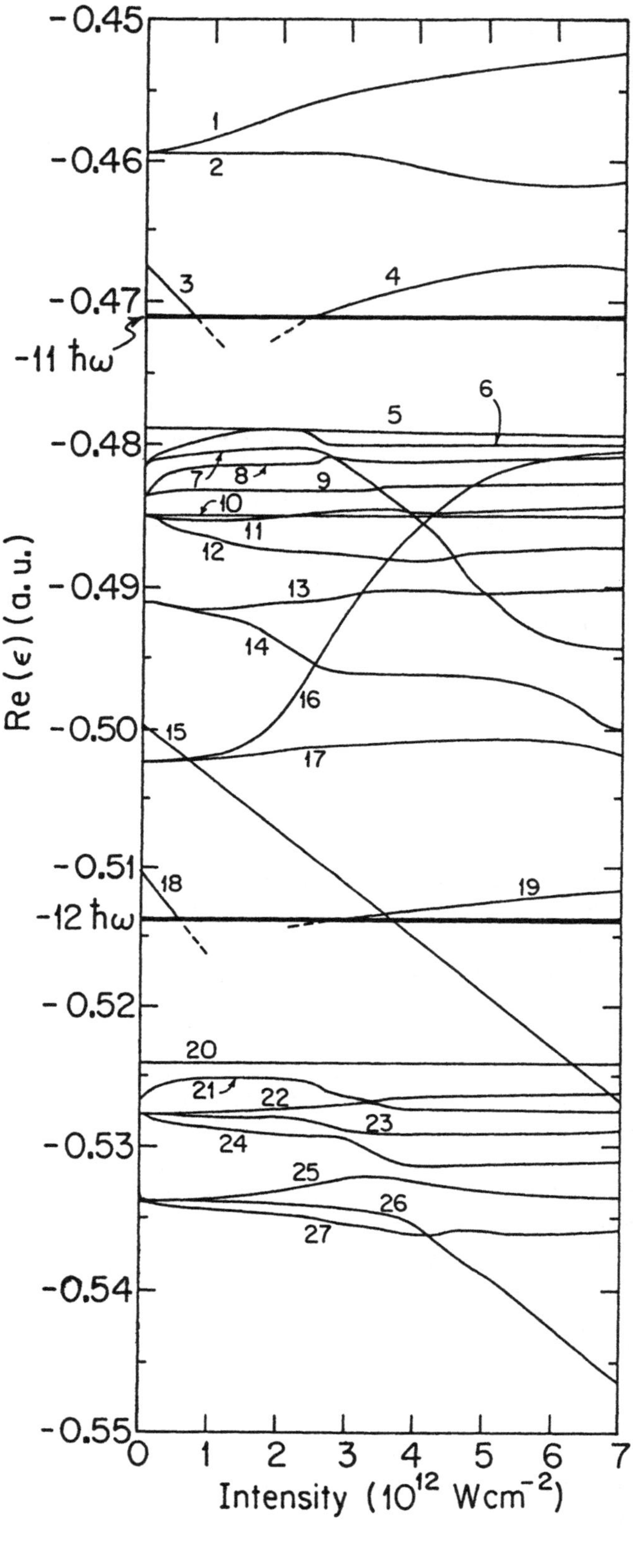

-0.45
-0.46
-0.47
-0.48
-0.49
-0.50
-0.51
-0.52
-0.53
-0.54
-0.55
-11 ħω
-12 ħω
Re (ε) (a.u.)
0
1
2
3
4
5
6
7
Intensity (10^12 Wcm^-2)
1
2
3
4
5
6
7
8
9
10
11
12
13
14
15
16
17
18
19
20
21
22
23
24
25
26
27

are associated with intermediate resonances, have been studied[10,18] under the assumption that the field is sufficiently weak that, at resonance, the atom can be viewed as consisting of just two bound states, p and q, with "uncoupled" widths γ_p and γ_q. The real parts of the quasienergies exhibit a true crossing when $|\gamma_p - \gamma_q| \ll 2\hbar\Omega_0$ (where Ω_0 is the Rabi frequency), but an avoided crossing with a gap of magnitude $\hbar\Omega_0$ when $|\gamma_p - \gamma_q| \gg 2\hbar\Omega_0$. It is worth noting that true crossings do not always lead to enhancements in the photoionization rate. For example, there is a 3-photon resonance between the 1s and the 2p states for linearly polarized 355 nm incident radiation, at $I \simeq 1.15 \times 10^{13}$ W/cm^2, associated with a real crossing in the real parts of the corresponding quasienergy curves. The rate for photoionization of the ground state, as a function of the intensity for a fixed wavelength, is *not* enhanced at this resonance (however, this rate is enhanced when the intensity is fixed but the wavelength varies through the resonance).

We calculated the quasienergies within the "velocity" gauge defined by Eq. (1). The continuum threshold does not shift in this gauge. Thus we might expect the shifts of the high Rydberg states to be small. This is indeed the case for many of these states ; see, for example, ϵ_5, ϵ_{8-11}, ϵ_{17}, ϵ_{20}, and ϵ_{22}, which stay rather flat as I varies. However, some of the Rydberg states undergo large shifts, often because of strong coupling to the 2s or 2p states.

For example, as I increases, ϵ_3, the eigenvalue originating from the 2s level, shifts downwards by roughly the ponderomotive energy P ($P = 2\pi e^2 I/\mu c\omega^2$ is proportional to the intensity) and passes below a multiphoton ionization threshold at about 0.7×10^{12} W/cm^2 ; the number of 1064 nm photons required to ionize an atom which is initially in the 2s state increases from 3 to 4 as this threshold is passed. At intensities just above and just below this threshold intensity, ϵ_3 corresponds to different poles, on different sheets of the Riemann surface. Not far below the threshold, ϵ_3 undergoes an avoided crossing with an eigenvalue curve originating from a very high Rydberg level. We do not know which Rydberg level this is ; in fact, we cannot precisely calculate the eigenvalues just below a multiphoton ionization threshold because an infinite number of Rydberg states accumulate there and give rise to a myriad of true and avoided crossings. However, when two eigenvalue curves undergo an avoided crossing, the characters of the corresponding Floquet states interchange. Thus, at the first avoided crossing encountered by ϵ_3, the other curve (which originates from a very high Rydberg level) acquires 2s character, and at intensities above the crossing intensity this second curve shifts downwards until, almost immediately, it encouters an avoided crossing with another curve originating from a high Rydberg level, and so on. In Fig. 3, the eigenvalue ϵ_6, which originates from a level with principal quantum number 7, has a noticeable bend at $I = 2.3 \times 10^{12}$ W/cm^2 which in fact signifies almost simultaneous avoided crossings with a curve originating from a higher Rydberg level, and with ϵ_7, another curve originating from a level with principal quantum number 7. The 2s character acquired by ϵ_6 is almost immediately transferred to ϵ_7 as I increases, and ϵ_7 shifts downwards without interruption as it undergoes true crossings with ϵ_{8-13}, until the avoided crossing with ϵ_{14}. One can easily trace, through the avoided crossings, the diabatic curve which at each intensity (far from a crossing) corresponds to the Floquet state which has predominantly 2s character. This diabatic curve is more or less the straight line $E_{2s} - 8\hbar\omega - P$. Similarly, the diabatic curve which corresponds to the Floquet state that is predominantly 2p in character starts as ϵ_{18}, but after passing a threshold and many avoided crossings becomes ϵ_{26} ; it is approximately $E_{2p} - 9\hbar\omega - P$.

The curve 15 in Fig. 3 corresponds to the 1s state when $I \sim 0$. We see that Re(ϵ_{15}), which is approximately $E_{1s} - P$, exhibits numerous true crossings, but no avoided crossings with other eigenvalue curves. (Of course it is not possible to distinguish a true crossing from an avoided crossing with a very small gap.) This is because at resonances between states 15 and j the difference of the "uncoupled" widths γ_{15} and γ_j is larger than $\hbar\Omega_0$. Since a large number of 1064 nm photons are required to either ionize or (on resonance) excite H(1s), the width γ_{15} and (on resonance) the Rabi frequency Ω_0 are both very small, while since few photons are required to ionize from an excited state j the width γ_j is relatively large, at least if j is not too highly excited. Hence an electron which starts in the 1s state will remain in the same Floquet state, whose eigenvalue is ϵ_{15} ; the coupling to other Floquet states, at true crossings, is small, and the electron will simply jump across the gap of an avoided crossing if this gap is very small. Therefore, ionization from H(1s) at long wavelengths (eg. 1064 nm), by a pulse whose intensity envelope $I(t)$ varies slowly on the time scale of a cycle, can be described by the width of a single eigenvalue. The total ionization probability is $1 - \exp(-\int dt\, \Gamma_i/\hbar)$, where the width Γ_i (with i=15) depends implicitly on t through $I(t)$, and where the integration limits are $\pm\infty$.

In Fig. 4 we show $\Gamma_{15}/\hbar$ versus I over the range $5 \times 10^{12} \leq I \leq 3 \times 10^{13}$ W/cm^2. The resonance peaks of Γ_{15} correspond to the (true) crossings of Re(ϵ_{15}) with other eigenvalue curves. The first group of resonances is associated with the group of crossings of Re(ϵ_{15}) with Re(ϵ_j) (with j=20-25, 27, and 1) in the intensity range $6 - 10 \times 10^{12}$ W/cm^2. These resonances involve intermediate 11- or 12-photon transitions from the 1s state to highly excited states. The middle group of resonances involves intermediate 12- or 13-photon transitions from the 1s state to excited states occuring in the intensity range $1.5 - 2.0 \times 10^{13}$ W/cm^2 (not covered in Fig. 3) where Re(ϵ_{15}) crosses the group of curves Re(ϵ_{5-14}) and Re(ϵ_{16-17}) when this group is displaced downwards by $2\hbar\omega$. The third group of resonances, occuring at still larger intensities, is a replication, though an imperfect one, of the first group ; the energy difference between ϵ_{15} and the ϵ_j of the first group has shifted by $2\hbar\omega$, and consequently N_j has increased by two so that the resonances now involve intermediate 13- or 14-photon transitions to the same states as in the first group. Note that although Re(ϵ_{15}) crosses many eigenvalue curves immediately below a multiphoton ionization threshold, the rate of excitation to these very high Rydberg states is very small, and so Γ_{15} exhibits little or no enhancement just below a threshold.

In Fig. 5 we show the total ionization yield for H(1s) irradiated by a realistic pulse whose intensity distribution is (ρ and z are cylindrical coordinates)

$$I(\rho, z, t) = I_0[R^2/r^2(z)]\exp[-2\rho^2/r(z)^2]\ \exp(-t^2/t_p^2) \tag{25}$$

where $t_p = 200$ psec, where $I_0 = 2.5 \times 10^{13}$ W/cm^2, where $R = 20\ \mu$m is the spot size at the focus, and where, with the wavelength $\lambda = 1064$ nm, $[r(z)]^2 = R^2[1 + (\lambda z/\pi R^2)^2]$. The figure represents the yield in ions as a function of the intensity at which they are produced, calculated from the nonperturbative rates of Fig. 4. Saturation effects are not apparent for pulses of this duration (they manifest themselves for $t_p > 1$ ns). The resonance peaks are very prominent in the yield, whether or not we average over the spatial dimensions of the pulse. The contribution of these peaks, between $I = 1.55 \times 10^{13}$ and $I = 2.0 \times 10^{13}$ W/cm^2, is about 30 % of the total yield produced by the pulse. Since we have calculated the total yield, rather than the partial N-photon ionization yields, we cannot confirm the substructures seen in the peaks of the emergent electron energy distribution by Freeman et al[19]. However, our results do support the claim that levels shifting in and out of resonance strongly enhance the ionization yield.

We also show in Fig. 4 the partial ionization rates, calculated within lowest order perturbation theory, for $(12 + S)$-photon ionization, with $S = 0$ and 1. Clearly, these

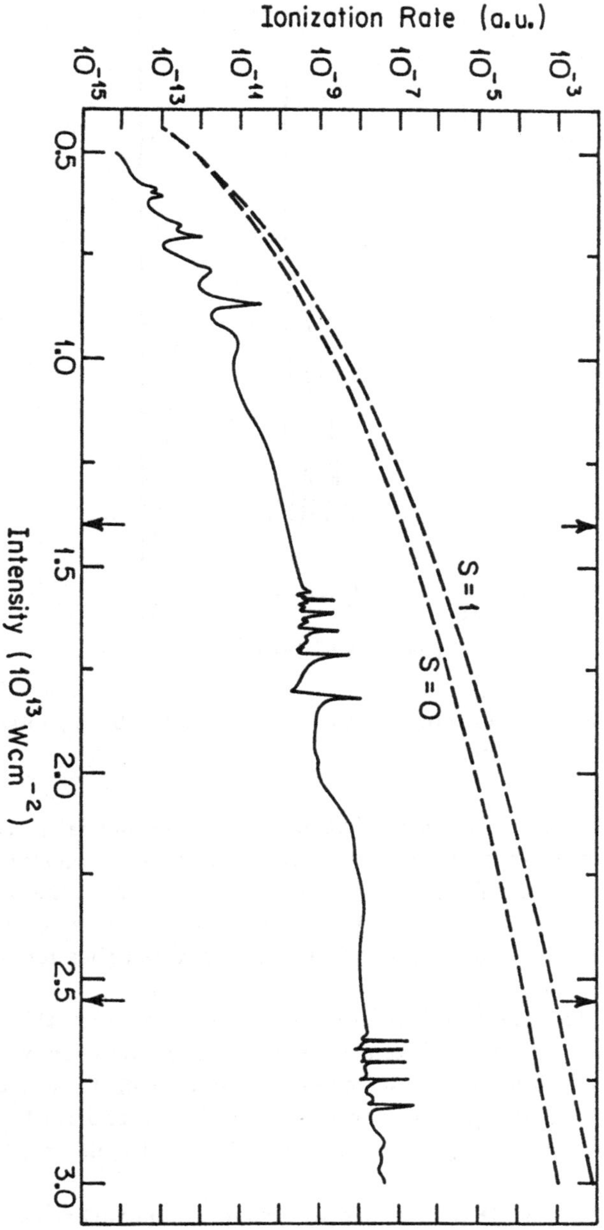

Fig. 4. Total ionization rate vs intensity for H(1s) irradiated by linearly polarized light of wavelength 1064 nm. Dashed curves are partial rates for (12+S)-photon ionization obtained within lowest order perturbation theory. The arrows indicate the intensities at the thirteen- and fourteen-photon ionization thresholds.

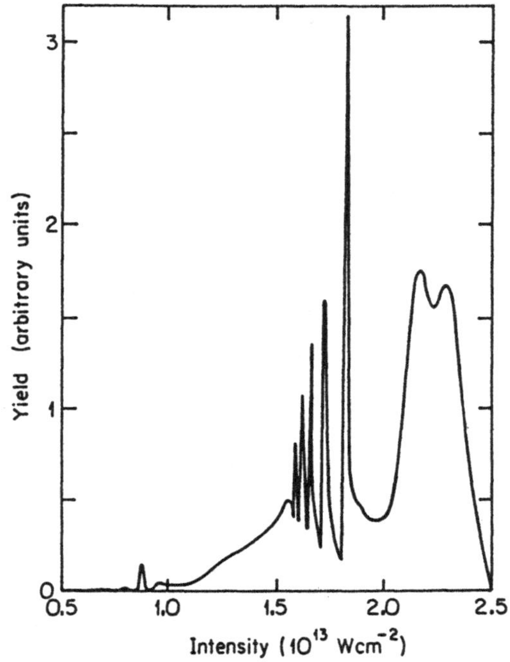

Fig. 5. Total ionization yield vs intensity for H(1s) irradiated by a gaussian pulse of peak intensity 2.5×10^{13} W/cm^2. See text.

rates are grossly overestimated by perturbation theory (the partial rates should sum to the total rate). In Fig. 6 we show the behaviour of the total ionization rate from the ground state, versus wavelength, for a fixed intensity of 5×10^{12} W/cm^2 ; the perturbative rates are too large at long wavelengths, though they are accurate at short wavelengths. As a consequence, perturbation theory predicts saturation intensities for long wavelengths which are much smaller than the nonperturbative estimates.

It is worth noting that significant differences may exist between the results for different isotopes due to the different reduced mass, even in the absence of resonance, when the order of the process is large. As an example, we give in Table I the perturbative photoionization rates for hydrogen and deuterium, at 355 nm and 1064 nm (rates for 265 nm and 532 nm can be found in Refs. 16 and 17). The results displayed in Figs. 3-6 are for hydrogen.

Though it is not impossible in practice to obtain nonperturbative differential rates for multiphoton ionization from Eqs. (19) or (20) by using the Sturmian methods briefly described at the end of Section II, it is much more difficult to do this than to obtain nonperturbative total rates, particularly in the case of high-order processes. We illustrate[14] in Fig. 7 the departure from perturbative theory in the angular distributions of the photoelectron ejected in 3- and 4-photon ionization of H(1s) by 265 nm linearly polarized light of 5×10^{13} W/cm^2 ; this intensity is not far below the threshold intensity for which the ionization requires 4 photons to occur. The Bessel function in the denominator of Eq. (19) has hardly a noticeable effect on the results shown in Fig. 7.

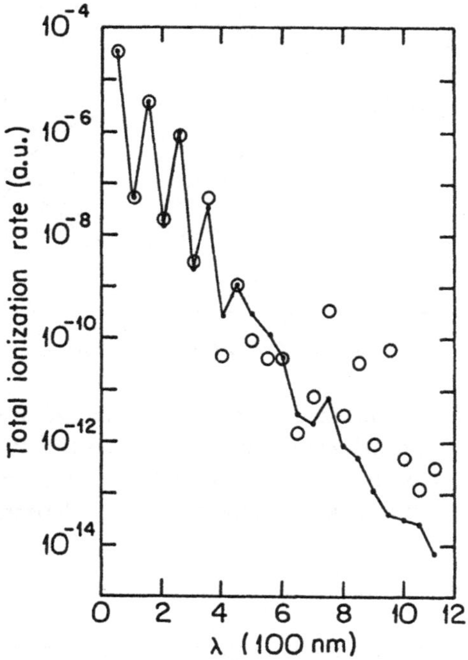

Fig. 6. Total rate for ionization of H(1s) by linearly polarized light of various wavelengths, with the intensity fixed at 5×10^{12} W/cm^2. We have connected the nonperturbative estimates (the dots) by straight lines to distinguish them from the circles, which represent the partial rate for ionization into the S=0 channel obtained within lowest order perturbation theory.

	355 nm ($N_0 = 4$)		1064 nm ($N_0 = 12$)			1064 nm	
S	(a) H	(a) D	(a) H	(a) D	N	(b) H	(b) D
0	3.36(+10)	3.42(+10)	8.68(+7)	9.49(+7)	3	1.28(0)	1.27(0)
1	1.67(+9)	1.69(+9)	1.91(+8)	2.13(+8)	5	3.20(-3)	3.18(-3)
2	3.81(+7)	3.86(+7)	2.36(+8)	2.64(+8)	7	2.45(-4)	2.41(-4)
3	7.64(+5)	7.77(+5)	2.23(+8)	2.51(+8)	9	2.94(-2)	2.93(-2)
4	1.56(+4)	1.59(+4)	1.80(+8)	2.03(+8)	11	2.94(0)	3.28(0)
5	3.15(+2)	3.22(+2)	1.27(+8)	1.43(+8)	13	3.87(0)	4.42(0)
6	6.02(+0)	6.15(+0)	7.85(+7)	8.92(+7)	15	2.33(-1)	2.77(-1)
7	1.06(-1)	1.08(-1)	4.34(+7)	4.95(+7)	17	2.56(-3)	3.29(-3)
8	1.69(-3)	1.72(-3)	2.16(+7)	2.48(+7)	19	9.60(-6)	1.36(-5)

Table I. Total rates (in 1/sec) (a) for N-photon ionization of H(1s) or D(1s) by 355 or 1064 nm light, with $N = N_0 + S$, and (b) for N-th order harmonic generation by 1064 nm light incident on H(1s) or D(1s), calculated in lowest order perturbation theory. The intensity is 1×10^{13} W/cm^2, and the polarization is linear.

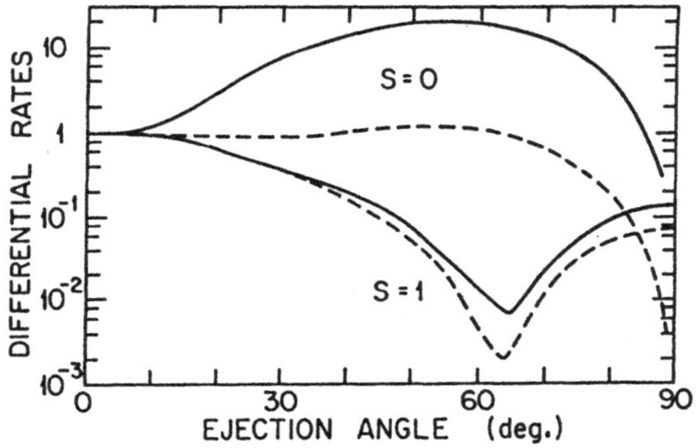

Fig. 7. Angular distribution, normalized to unity at zero angle, for (3+S)-photon ionization by linearly polarized light of wavelength 265 nm and intensity 5×10^{13} W/cm^2. The solid and dashed curves are the results obtained from a nonperturbative calculation[14] and from perturbation theory[20]. The angle of ejection is measured relative to the polarization axis.

IIIB Harmonic generation

The real atomic quantum-mechanical dipole moment induced by the laser field is

$$\mathbf{d}(t) = \langle \Psi(t)|e\mathbf{x}|\Psi(t)\rangle \tag{26}$$

where e is the electron charge. If we make the Floquet ansatz, we obtain

$$\mathbf{d}(t) = \exp(-\Gamma t/\hbar)\{\mathbf{d}_0 + \sum_{N>0} \mathrm{Re}[(2\mathbf{d}_N \exp(-iN\omega t)]\} \tag{27}$$

where

$$\mathbf{d}_N = \sum_n \langle F_{n-N}|e\mathbf{x}|F_n\rangle. \tag{28}$$

From general results of Electromagnetism, the rate for (spontaneous) emission of photons of frequency $\Omega = N\omega$ and polarization ϵ into the solid angle $d\hat{\mathbf{n}}$ is given by putting $\mathbf{D} = 2\mathbf{d}_N$ in

$$\frac{dR}{d\hat{\mathbf{n}}}(\Omega,\epsilon) = \frac{\Omega^3}{8\pi\hbar c^3}|\epsilon^* \cdot \mathbf{D}|^2. \tag{29}$$

Note that this rate depends on the normalization of the Floquet wavefunction, but not on the gauge used to described the laser field or the spontaneous emission. If the field-free initial state has a definite parity, all $\mathbf{d}_N$ with N even vanish, and only harmonics of odd orders are generated.

In Fig. 8 we show rates, both perturbative and nonperturbative, versus the order of the harmonic at several different intensities for linearly polarized 1064 nm light incident on H(1s) (as illustrated by Table I, the rates for high-order harmonic generation can be sensitive to reduced mass effects at long wavelengths). We see that perturbation theory yields a behaviour with respect to order that is reasonable at low orders, but breaks down at high orders, more seriously for higher intensities I. The nonperturbative estimates of the rates do not follow a power law in I as the intensity varies for a fixed order, like the perturbative rates. At sufficiently large intensity, they first decrease sharply as the order increases, but as the order increases still further, they increase, roughly level out into a

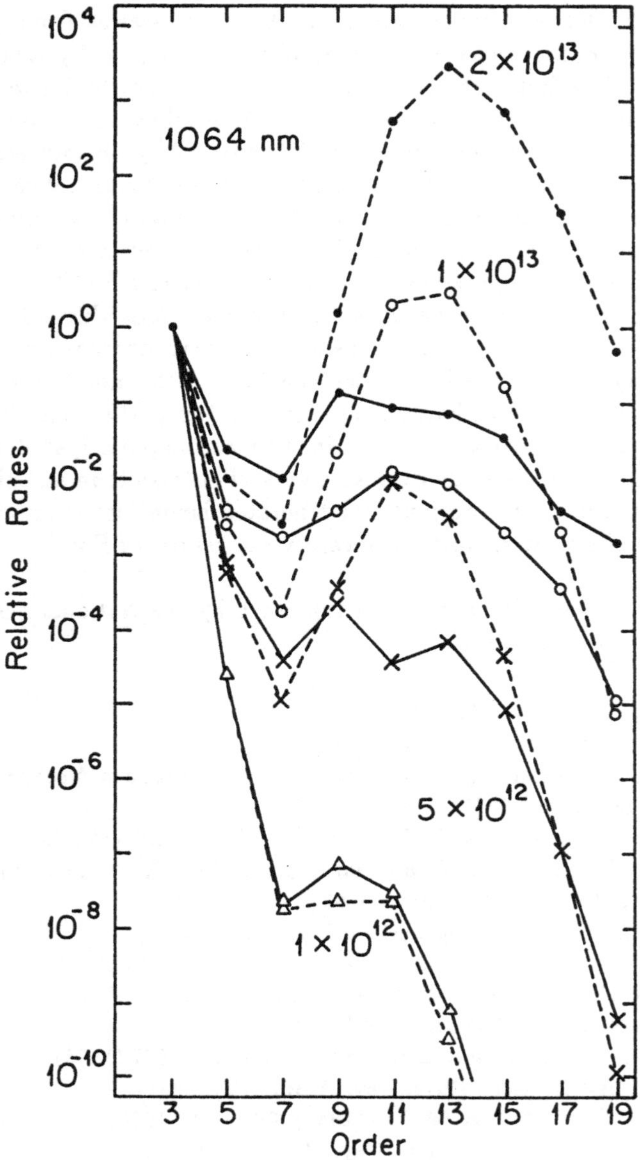

Fig. 8. Rates for harmonic generation of various orders by linearly polarized light of wavelength 1064 nm incident on H(1s), normalized to 1 for the third harmonic. These rates are for emission into the direction of incident propagation, and are summed over polarizations of emitted light. Dashed and solid curves are results obtained in lowest order perturbation theory and our nonperturbative approximation, respectively. $\triangle$: 1×10^{12} W/cm^2 ; $\times$: 5×10^{12} W/cm^2 ; $\circ$: 1×10^{13} W/cm^2 ; $\bullet$: 2×10^{13} W/cm^2.

plateau, and finally decrease beyond an order which increases as I increases. In contrast, the perturbative estimates peak at an order which is roughly N_0, and decrease rapidly beyond this order. The fact that, on the basis of our nonperturbative results, the strength of the 9-th harmonic is larger than that of the 7-th, shows that one cannot conclude that the breakdown of perturbation theory is signalled by an increase in the strength of the harmonic from one order to the next. The difference between the qualitative behaviours of the perturbative and nonperturbative results is less and less striking, and the nonperturbative plateau vanishes, as the wavelength decreases for a fixed intensity.

The appearance of an approximate plateau in the strength of the harmonics at 1064 nm was observed in a recent experiment[21]. However, it is difficult to make a direct comparison, not only because the rare gases rather than hydrogen were studied in this experiment, but also because we do not take into account the modification usually referred to as "phase matching". (Since harmonic generation leaves the atoms in the same initial state, it is impossible to know which atoms in the medium participated, and therefore the signal must be summed coherently over all atoms. This results in an enhancement of the signal when the phases of the incident and generated beams are appropriately matched.) This macroscopic effect can distort significantly the results of Fig. 8.

We gratefully acknowledge support of this work by the National Science Foundation under Grant No. PHY-8713196.

1 J. H. Shirley, Phys. Rev. **138**, B979 (1965).
2 S.-I Chu, Adv. At. Mol. Phys. **21**, 197 (1985).
3 See e.g., C. Cohen-Tannoudji, B. Diu, and F. Laloë, *Quantum Mechanics* (Hermann-Wiley, Paris, 1977), Vol. 1, p. 315.
4 R. M. Potvliege and R. Shakeshaft, Phys. Rev. A **38**, 6190 (1988).
5 A. Maquet, S.-I Chu, and W. P. Reinhardt, Phys. Rev. A **27**, 2946 (1983).
6 S.-I Chu and J. Cooper, Phys. Rev. A **32**, 2769 (1985).
7 M. Pont and M. Gavrila, Phys. Lett. A **123**, 469 (1987).
8 A. Giusti-Suzor and P. Zoller, Phys. Rev. A **36**, 5178 (1987).
9 L. Dimou and F. H. M. Faisal, Phys. Rev. Lett. **59**, 872 (1987).
10 R. M. Potvliege and R. Shakeshaft, to be published.
11 R. M. Potvliege and R. Shakeshaft, Phys. Rev. A **38**, 4597 (1988).
12 R. Shakeshaft and X. Tang, Phys. Rev. A **36**, 3193 (1987).
13 R. Shakeshaft, Phys. Rev. A **34**, 244 (1986); **34**, 5119 (1986).
14 R. M. Potvliege and R. Shakeshaft, Phys. Rev. A **38**, 1098 (1988).
15 S. Klarsfeld and A. Maquet, Phys. Lett. **78A**, 40 (1980), and references therein.
16 R. M. Potvliege and R. Shakeshaft, Phys. Rev. A **39**, 1545 (1989).
17 R. M. Potvliege and R. Shakeshaft, Z. Phys. D **11**, 93 (1989).
18 Y. Gontier and M. Trahin, Phys. Rev. A **19**, 264 (1979) ; C. R. Holt, M. G. Raymer, and W. P. Reinhardt, Phys. Rev. A **27**, 2971 (1983).
19 R. R. Freeman, P. H. Bucksbaum, H. Milchberg, S. Darack, D. Schumacher, and M. E. Geusic, Phys. Rev. Lett. **59**, 1092 (1987).
20 Y. Gontier, N. K. Rahman, and M. Trahin, Phys. Rev. A **34**, 1112 (1986).
21 M. Ferray, A. L'Huillier, X. F. Li, L. A. Lompré, G. Mainfray, and C. Manus, J. Phys. B **21**, L31 (1988) ; X. F. Li, A. L'Huillier, M. Ferray, L. A. Lompré, and G. Mainfray, unpublished.

PART II: Atoms in Intense Laser Fields

HARMONIC GENERATION IN RARE GASES AT HIGH LASER INTENSITY

L.-A. Lompré, A. L'Huillier and G. Mainfray

Service de Physique des Atomes et des Surfaces,
C.E.N. Saclay, 91191 Gif-sur-Yvette, cédex, FRANCE.

ABSTRACT

The purpose of the present paper is to discuss experimental measurements of VUV light emission processes in a 15 Torr rare gas medium exposed to a strong 1064 nm laser field. Apart from a small number of lines which correspond to discrete transitions, we essentially observe the odd harmonics of the laser field, up to very high order. At a 3×10^{13} Wcm^{-2} laser intensity, the highest-order harmonics observed are the 33th in Ar, the 29th in Kr and the 21st in Xe. The study of the variation of the conversion efficiency for some of the harmonics as a function of the incident laser intensity leads to the conclusion that the harmonics are mainly produced in an atomic medium. After a characterization of the phase matching conditions in our experiment, harmonic distributions related only to the atomic response will be tentatively presented showing very interesting features.

1. INTRODUCTION.

An atom exposed to a strong laser field can absorb a large number of photons through high-order nonlinear processes. This leads to the production of multiply charged ions[1]. Consequently electrons are also created but with energies which can be much larger than the energy expected for a N_0-photon absorption, N_0 being the minimum number of photons required to ionize the considered atom. This effect is explained in terms of above-threshold ionization (ATI)[1]. But the energy absorbed by the atomic system may be also transfered back to the photon field under the form of energetic photons with energy $Nh\nu$, $h\nu$ being the laser photon energy, and N the number of photons absorbed. Ion and electron experiments are performed at low pressure, typically 10^{-5} Torr, about 3×10^{11} atoms/cm^3, in order to minimize collective effects. Due to the poor collection efficiency of the photon detection, photon experiments require to be performed at a much higher atomic density, typically 10^{17} to 10^{18} atoms/cm^3, about 10 to 20 Torr. At such a high pressure, the laser light does not interact anymore with isolated atoms, but with a collective assembly of atoms. The emitted radiation may be due to fluorescence from excited states of atoms or ions produced either directly by the laser or by various plasma recombination mechanisms. The additional processes due to the high pressure turn out to be dominant effects. Provided the medium has the required pressure and geometrical properties for proper phase matching of the generated radiation, one may expect production of high-order harmonics of the laser field[2,3].

With the exception of two experiments using a 1064 nm[4] and a 1315 nm[5] laser radiation, most of the previous experiments on harmonic generation in a gaseous medium use short incident wavelengths in order to obtain the shortest generated wavelengths[6-8]. A number of experiments have been performed by using the fourth harmonic (266 nm)[6,7] of the Nd-YAG laser, or the fundamental of an excimer, 308 nm (XeCl)[8] or 248 nm (KrF)[3]. McPherson et al have recently reported up to the 17th harmonic in Ne using a 248 nm excimer laser at an intensity of 10^{16} Wcm^{-2} [3]. The conversion seems to be strongly dependent on the medium.

The purpose of this paper is to describe experimental measurements of VUV light emission (355 - 30 nm) and in particular harmonic generation in a rare gas medium irradiated by an intense 1064 nm laser field. In a 10^{13} Wcm^{-2} intensity range, a 15 Torr Ar gas jet, odd harmonics of the laser field are observed, from the 3rd up to the 33rd harmonic. After a brief description of the experimental setup and a careful analysis of the phase matching conditions in our experiment, harmonic distributions related to the atomic response alone will be tentatively deduced from the experimental results.

2. EXPERIMENTAL ARRANGEMENT.

The experimental set-up shown in Fig.1 has been described in details elsewhere and only a brief description will be given here[2,9]. The laser used is a 30 ps mode-locked Nd-YAG laser operated at 10 Hz. It is focused into the vacuum chamber by a 200 mm focal length lens. The gaseous target is provided by a pulsed gas jet. At 0.5 mm from the nozzle, the gas density in the jet is about 15 Torr, for a backing pressure of 150 Torr, and a nozzle

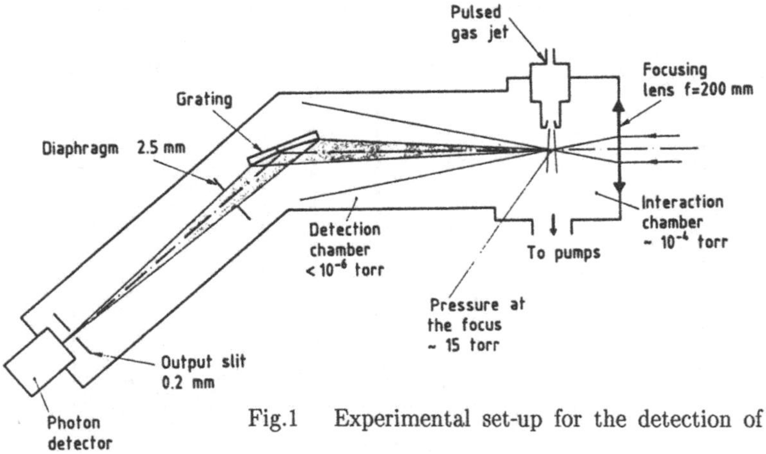

Fig.1 Experimental set-up for the detection of VUV light

diameter of 1 mm. The low ratio between the backing pressure and the pressure measured at the output of the gas jet is very important in order to minimize the contribution of dimers. The complete description of the gas jet and the method to measure the pressure in the gas jet has been described elsewhere[10]. The use of a gas jet limits the reabsorption of the VUV light and enables the detection of very short wavelengths (no cutoff by a window as in a cell geometry). The VUV radiation emitted along the laser axis is analyzed by a monochromator. It is composed of a toroidal grating which forms on the output slit an aberration free image of the interaction region for the spectral range of the grating. In order to cover the complete spectral range, two different gratings have been used : (I) from 50 to 355 nm, (II) from 10 to 50 nm. The VUV radiation is then detecting by a photomultiplier or a windowless electron multiplier depending on the spectral range investigated.

3. EXPERIMENTAL RESULTS.

Figure 2 shows typical spectra between 30 and 160 nm in Xe and Ar obtained using the grating I, an electron mutiplier as photon detector and for a gas pressure of 15 Torr. The laser intensity is the same for the two scans and is equal to 4.5×10^{13} Wcm^{-2}. Several major points come out at first sight : (1) only odd harmonics are observed. That is due to the fact that the harmonics are created in an isotropic gaseous medium with inversion symmetry which excludes even harmonics. (2) the maximum number of harmonics created at the a)

b)

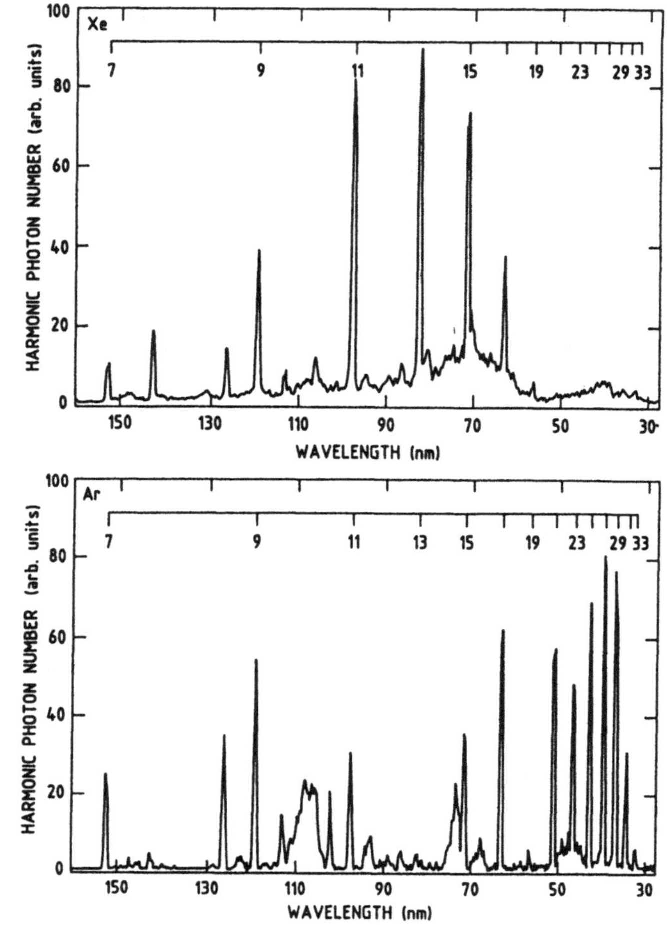

Fig.2 Photon spectra obtained at 1064 nm, 3×10^{13} Wcm^{-2}, 15 Torr of a) Xe and b)Ar

maximum laser intensity is strongly dependent on the rare gas investigated; up to the 21st harmonic in Xe (Fig.2a) and up to the 33rd harmonic in Ar (Fig.2b). (3) a continuous background appears in Xe, which is much reduced in Ar. It must be mentioned at this point that there is no convincing explanation about the origin of this background. (4) Other lines than the first and second orders of the grating of the harmonics can be observed : they correspond to transitions 6s → 5p (147 nm) and 6s' → 5p (129.5 nm) in neutral Xe (Fig.2a), or 4s → 3p in neutral Ar, and 4d → 3p in ionic Ar (Fig.2b)[9]. Over the whole spectrum extent (10-355 nm), a small number of lines are observed, corresponding to transitions from excited states to the fundamental of the neutral atom and also of the singly

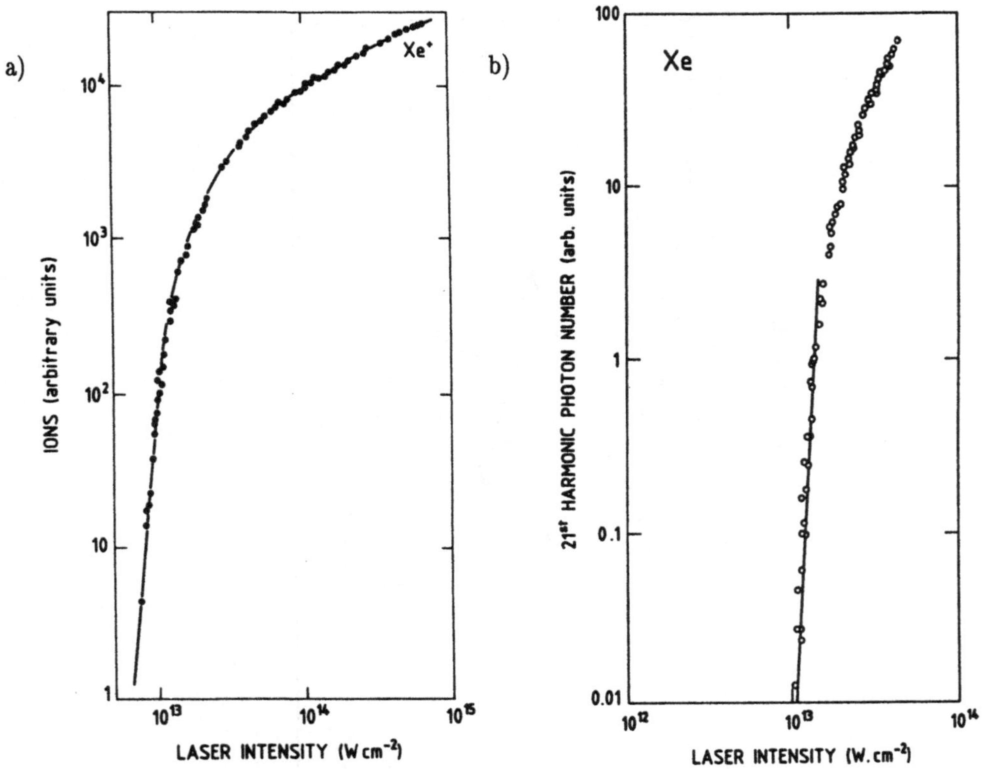

Fig.3 Intensity dependence of a) number of ions b) 21th harmonic number of photons in Xe

charged ion for our maximum laser intensity.

Before going further into the description of light emission measurements, it is worthwhile recalling a typical multiphoton ionization result[11]. Figure 3a shows the intensity dependence of the number of singly charged Xe ions at 10^{-4} Torr. At low laser intensity, the experimental points lie on a straight line with a slope equal to the minimum number of photon N_0 necessary to ionize the atom. Then, for higher laser intensities, points depart from this straight line. There is a saturation of the interaction volume, and the ions created for

	Xe	Kr	Ar	Ne	He
$I_s(W.cm^{-2})$	$1.2\ 10^{13}$	$2.5\ 10^{13}$	$3.\ 10^{13}$	$1.9\ 10^{14}$	$2.8\ 10^{14}$

Table 1 Saturation intensities for the rare gases.

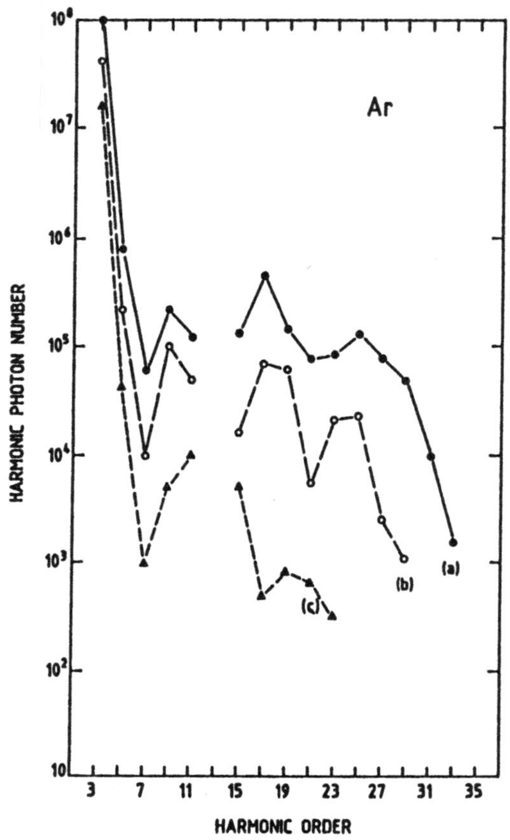

Fig.4 Number of photons at each harmonic at three different laser intensities.

high intensities are produced in an expanded volume. The intensity for which this departure happens, called saturation intenisty I_S, depends on the atom. Table 1 summarizes the saturation intensities for the five rare gases, for a pulse duration of 30 ps.

Coming back to the photon detection results, Fig.3b shows an intensity dependence of the 21st harmonic in Xe, taken as an example. The behaviour of the harmonic is very similar to the ion behaviour (Fig.3a). A saturation is observed at the same laser intensity. The two results have however been obtained at different pressures. This observation is very important because it shows that the harmonic production is strongly influenced by the presence of ions in the interaction volume. It must be pointed out that this observation has been made for all the harmonics in all the rare gases studied. It leads to two important

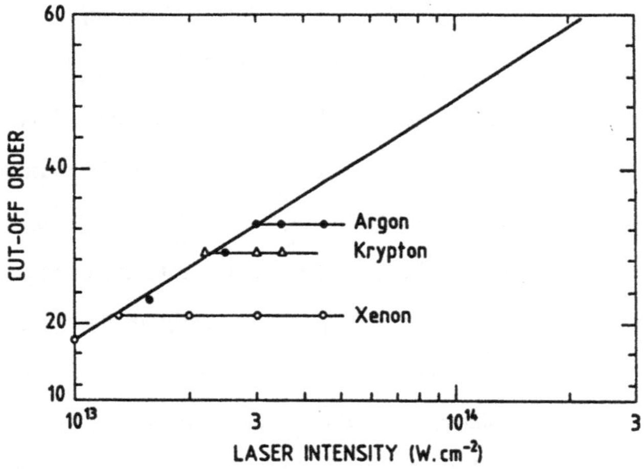

Fig. 5 Maximun number of harmonics versus laser intensity for three different rare gases

conclusions : (1) the harmonic production efficiency is much smaller in an ionic medium than in an atomic vapor. In other words, the harmonic production occurs preferentially in the neutral medium. (2) For comparison with theoretical calculations in atoms, data must be taken for laser intensities below or close to the saturation intensiy, but in any case for laser intensities larger than I_S.

Let us focused our attention on Ar. Figure 4 shows the number of photons produced at each harmonic at three different laser intensities, a) $3x10^{13}$ Wcm^{-2} (= I_S), b) $2.2x10^{13}$ Wcm^2 and c) $1.6x10^{13}$ Wcm^{-2}. 33rd harmonic generation, the highest-order conversion process ever observed in our experiment, corresponds to a radiation emitted at 32.3 nm. The vertical scale has been calibrated in absolute value by estimating the collection efficiency, in particular by taking into account the spectral responses of the grating and the photon detectors for the different harmonics. The experimental uncertainty is estimated to a factor of 5. Most of the modulations in Fig.4 are therefore not significant. For the highest laser intensity, three different parts can be observed. After a steep decrease from the 3rd to the 7th, the harmonic distribution presents a kind of plateau from the 9th to the 27th harmonic, with a missing 13th harmonic. Then it decreases again rather steeply down to the 33th harmonic. The ionization rates corresponding to these intensities are approximately 1, 10^{-2} and 10^{-4}. These harmonic distributions shows that most of the harmonics are created mostly in an atomic medium, or at an intensity such that the medium is not or weakly ionized. Similar distributions were obtained in Kr and Ar for different laser intensities. As in Ar, the distribution in Kr presents an anomaly for the 11th harmonic. There are also abrupt cutoff in both rare gases at the 21st harmonic in Xe and at the 29th harmonic in Kr. This cutoff has been plotted versus laser intensity in Fig.5 for the different rare gases. It appears clearly that, for each rare gas for laser intensities below the saturation intensity (see table 1) experimental cutoff points lie on the same line, independently of the rare gas considered : the maximum number of harmonics which can be generated is only intensity dependent. Of course, the conversion efficiency depends on the atom. When the laser intensity is greater than the saturation intensity I_S, the maximum harmonic order stops increasing. Recalling that when the intensity reaches I_S there are not any more atoms in the interaction volume but only ions, this observation confirms that the harmonics are mainly created in an atomic medium and not in an ionic medium. Let us point out that, in the case of Ne and He which have higher saturation intensity about $2x10^{14}$ and $3x10^{14}$ Wcm^{-2} respectively, an extrapolation of the curve showed in Fig.4 predicts the creation of the 58th and the 64th harmonics of the incident laser radiation at 1064 nm in Ne and He. This would correspond to generated radiations at 18.3 and 16.8 nm. But are these radiations observable because of the poor efficiency we may expect compare to Xe, Kr and Ar....

4. DISCUSSION.

The coherent radiation field is created from the laser induced polarization of the medium. This is a collective process due to the response of the whole medium to the incident field : it approximately varies as the square of the atomic density. In order to extract some informations on the microscopic response of the atomic medium, it is necessary to briefly recall the main equations which govern nonlinear optical phenomena [12,13].

An electromagnetic field propagating in a dielectric medium characterized by an electronic polarization must satisfy the Maxwell propagation equation. If the incident laser field is assumed to be monochromatic, and considering only the lowest-order in the interaction with the radiation field, then, for a gaussian beam focused into a nonlinear medium of finite length, the intensity of the qth harmonic field is [12]:

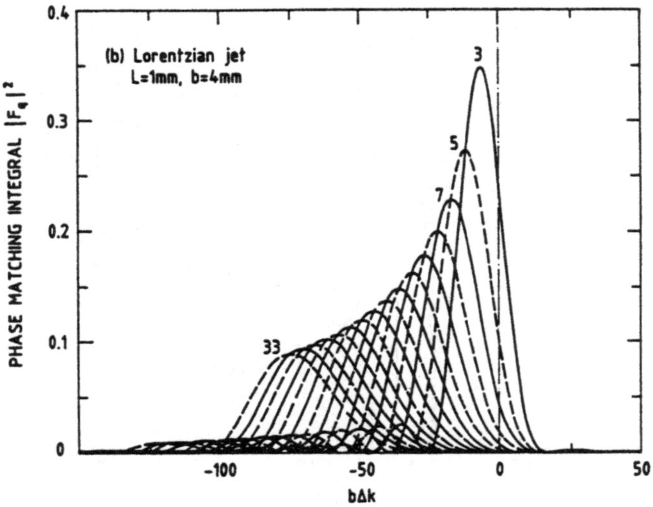

Fig.6 Phase matching integral as a function of the dimentionless parameter bΔk.

$$I_q(0,0,0) = (2^{q+1}\ b^2\ q^2\ \pi^{q+3}\ /\ n_1\ n_q\ \lambda^2\ c^{q-1})\quad I^q\ |F_q|^2\ |N_a\ \chi^{(q)}|^2$$

Finally, the total number of photons of the qth harmonic field generated at each laser pulse, which is measured in the experiment, is given by :

$$N_q = \int 2\pi r\ I_q(r,z,t)/h\omega_q\ dr\ dt\ = (b\ \lambda^2\ \tau/\ q^{5/2}\ h\ c)\ I_q$$

In these expressions, h is the Planck constant, λ is the incident wavelength, c is the speed of the light, n_1 and n_q are the refractive index of the nonlinear medium at ω and $q\omega$ respectively, N_a the atomic density and τ the pulse duration (the area of the temporal distribution). b is the laser confocal parameter closely related to the focalisation of the laser. The qth order susceptibilty $\chi^{(q)}$ characterizes the response of an individual atom to the radiation field. It is the quantity which is of interest in the next few paragraphs. F_q is a phase matching factor which depends on the phase mismatch between the harmonic and its driving polarization. It depends on the macroscopic properties of the medium (geometry, pressure, dispersion). The interplay between microscopic and macroscopic properties of the nonlinear medium in the intense field makes the interpretation of the experimental results rather difficult. The conversion efficiency is influenced by both the atomic and the collective response of the gaseous medium.

Let us now examine the phase matching factor F_q. The length of the nonlinear medium

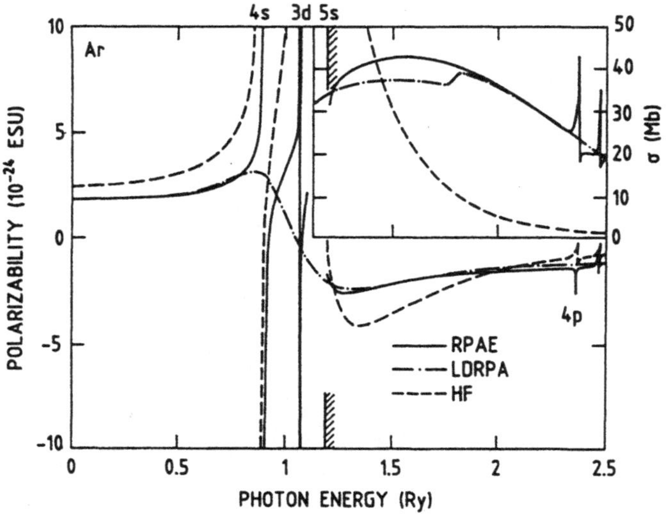

Fig.7 Photon energy dependence of the polarizabilty of Ar. (Ref. 16).

(L = 1 mm) is such that the radiation beam remains approximately collimated throughout the medium (b = 4.2 mm for the 200 mm focal length lens used in the experiment). The qth harmonic field as well as the incident field have gaussian profiles with the same confocal parameter. The gas jet have been modelized in a realistic way, by using a lorentzian distribution with a 1 mm width at half maximum for representing the non-uniform atomic density in the gas jet[14]. Figure 6 shows $|F_q|^2$ for q odd varying from 3 to 33, as a function of the dimentionless parameter $b\Delta k$. Δk is the phase mismatch between the generated field and the incident field. To complete this discussion, the phase mismatch must be estimated in the experimental conditions.

$$\Delta k = k_q - qk_1 = (n_q - n_1)\, q\omega/c = 2\pi\, q\omega/c\, N_a[\alpha(q\,\omega) - \alpha(\omega)]$$

α is the frequency-dependent polarizability. As for the susceptibilities, there is very little information concerning refractive indices or dynamical polarizabilities for rare gases. Figure 7 shows the polarizability in Ar versus the photon energy calculated using many-body perturbation theory and the random phase approximation with exchange[15,16]. In fact, three regions would have to be taken into consideration, (1) below the first excited state : Δk is positive and small, (2) near the resonance : Δk could be very large or very small for one or two harmonics in the vicinity of the ionisation threshold. This could explain the non-a)

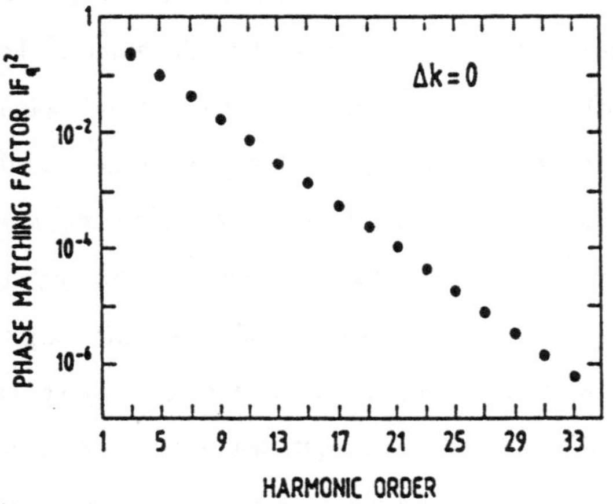

b)

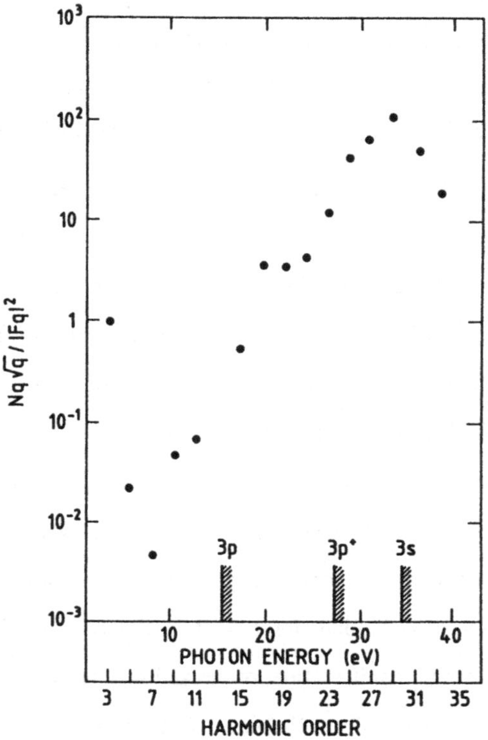

Fig.8 a) Phase matching factor as a function of the harmonic order for $\Delta k = 0$

b) Variation of $|\chi^{(q)}|^2 \, I^q$ as a function of the harmonic order

observation of the 13th harmonic in Ar (Fig.4) (3) above the ionization threshold : n_q becomes imaginary : the real part is negative and small because of the low pressure, the imaginary part is responsible for the absorption of the generated radiation due to photoionisation. In summary, apart possibly from particular frequencies corrresponding to discrete excitations, the nonlinear medium in our experimental conditions can be considered to be dispersionless ($b\Delta k \approx 0$). Figure 8a) shows the variations of $|F_q|^2$ for $\Delta k = 0$ versus the harmonic order, under these conditions. Taking into account this factor, it is possible to deduce $N_q \, q^{1/2} / |F_q|^2$ as a function of the harmonic order, which is directly proportional to $|P_q^{NL}|^2$. P_q^{NL} is the nonlinear polarization at the qth harmonic involving only the incident laser field E_1. Figure 8b) shows the result in Ar (Fig.4) at $3x10^{13}$ Wcm^{-2}, normalized to 1 for the third harmonic. This quantity should be directly comparable to single-atom calculations[17-21]. It is however based upon a I^q dependence of qth harmonic generation which may not be valid for very high orders and intensities. 21th harmonic intensity dependence, with a slope of 20 ± 2 for laser intensities $\leq I_S$ (Fig.3b)), seems to indicate that this hypothesis is still valid at such order and intensity. This result shows that, for low harmonic order, there is a steep decrease of the nonlinear susceptibilities. This could be easily understand by the fact that the nonlinear susceptibilities are very similar to multiphoton ionization : the main difference is that the final state is the ground state of the atom in the first case, a continuum in the second case. On the basis of what is known in multiphoton ionization and also calculations derived from the lowest perturbation theory (some performed in H[21,22]), one may expect that the quantities $|\chi^{(q)}|^2 I^q$ would decrease rapidly for low orders, then more and more slowly. But it is not exactly what it is observed for higher orders : the nonlinear polarization increases with the order to reach a value 10 times larger for q = 27 than for q = 3 in the Ar case. Then it drops dramatically at the cutoff order. Is this a manifestation of breakdown of perturbation theory ?. The question is still open and will hopefully stimulate further experimental and theoretical investigation in order to understand this behavior.

The authors would like to thank Dr. C. Manus for many stimulating discussions, M. Ferray, X.F. Li and A. Sanchez for their help in the experiments.

REFERENCES

1. see articles in "Multiphoton Processes", Proc. of the 4th Int. Conf. on Multiphoton Processes, JILA, Boulder, July 1987, Eds. S.J. Smith and P.L. Knight, Cambridge University Press, 1988.
2. M. Ferray, A. L'Huillier, X.F. Li, L.A. Lompré, G. Mainfray and C. Manus, J. Phys. B 21, L31 (1988).
3. A. McPherson, G. Gibson, H. Jara, U. Johann, T.S. Luk, I. McIntyre, K. Boyer and C.K. Rhodes, J. Opt. Soc. Am. B 4, 595 (1987).
4. M.G. Groseva, D.I. Metchkov, V.M. Mitev, L.I. Pavlov and K.V. Stamenov, Opt. Commun. 23, 77 (1977).

5. J. Wildenauer J. Appl. Phys. **62**, 41 (1988).

6. J. Bokor, P.H. Bucksbaum and R.R. Freeman, Opt. Lett. **8**, 217 (1983).

7. J. Reintjes, C.Y. She and R. Reckardt, IEEE J. Quatum Electron. QE-**14**, 581 (1978).

8. J. Reintjes, L.L. Tankersley and R. Christensen, Opt. Commun. **39**, 334 (1981).

9. X.F. Li, A. L'Huillier, M. Ferray, L.A. Lompré and G. Mainfray, Phys. Rev. to be published.

10. L.A. Lompré, M. Ferray, A. L'Huillier, X.F. Li and G. Mainfray, J. Appl. Phys. **63**, 1791 (1988).

11. A. L'Huillier, L.A. Lompré, G. Mainfray and C. Manus, J. Phys B **16**, 1363 (1983).

12. J.F. Reintjes, in "Nonlinear Optical Parameter Processes in Liquids and Gases", Academic Press, Inc. (1984).

13. N. Bloembergen, in "Nonlinear Optics", Benjamin, New-York (1965).

14. A. Lago, G. Hilber and R. Wallestein, Phys. Rev. A **36**, 3827 (1987).

15. A. L'Huillier and G. Wendin, Phys. Rev. A **36**, 4747 (1987).

16. A. L'Huillier, to be published.

17. L. Pan, K.T. Taylor and C.W. Clark, private communication.

18. K. Kulander and B.W. Shore, Phys.Rev. Lett. **62**, 524 (1989).

19. J.H. Eberly, Q. Su and J. Javanainen, private communication (preprint).

20. R.M. Potvliege and R. Shakeshaft, private communication.

21. Bo Gao and A.F. Starace, to be published.

22. Y. Gontier and M. Trahin, IEEE, J. Quantum Electron. QE-**18** ,1137 (1982).

IONIZATION OF ATOMS BY AN INTENSE CO_2 LASER

S.L. Chin and W. Xiong
Laboratoire de Recherches en Optique et Laser (LROL)
Département de Physique, Université Laval
Québec (Québec) G1K 7P4 Canada

(I) Introduction

One of the criteria for defining an intense laser field during its interaction with atoms is the Keldysh parameter γ,

$$\gamma^2 \equiv \frac{\text{ionization potential of the atom}}{\text{ponderomotive potential}}$$

$$\equiv \frac{I_p}{U_p}$$

where the ponderomotive potential of an electron (charge e, mass m) in a laser field is

$$U_p = \frac{e^2 E_o^2}{4m\omega^2}$$

ω and E_o are the angular frequency and the peak electric field of the laser radiation respectively. One might define the field to be strong when

$$U_p \gg I_p$$

Under such a condition, the Coulomb field binding an electron to the nucleus is much smaller than the laser field. The previous equation also show that the criterion is frequency dependent. For the same laser intensity, it is easier to achieve the same small value of γ using a longer wavelength (lower frequency) laser than a shorter wavelength (higher frequency) one. Experiments have been done during the past few years by Chin's group (Chin et al, 1983, 1985; Yergeau et al, 1987; Xiong et al, 1988) using an intense CO_2 laser to ionize rare gas atoms with γ^2 for Xe for example, as small as about 0.006. Both ions and electron energy spectra have been measured. The original idea of the first few experiments was to study laser induced tunnel ionization (Keldysh, 1964; Reiss, 1980).

Various models and calculations have since been used to fit the ion data with different degrees of satisfaction (Amosov et al, 1986; Yergeau et al, 1987; Faisal, 1987; Geltman, 1988). Our electron measurement had added complementary data not only to the ionization process, but also to the phenomenon of accelerationg electrons by a strong laser field through ATI (above threshold ionization) and the ponderomotive acceleration.

This paper presents some new fittings of some of the CO_2 laser ionization data based on a modified theory of tunnel ionization (Perlomov et al, 1966; Ammosov et al, 1986) and the assumption that ATI is present in the ionization process, together with ponderomotive acceleration. The ATI process is necessary in order to have a quantitative fit of the electron data. We only use Xe as an example and we believe that similar good fittings should also be found in other atoms.

(II) Modified tunnelling theory

In their treatment, Perlomov et al (1966) obtained the d.c. tunnel ionization rate:

$$W_{stat}(E) = \omega_o |C_{n^*\ell}|^2 (2\ell + 1) \frac{(\ell + |m|)!}{2^{|m|}(|m|)!(\ell - |m|)!} \times \left(\frac{2F_o}{E}\right)^{2\lambda - |m| - 1} \exp\left(-\frac{2F_o}{3E}\right) \quad (1)$$

where ℓ is the orbital angular momentum of the electron in the unperturbed atom and m its projection along the direction of the field. F_o is the typical interatomic field and E the strength of the laser field. Here we fix $\ell = 1$ and $|m| = 0$ in our calculation since the contributions from higher ℓ and m to W_{stat} are relatively small (Ammosov et al, 1986). $C_{n^*\ell}$ is a constant determined by the atoms used. In our calculation, we use the value deduced from the arguement of Ammosov et al (1986) as explained in the following. Finally, ω_o is the binding energy of the electron in atomic units.

In Eq.(1), taking E as an alternating field, one can average the ionization rate over one cycle of an alternating electric field with amplitude E_o and frequency ω. Under the quasi-classical limit, one obtains the tunnel ionization rate of Ammosov et al (1986):

$$W = \left(\frac{3e}{\pi}\right)^{1/2} \frac{Z^2}{n^{*\,4.5}} \left(\frac{4eZ^3}{E_o n^{*4}}\right)^{2n^*-1.5} \exp\left(-\frac{2Z^3}{3n^{*3}E_o}\right) \tag{2}$$

where e is the base of natural logarithm, Z is the charge of the atomic residue, n^* the effective quantum number of the fundamental state. The value of $C_{n^*\ell}$ in Eq.(1) is given by (Ammosov et al, 1986).

$$C_{n^*\ell} = \left(\frac{2e}{n^*}\right)^{n^*} \frac{1}{(2\pi n^*)^{1/2}} \tag{3}$$

The above formulae have been used by Ammosov et al (1986) to calculate the saturation intensities which agree rather well with the results of CO_2 laser ionization of rare gases of Yergeau et al (1987).

Using Ammosov's tunnel ionization rate Eq.(2), we performed the spatial an temporal integration, and obtained the total ion yield N_{ion} for a laser pulse with peak intensity I_o:

$$N_{ion} = \iiint_{-\infty}^{+\infty} d^3r\{1 - \exp\left[-\int_{-\infty}^{+\infty}dt\ W(I_o\exp(-(r/ro)^2-(t/\tau)^2))\right]\} \tag{4}$$

where ro is the focal radius and τ is the pulse duration.

When we plot N_{ion} against I_o, the fits of the plots to the Xe^+ and

Xe^{++} measurement of Yergeau et al (1987) at $\lambda = 10.55\mu m$ are good in orientation but shifted by a factor of 2.02 in the intensity scale. In Ammosov's formula, there is no adjustable parameter, so this discrepancy could result from two things. Either Ammosov's tunnel-like theory does not apply to the ionization of Xe by a CO_2 laser pulse or Yergeau et al's calibration of the laser intensity was off by a factor of 2. The latter is possible in view of the imprecision of the spot size measurement. Furthermore, the second explanation will be confirmed later on when we calculate the electron energy spectrum since the acceleration of electrons depends on the absolute ionization rate of atoms. Fig. 1 shows a very good fit when the intensity scale of the experimental points is down shifted by a factor of 2.02.

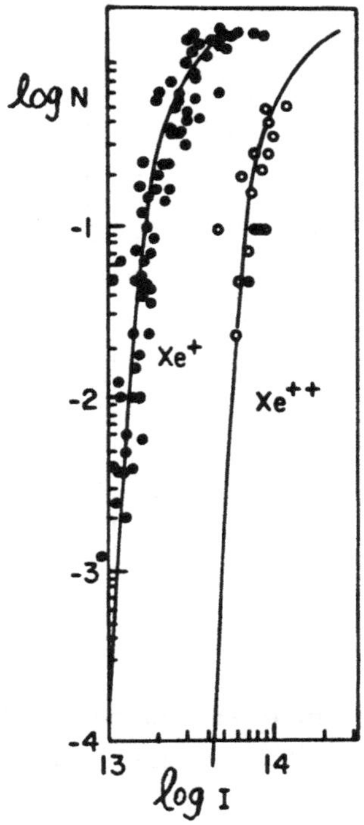

Figure 1: The fit of tunnel-like ionization rates of Ammosov et al (1986) to data of CO_2 ionization of Xe by Yergeau et al (1987). The experimental data were downshifted by a factor of 2.02 in the laser intensity scale. The horizontal scale is in W/cm^2.

(III) Ponderomotive acceleration

The ponderomotive acceleration originates from the intensity gradient of the laser field and can be calculated by the classical Newtonian mechanics equation

$$m \frac{d^2 r}{dt^2} = - q E_o(r,t) \cos\omega t \qquad (5)$$

$E_o(r,t)$ is the temporal profile of the laser pulse at the point r, q and m, the electron charge and mass and ω, the radiation frequency. We assume that once the electron tunnels out of the atom (ionized), it sees the classical laser field at a certain moment within an optical cycle (Fig. 2). This optical cycle can be any one within the laser profile.

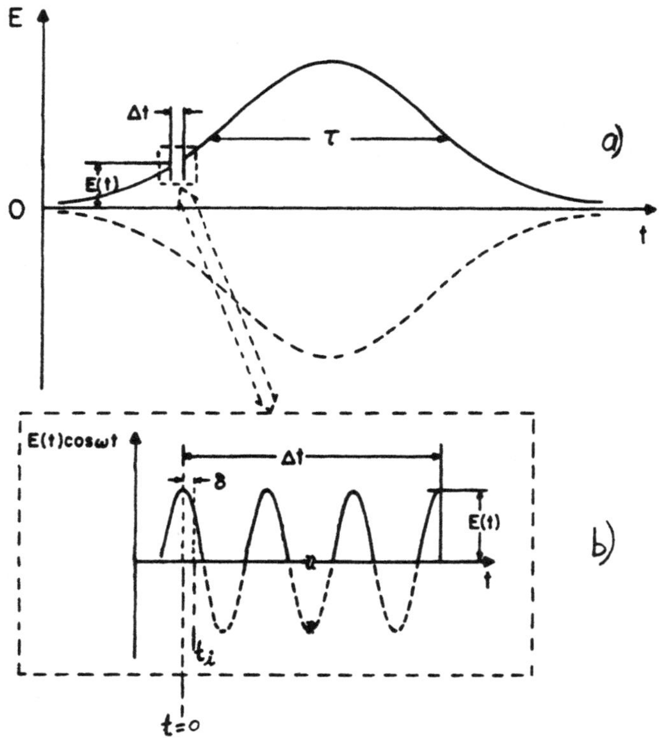

Figure 2: Schematic drawing of the classical field oscillation of a laser pulse as a function of time. (a) The envelop is $E_i=E_o(r,t)$. (b) The enlarged section Δt contains many cycles of oscillations ($2\pi/\omega<<\Delta t<<\tau$). An atom is assumed ionized at $t = \delta/2\pi$ by the field $E_o(r,t)\cos\delta$, where $0<\delta<2\pi$.

Within one optical cycle, we take $E_o(r,t)$ as a constant and let the initial kinetic energy of the electron be zero, the reason of which will be discussed later. Taking this as our initial condition, we can solve Eq.(5) easily. For an electron ionized at time $t_i = \delta/\omega$

$(0<\delta<2\pi)$, it will pick up a kinetic energy U_d by the end of this optical cycle $t = 2\pi/\omega$ (See Fig. 2(b)).

$$U_d = \frac{q^2E_o^2}{2m\omega^2}\sin^2\delta \tag{6}$$

This is the acceleration within the first optical cycle. Starting from $t = 2\pi/\omega$, the electron, undergoes a "standard" PMF acceleration process, i.e. gain a kinetic energy U_p (Kibble, 1966) through the intensity gradient of the laser.

$$U_p = q^2E_o^2 / 4m\omega^2 \tag{7}$$

Here, we assume that the electron flies out of the focal region within a time Δt short compared to the laser pulse duration τ but long compared to an optical cycle (Fig. 2). Thus, E_o is assumed constant during the acceleration process and the ionization rate is assumed instantaneous, i.e. W_{stat}.

Also, if the system is conservative, an electron ionized by the laser pulse at $t = \delta/\omega$ will gain a kinetic energy which is the sum of eq.(6) and (7),

$$U_\alpha = \frac{q^2E_o^2}{4m\omega^2}\left(1 + 2\sin^2\delta\right) \tag{8}$$

A plot of W_{stat} against U_α gives an energy spectrum of electrons at a certain laser field E_i where $E_i = E_o(r,t)$. In a laser pulse, the laser field E_i increases from zero to a maximum, then drops back to zero (Fig. 2). For each value of E_i, there is an energy spectrum. The superposition of all spectra (weighted with respect to population depletion) gives the energy spectrum of electrons.

As argued by Yergeau et al (1987), due to the very high nonlinearity of the ionization process, the ionization rate is significant only at an intensity near saturation. We define an effective saturation intensity (I_s) by

$$W(I_s)\tau = 1 \tag{9}$$

where $W(I_s)$ is given by eq.(2).

Thus, in our calculation, we first solve for I_s using eq.(9) and obtain the corresponding time t_s at which the laser intensity has risen to I_s. This effective saturation intensity is used in the integration over the focal volume.

IV. Comparision with experimental results

For a peak laser intensity I_o, at which only single ionization takes place, which is 5×10^{13} W/cm^2 (defined here as the low energy case), the electron spectrum is plotted and we obtain the dashed line in Fig. 3a. When the laser intensity is increased so that both single and double ionizations occur (10×10^{13} W/cm^2), we obtain two Xe peaks (Solid line in Fig. 3b, "high energy case"). The points in Fig. 3a and Fig. 3b are experimental data published in our last paper (Xiong et al, 1988).

In Fig. 3a, we can see that the fit is very good for single ionization. In Fig. 3b, the second peak corresponding to stepwise double ionization also has a good fit; however, the first peak corresponding to single ionization doesn't fit well. The rising part of the spectrum has a good fit with the experimental result, but there is a considerable discrepancy at the falling part. The experimental curve extends further than the theoretical one. It's hard to explain such a big discrepancy as experimental errors. In the following, we remove this difference by taking into account the ATI process.

The quantum process of ATI in the visible, near I.R. and U.V. can be considered as a direct coupling of the ground state of the atom to the continuum by the absorption of H photons more than the N strictly necessary for ionization. The cross sections for such processes have been calculated by a number of authors both for hydrogen and alkali atoms (Zernik et al, 1965; Karule, 1978; Klarsfeld et al, 1980; Aymar et al, 1981; Gontier et al, 1981). Their values are quite "normal" in the sense that we can study their behaviour as if the transition involves a direct non-resonant absorption of (N + H) photons in the discrete part of the spectrum. We extend this idea to the present case and assume that an electron can absorb H more photons during its ionization; i.e., an electron in such a case has to overcome an ionization potential of $(I_p)_H = (I_p)_o + H\hbar\omega$ where $(I_p)_o$ is the unperturbed ionization potential and $H = 0, 1, 2, 3 \ldots$ In what follows, we loosely use the notations ATI and MPI to represent ionizations for $H \neq 0$ and $H = 0$, respectively.

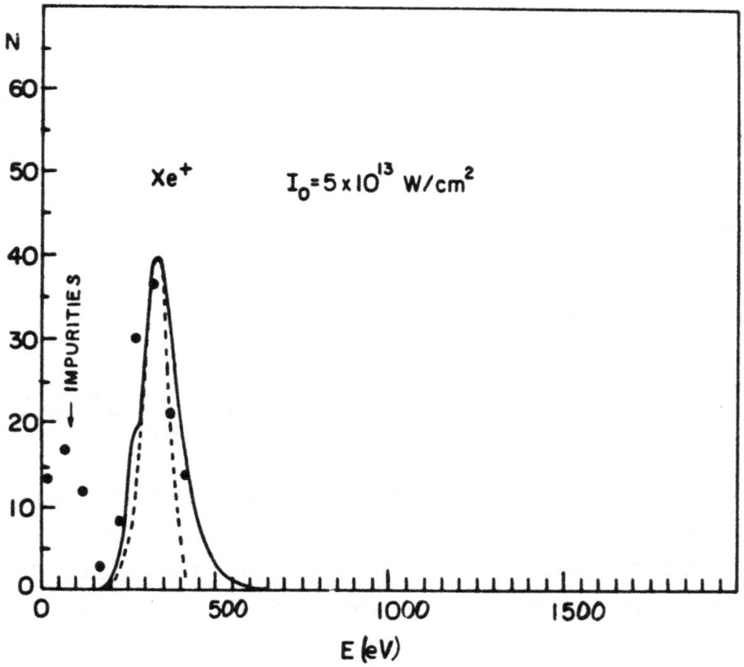

Figure 3a: The energy spectrum of electrons ejected from the single ionization of Xe. Peak laser intensity is 5×10^{13} W/cm^2. The solid line and dashed lines are the fits with and without ATI consideration, both of them fit the experimental data rather well.

Again, due to the very high nonlinearity of the ionization process, the ionization through the Hth ATI channel is assumed to take place mainly around an effective saturation intensity $(I_s)_H$ defined by:

$$W_H((I_s)_H)\tau = 1 \qquad (10)$$

where W_H is the ionization rate of the Hth ATI channel using eq. 2 for an atom with ionization potential $(I_p)_H$.

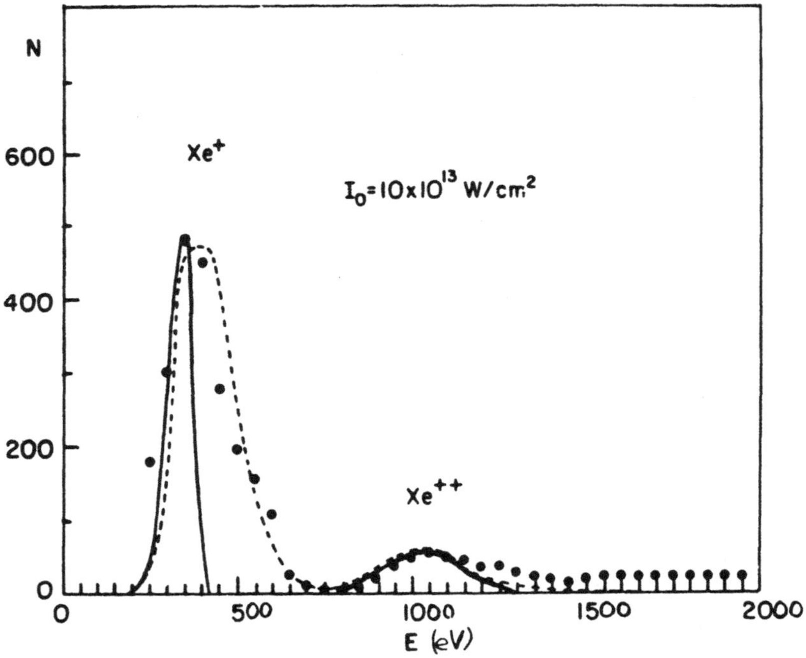

Figure 3b: The energy spectrum of electrons ejected from single and double ionizations of Xe at a peak laser intensity of $10x10^{13}$ W/cm^2. The dashed line and solid line is the theoretical fits with and without assuming ATI processes during ionization.

We can solve Eq.(10) for $(I_s)_H$ and obtain the corresponding t_H at which the laser intensity has risen to $(I_s)_H$. Using this effective saturation intensity in the integration over the focal volume, and the values of H from zero to 25 (the electron number is practically zero at H > 25), for electrons coming from Xe --> Xe^+, we obtain the result shown as the dashed line in Fig. 3b. This fit extending over the second electron spectrum (step-wise double ionization with H=0 only because the electron number is insignificant at H > 0), overlaps with the second solid line. One can see that the fit is very good. This good fit also indicates that the maximum ATI energy of an electron of any significance in our calculation is only 25 $\hbar\omega \cong$ 3eV (H = 25) in the case of ejecting the first electron, much smaller than the ponderomotive energy. That's why we could ignore the initial energy of electrons when we calculate the PMF acceleration.

V. Discussion

By down shifting the experimentally measured laser intensity by a factor of 2, we obtained very good fits between theoretical and experimental results for both ion yield and electron acceleration. The fit of electron energy spectrum confirms the fact that due to the long wavelength of the CO_2 laser, the absence (or "suppression") of low energy electrons up to 200 eV can happen even at a moderate intensity ($5 \times 10^{13}/cm^2$). Based on this fact, we can say that the difference between the results of theoretical calculations and Yergeau et al (1987) is due to the error in the experimental intensity calibration. In fact, if we fit a curve through the unshifted experimental points of Yergeau et al (1987) and use it as the dependence of ionization rate on the laser intensity in calculating the energy spectrum of electrons, the fit will be off completely! We have a spectrum centered at a much higher energy.

In our calculation, we have made the following assumptions:

(a) Atoms are first ionized through MPI and ATI processes and then accelerated by the PMF.

(b) When the electron is born (ionized), it finds itself already in the strong radiation field which provides it with the drift energy U_d (eq. 6) and the ponderomotive energy U_p (eq. 7).

(c) The laser pulse is assumed to be much longer than the time (Δt) it takes an electron to fly out the focal volume. That is to say, once born, the electron will only pick up the U_d and U_p at that moment as it kinetic energy. Any subsequent change in the laser intensity gradient doesn't affect the electron.

We did not apply the idea of "ponderomotive shift" of the ionization potential and high lying Rydberg states as proposed by many authors (e.g. Muller et al, 1983, Freeman et al, 1986) in the cases of visible and near I.R. photons. In our case, this shift could be as large as a few hundreds volts, much larger than the ionization potential of Xe atoms. This would not fit our results if the ionization potential becomes so large.

We like to stress that our idea doesn't necessarily completely contradict that of Muller et al (1982) and Freeman et al (1986) concerning "ponderomotive shift" of the ionization potential. Our definition of ATI in this paper, by assuming that the ionization potential changes from $(I_p)_o$ to $(I_p)_H$ is essentially equivalent to a shift of the unperturbed ionization potential $(I_p)_o$. The ATI probability depends on the laser intensity and frequency; i.e. the physics of interaction. It just happens that we have used a long wavelength and intense CO_2 laser so that the ponderomotive potential is very high as compared to to those in other work using intense near i.r., visible and u.v. lasers. In our case, the interaction physics is dominated by tunnelling while the others are dominated by multiphoton ionization (at least in the creation of the first electron). However, when the electron "emerges" from the atom, it should see the radiation field, and thus experience the ponderomotive potential (intensity gradient) whatever the laser wavelength is. Thus we can imagine that when the electron "emerges" from the atom into the laser field, it possesses already an initial kinetic energy of $H\hbar\omega$. If the laser pulse is long and the system is conservative, the electron will pick up an extra kinetic energy equal to U_α (eq. 8) before it is detected. If the laser pulse is very short (Freeman et al, 1986), it will not pick up any significant kinetic energy from the field gradient and what it possesses in the final state is essentially its initial kinetic energy.

It is perhaps a coincidence that in the experiments of Muller et al (1983), Freeman et al (1986) the ponderomotive potential was not much different from the ATI energy ($H\hbar\omega$) so that it appeared that the electron had to overcome the ponderomotive potential (or equivalently ponderomotive shift) before it could be ionized.

(VI). Conclusion

By taking into account the temporal profile of the laser pulse and the participation of ATI in the ionization process, we succeeded in describing the acceleration of electrons by PMF in the tunnel ionization of Xe by a CO_2 laser pulse.

Acknowledgements

This work was supported in part by The Natural Science and Engineering Research Council of Canada and in part by Le Fonds FCAR de la province de Québec. The stimulating discussions with Dr. P.B. Corkum of the National Research Council of Canada, in particular, about the influence of the initial phase angle δ on the acceleration of electrons within the first optical cycle are greatly appreciated. The fruitful discussion with Dr. F. Yergeau and the helpful communication with Prof. N.B. Delone are also acknowledged.

References

M.V. Assomov, N.B. Delone and V.P. Krainov, 1986, JETP **21**, 2008; 1986, English translation in Sov. Phys. JETP **64**, 1191.

M. Aymar and M. Crance, 1981, J. Phys. B: At. & Mol. Phys., **14**, 3585.

S.L. Chin, F. Yergeau, P. Lavigne, 1985, J. Phys. B: At. & Mol. Phys. **18**, L213.

F.H.M. Faisal, 1987, J. Phys. B: At. & Mol. Phys. **20**, L299.

R.R. Freeman, T.J. McIlrath, P.H. Bucksbaum, and M. Bashkansky, 1986, Phys. Rev. Lett. **57**, 3156.

S. Geltman and J. Zakrzewski, 1988, J. Phys. B: At. & Mol. Phys. **21**, 47.

Y. Gontier and M. Trahin, 1981, J. Phys. B.: At. & Mol. Phys. **13**, 4381.

E. Karule, 1978, J. Phys. B: At. & Mol. Phys. **11**, **441**.

L.V. Keldysh, 1964, JETP **47**, 1945; 1965, English Translation in Sov. Phys. JETP **20**, 1307.

T.W.B. Kibble, 1966, Phys. Rev. **150**, 1060.

S. Klarsfeld and A. Maquet, 1980, Phys. Lett. **78A**, 40.

H.G. Muller, A. Tip and M.J. Van Der Wiel, 1983, J. Phys. B: At. & Mol., **16**, L679.

A.M. Perelomov, V.S. Popov and M.V. Terent'ev, 1966, JETP 50, 1393, 1966 English translation in Sov. Phys. JETP **23**, 924.

G. Petite, F. Fabre, P. Agostini, M. Crance and M. Aymer, 1984, Phys. Rev. **A 29**, 2677.

G. Petite, P. Agostini, 1987, "Multiphoton processes", Proceedings of the 4th International Conference on Multiphoton Processes, Ed. by S.J. Smith & P.L. Knight.

H.R. Reiss, 1980, Phys. Rev. A **22**, 1786.

W. Xiong, F. Yergeau, S.L. Chin and P. Lavigne, 1988, J. Phys. B: At. & Mol. **21**, L159-L164.

F. Yergeau, S.L. Chin and P. Lavigne, 1987, J. Phys. B: At. & Mol. Phys., **20**, 723.

W. Zernik and R.W. Klopfenstein, 1965, J. Math. Phys. **6**, 262.

STUDIES OF STRONG–FIELD EFFECTS IN MULTIPHOTON SUBPICOSECOND EXCITED PLASMAS: SOFT X–RAY FLUORESCENCE AND PROPAGATION

A. McPherson, T. S. Luk, G. Gibson,
J. C. Solem, K. Boyer and C. K. Rhodes
Laboratory for Atomic, Molecular, and Radiation Physics
Department of Physics, University of Illinois at Chicago
P. O. Box 4348, Chicago, Illinois 60680, USA

ABSTRACT

Studies concerning two aspects of the strong–field multiphoton response of matter are discussed, namely, (1) the generation of excited states and (2) the propagation of extremely intense radiation in plasmas. Fluorescence in the 7 – 79 nm region arising from excited ionic states produced in gas targets composed of the rare gases and N_2 is examined. Emissions from complex transition arrays stemming from open nd configurations in Kr and Xe are observed. Strong–field conditions are expected to produce significantly altered characteristics of propagation in plasmas which include the possibility of a new mode of channeled propagation.

I. INTRODUCTION

The response of materials to intense short–pulse radiation is an active area of inquiry. Two subjects of marked interest, in which multiquantum and strong–field processes can play a significant role, are (1) the mechanisms leading to the production of electronically excited states, particularly levels radiating in the soft x–ray region, and (2) the characteristics of electromagnetic propagation in plasmas experiencing very high rates of multiphoton ionization. Roughly speaking, we are interested in the physical regime involving intensities $\gtrless 10^{18}$ W/cm^2 so that the peak field strength E is significantly above an atomic unit (e/a_0^2). The experimental picture is now such that this regime can be systematically studied and we report on recent results concerning both of these areas.

II. DISCUSSION OF RESULTS

A. Ion Production Studies with the 248–nm KrF* Source

The presently operating subpicosecond (~ 600 fs) KrF* (248 nm) system, which incorporates a large aperture (100 cm^2) power amplifier, is illustrated in Fig. (1). The full operational characteristics are described in more detail elsewhere.[1] Experiments performed on Ne examining collision–free multiphoton ion production using a time–of–

SUBPICOSECOND KrF* LASER SYSTEM
at the University of Illinois at Chicago

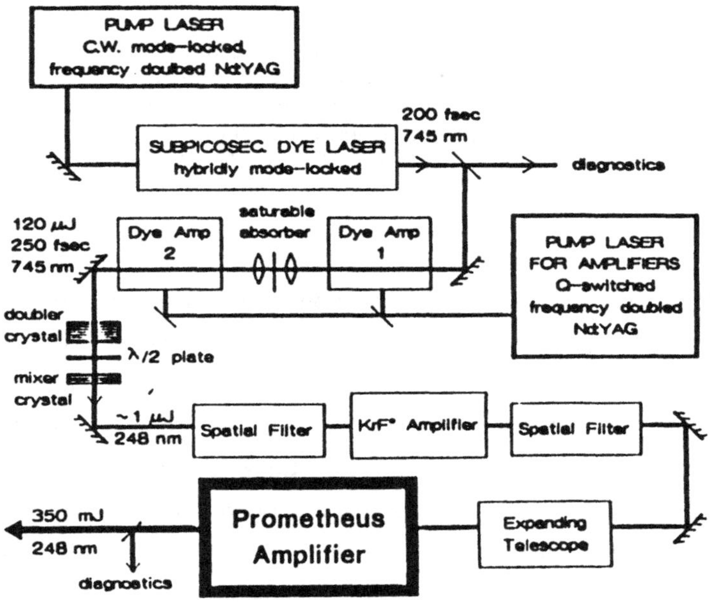

Fig. (1): Schematic of high brightness subpicosecond source incorporating a large aperture (100 cm^2) power amplifier (Prometheus).

flight technique[2] with this source have demonstrated the strong production of He–like Ne^{8+}, a significantly higher stage of ionization for this atomic system than had been previously[3] observed (Ne^{4+}) with an intensity estimated to be ~ 10^{16} W/cm^2. Although an accurate measurement of the peak intensity generated with the present source has not been concluded, the implication of this result is that a considerably higher intensity is now being produced. The total energy[4] needed to ionize the neutral Ne atom to the Ne^{8+} state is ~ 1 keV.

It is noted that the production of charged ions in this manner, provided that the density is sufficiently low so that space–charge influences are minimized,[5] is associated with a very small volume of phase space. The ions are spatially localized to within a rather small focal volume, perhaps on the order of ~ 10^{-9} cm^3 or less for the higher charge states, and the ions possess low kinetic energies, on the order of thermal values. Therefore, this technique of ionization might be used as a method to load ion traps with highly charged ions.

B. Studies of Soft X–Ray Fluorescence

The fluorescence spectra in the range spanning ~ 7 – 79 nm have been recorded for all the rare gases and N_2 using an experimental configuration previously described,[6] but modified to improve the short wavelength detection and the signal-to-noise. On the basis of the new results for multiphoton ionization of Ne, we expected the production of appreciably higher charge states in the heavier rare gases, such as Kr and Xe, by stripping well down into the 3d and 4d shells, respectively, of these materials.

Overall, there is considerable spectral evidence for the generation of more highly charged species than seen in past studies through the appearance of features at shorter wavelengths. For example, numerous new transitions were present in N_2, Ne, Kr, and Xe. Specifically, in Kr, radiation was observed down to 7.48 nm, a value essentially at the short wavelength limit of the instrument used. The spectral data are consistent with the identification[7] of emissions from Kr^{8+}, Kr^{9+}, Kr^{10+}, and Kr^{11+} in the wavelength region covering 7.3 – 11.8 nm. In general, the transition arrays associated with these ions in this spectral region are quite large, a feature complicating the identifications. Although the identifications must be regarded as tentative and are almost certainly incomplete, it appears that the types of transitions compiled in Table I are produced. The gas density of Kr generated by the pulsed valve was estimated to be ~ 10^{19} cm^{-3} for these measurements.

Table I: Compilation of tentatively identified transition arrays observed in Kr in the 7.3 – 11.8 nm region.

Ion	Transition Array
Kr^{8+}	$3d^9 4p \rightarrow 3d^{10}$
Kr^{9+}	$3d^8 4p \rightarrow 3d^9$
Kr^{9+}	$3p^5 3d^{10} \rightarrow 3p^6 3d^9$
Kr^{10+}	$3d^7 4p \rightarrow 3d^8$
Kr^{11+}	$3d^6 4p \rightarrow 3d^7$
Kr^{11+}	$3d^5 4s4p \rightarrow 3d^6 4s$

The shorter wavelength emissions seen in Kr and Xe are not necessarily weak components of the radiated energy. The Kr^{8+} features at ~ 11.6 nm and the Kr^{9+} array at ~ 10.3 nm are extremely strong emissions, being among the brightest observed throughout the entire region studied.

The most intense signals observed arise from Xe, specifically on the Xe^{8+} $4d^9 5p \rightarrow 4d^{10}$ transition[8] at 16.53 nm. The existence of such strong features, from this transition and other comparably intense arrays, is permitting the density dependence of these spectra to be examined. Although a full study has not been completed, we can state the following experimental observations. Identifiable features from Xe^{7+}, Xe^{8+}, and Xe^{9+} in the spectra exhibit complicated behaviors on density in the region covering ~ 5 x 10^{17} – 10^{19} cm^{-3}. The relative strengths of the emissions are also found to be quite sensitive to the position of the focus in the gas jet, a result that may indicate modifications in the propagation of the intense incident radiation. Of course, even for gas targets with a spatial size of ~ 1 mm, absorption of the emitted radiation by the target material is an important factor influencing the observed line strength.

Preliminary studies of emission from N_2 have also been performed. A strong and well developed Li-like spectrum[9,10] of N^{4+} is observed. As shown in Fig. (2), both the $1s^2np \rightarrow 1s^22s$ and $1s^2nd \rightarrow 1s^22p$ series clearly appear. Full examination of the spectrum also reveals the presence of f → d and s → p transitions. The pattern of observed transitions suggests that recombination from N^{5+} is playing an important role in the appearance of the Li-like spectrum. An analysis of the line strengths[11] of the 4d → 2p, 5d → 2p, and 4p → 2s transitions reveals that the observed relative intensities are quite close to those corresponding to a statistical distribution. This would be expected for the conditions of the plasma produced.[12]

If N^{5+} is present, emission from some of its excited states could also be generated. The position of the N^{5+} $1s3p(^3P^0) \rightarrow 1s2s(^3S)$ transition is indicated in Fig. (2) in connection with the small feature on the shoulder of the $1s^24p \rightarrow 1s^22s$ line at 162.4 Å. The presence of another feature at 122.4 Å in the spectrum also suggests emission from N^{5+}, in this case on the next higher member of that series, the $1s4p(^3P^0) \rightarrow 1s2s(^3S)$ transition. Additional studies will be needed to verify the presence of these N^{5+} lines.

C. Modes of Electromagnetic Propagation in the Strong–Field Regime

The new KrF* light source described in Section II.A enables the study of electromagnetic propagation under an unexplored regime of physical conditions. It appears[13–16] that a fundamentally new regime of electromagnetic propagation may develop in plasmas for subpicosecond radiation of sufficient intensity. In the high intensity case of interest, in which processes generating ionization dominate the coupling, multiphoton ionization is expected to produce a substantial reduction of both the linear and nonlinear refractive indices of the medium. For a suitably ionized plasma, this combination of the highly charged ions with the large electron density produced by the multiphoton coupling appears capable of producing an important new mode of radiative channeling.

Preliminary analysis[16] supports the presence of the following mechanism. For a sufficiently short laser pulse (~ 100 fs), the massive ions remain spatially fixed while the relatively mobile electrons are expelled by the ponderomotive force from the high

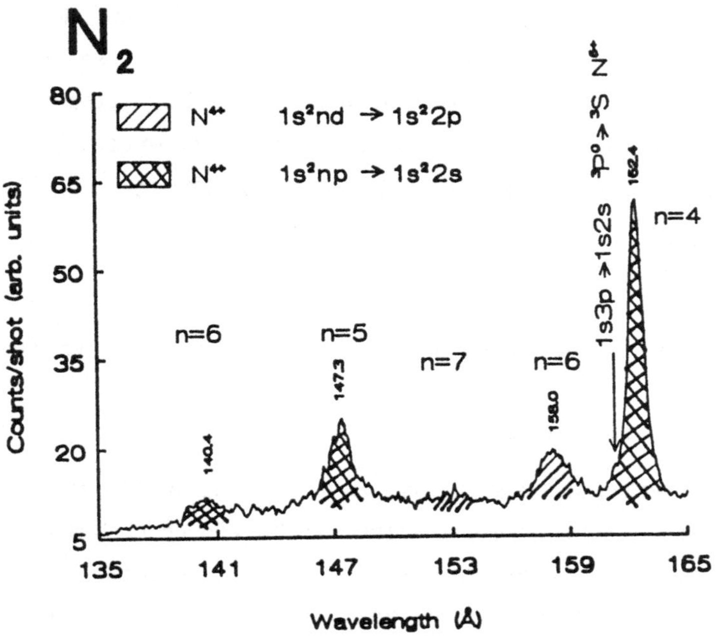

Fig. (2): A portion of the spectrum observed in N_2 illustrating the Li-like nd → 2p and np → 2s series. The position of the He-like (N^{5+}) $3p(^3P^o)$ → $2s(^3S)$ line is shown.

intensity zone. A state of equilibrium is then established between the ponderomotive and the electrostatic force densities owing to the charge displacement. Since the electrons, which embody a negative contribution to the index, are expelled, an on-axis region of relatively high refractive index is formed which supports the channeling of the radiation. This focusing mechanism is governed principally by the intensity and is not describable in terms of the conventional nonlinear index parameter[17] n_2. Since there is a strong tendency to locally reduce the plasma frequency ω_p in the region where the laser intensity is high, this process may enable propagation for appreciable distances in plasmas that would normally be considered as overdense.

A striking aspect of the analysis is the finding that an extremely strong frequency dependence favors the use of ultraviolet wavelengths in order to establish the conditions required for channeling. Scaling laws[16] indicate variations as rapid as ω^9.

The radial dependence of the refractive index n(r) governs the condition for channeling. In order to motivate the discussion, we employ a simple model and assume that the frequency of the radiation ω is above the plasma frequency ω_p. Since the ions are inertially confined for the short time (~ 100 fs) considered, the expulsion of the free electrons from the channel produces a reduced plasma frequency or, equivalently, an increase of the refractive index in the central region. It can be shown[16] that the contribution to the index from the ions under the strong-field condition is negligible.

An estimate of the form of the modified distribution of electronic charge density $\rho(r)$ arising from the presence of the intense radiation can be made in the following manner. The steady–state force balance between the radially outward ponderomotive force and the oppositely directed electron–ion attraction is represented in cylindrical symmetry for a completely ionized tenuous ($\omega >> \omega_p$) plasma by

$$\frac{2\pi e^2}{m\omega^2 c} \nabla I(r) + e^2 n_e \int \frac{\rho(r)\ (r - r')}{|r - r'|^3} d^3r' = 0 \quad . \tag{1}$$

For a gaussian intensity distribution $I(r)$, the exact solution for the charge density $\rho(r)$ is found to be

$$\rho(r) = 2B\ [1-(r/r_0)^2]\ e^{-(r/r_0)^2} \ , \tag{2}$$

in which $B = I_0(m\omega^2 c n_e r_0^2)^{-1}$ and I_0, m, c, n_e, and r_0 denote the peak intensity electron mass, speed of light, quiescent plasma density, and channel radius, respectively.

The condition necessary for focusing can now be estimated by describing the charge displacement by two regions such that $\rho(r<r_0\sqrt{2}) = \rho(0)$ and $\rho(r \geqslant r_0\sqrt{2}) = \rho(r_0\sqrt{2})$ and equating the angle of total internal reflection to the angle corresponding to the first minimum of diffraction. This procedure yields a critical intensity

$$I_c = \frac{\pi m\omega^2 c}{64(1 + e^{-2})r_e} \simeq (4.3 \times 10^{-2}) \frac{m\omega^2 c}{r_e} , \tag{3}$$

in which r_e denotes the classical electron radius. Significantly, this intensity is independent of the electron density n_e. For 248 nm radiation, $I_c \simeq 2.4 \times 10^{19}$ W/cm^2, a value approximately one half of the Compton intensity.

Energy deposition rates occurring in the formation of these channels can be large. The two principal mechanisms leading to energy deposition in the channel are (1) ionization and (2) the field energy associated with the charge displacement. The energy lost to ionization can be found by summing the ionization potentials[4] of electrons successively striped from the atom up to the ionization level produced by the field. For an example involving Ni–like Ho^{39+}, with $n_a = 2.3 \times 10^{18}$ cm^{-3} and a channel radius $r_0 \simeq 2c/\omega_p = 1.1$ µm, ionization produces a deposition of ~ 4.3 mJ/cm. However, for these conditions, the component corresponding to the charge–displacement accounts for ~ 160 mJ/cm, a substantially larger value.

Two questions concerning the stability of such channels have been examined. The first involves energetic stability associated with the splitting of a single channel into two or more having a reduced total electrostatic energy. The second considers stability against relativistic self–focusing in the walls of the channel, a mode observed in recent numerical simulations.[13] It has been shown[16] that neither of these questions leads to instability of the channel.

In light of the discussion given above, a series of simple experiments providing photographs of focused beams in air have been quite informative. In this set of experiments, the output beam was focused in ambient air with a lens (f/10). The result was a long luminescent streak associated with a very bright breakdown region. The incident energy in these studies was ~ 150 mJ, so the total peak power involved was approximately 250 gigawatts.

The observations made in these experiments consistently exhibited three salient features. They were (1) a marked reduction in the divergence of the transmitted radiation in comparison to that of the incident focused energy, (2) a substantial (~ 30% – 50%) absorption of energy in the focal region, and (3) the clear appearance of feathery spatially anisotropic structures or a wake in the emitted luminescence that was plainly outside the high intensity zone produced by the lens. Overall, the photographs showed that the spatial character of the propagating energy is strongly affected and that some energy is deposited in the air in regions well outside the directly produced focal region.

As a preliminary step toward an understanding of the observed behavior, we have analyzed the luminescent wake in the following manner. The transverse scale-length of the wake is ~ 2 – 3 mm. At least three possibilities for the transport and deposition of energy outside the high intensity zone exist; specifically, the energy could be carried by photons, electrons, or ions.

Since the range of energy transport is measured and the density of the medium is known, the following estimates can be made.

a. Photons

Soft x-rays in a sufficiently restricted range of wavelengths would have the observed range δ = 2 – 3 mm in air. If we approximate air as atomic nitrogen (two atoms per molecule) with a density of $\rho_N \sim 5.3 \times 10^{19}$ cm^{-3}, the absorption cross section σ_γ is given roughly by

$$\sigma_\gamma \simeq \frac{1}{\rho_N \delta} \simeq 0.6 - 0.9 \times 10^{-19} \text{ cm}^2. \tag{18}$$

We then take $\sigma_\gamma \simeq 10^5$ barns as an estimate. From available information on the photon absorption cross section[18] of N, the observed range is consistent with soft x-rays at ~ 300 eV or at ~ 900 eV. However, since the K-edge of nitrogen is ~ 400 eV, nitrogen is expected to be a strong radiator at neither.

b. Ions

Energetic ions could be the source of the energy causing the observed wake. Given the density of air of ~ 1 mg/cm^3, the corresponding ionic range would have to be ~ 0.2 – 0.3 mg/cm^2. It can be readily seen from data[19] on ion ranges in materials that the ion (N^+) kinetic energy necessary to produce the observed depth of penetration is ~ 1 – 2 MeV. For the conditions present, no reasonable mechanism seems capable of producing ions at this energy. Therefore, we believe that it is very unlikely that ions are the cause of the observed effect.

c. Electrons

Energetic electrons produced in the focal volume could generate the wake detected. If the origin of the luminescence arises from electrons expelled from the focal zone by the effect of the ponderomotive potential, then it may be possible to estimate the maximum intensity actually produced by measurement of the range of deposition in conjunction with the known range–energy[20] relationships for materials. Of course, only a rather small fraction of electrons would be expected to behave in this fashion, since the energy associated with the charge separation would otherwise be too great. Quantitatively, this latter aspect needs to be examined further.

The range–energy curve[20] for electrons absorbed in Al can be used to estimate the particle energy. The difference between Al and air for the same range (mg/cm^2) is within the uncertainty of this estimate, so we will take the liberty of neglecting it. In this case, the range of 0.2 – 0.3 mg/cm^2 corresponds to electrons in the 10–12 keV range. For a wavelength of 248 nm, a ponderomotive potential of ~ 10 keV corresponds to an intensity of ~ 1.7 x 10^{18} W/cm^2. An intensity of this level, although approximately eight–fold more than that expected to be produced by the f/10 optical system, could conceivably be produced by self–focusing action in small regions inside the focus at the power levels used in this experiment.

There is some additional experimental evidence from photoelectron studies[21] supporting the expectation that electrons having an energy characteristic of the ponderomotive potential will be produced in the focal volume. Figure (3) shows a time–of–flight electron spectrum for Ne observed with subpicosecond 248–nm radiation at a maximum intensity ~ 10^{16} W/cm^2. The peak value of the ponderomotive potential, $U(I,\omega)$ ~ 60 eV, with I denoting the intensity and ω representing the frequency, is indicated. The shaded region, which is taken to be between ~ $1/2U(I,\omega)$ and ~ $2U(I,\omega)$, simply illustrates that a significant fraction of the electrons produced have an energy scale in the approximate vicinity of $U(I,\omega)$. Furthermore, since the range depends

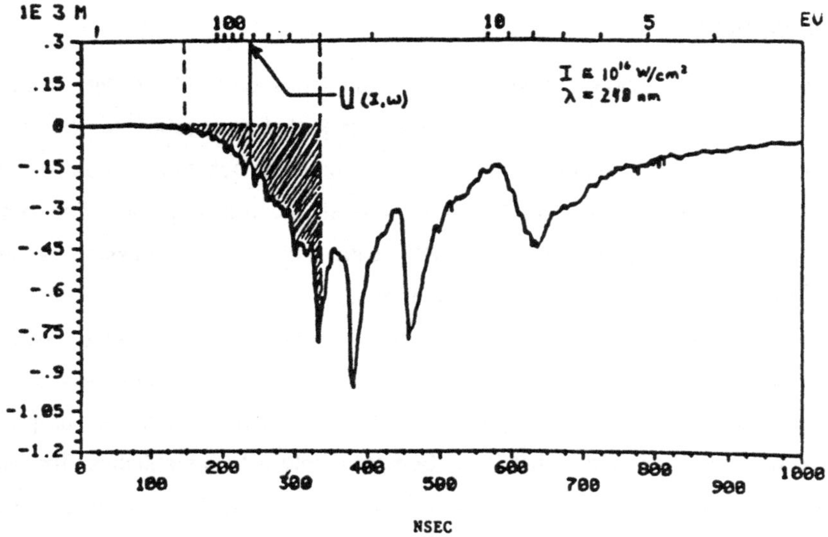

Fig. (3): Time–of–flight photoelectron energy spectrum of Ne obtained with subpicosecond 248–nm radiation at an intensity ~ 10^{16} W/cm^2. The peak ponderomotive potential U(I,ω) ~ 60 eV is shown. The shaded region corresponds to ~ 1/2U(I,ω) to ~ 2 U(I,ω). See text for discussion.

approximately quadratically[20] on the electron energy below ~ 100 keV, it is only the higher energy electrons that develop any appreciable range.

Finally, it is interesting to note that these general notions appear to qualitatively agree with streak camera data obtained in previous studies of optical breakdown in gases.[22] The velocity observed in that work was consistent[23] with the transport of energy outside the focal zone by energetic electrons.

The hypothesis that emerges from these estimates is that the observed wake is produced by electrons and that the limiting range gives an approximate measure of the magnitude of the ponderomotive potential. If this picture is valid, a measurement of the maximum range can furnish a <u>direct</u> indication of the peak intensity developed in the medium. Normally, the measurement of an intensity, particularly under dynamic conditions of propagation, is extraordinarily difficult. Consequently, this phenomenon may represent a rather simple diagnostic on the conditions that develop during propagation of extremely intense radiation through materials.

Although it is unlikely that these preliminary experiments generated intensities as high as the Compton intensity, qualitatively the theoretical discussion given above involves a physical mechanism with certain predicted tendencies. These anticipated properties are consistent with <u>all three</u> observed effects, namely, (1) the inclination to focus or collimate the radiation propagating in the plasma, (2) the significant energy deposition, possibly stemming from ionization and charge–displacement, and (3) the expulsion of energetic charged particles from the high intensity region.

III. CONCLUSIONS

Experiments have shown that intense subpicosecond ultraviolet irradiation of gaseous targets gives rise to the copious production of fluorescence in the 7 – 79 nm region. The heavier rare gases, Kr and Xe, exhibit a complex series of emissions arising from excitations of the n = 3 and n = 4 shells, respectively. The data on Kr are consistent with the production of Kr^{11+} with subsequent emission arising from both the $3d^6 4p \rightarrow 3d^7$ and $3d^5 4s4p \rightarrow 3d^6 4s$ arrays.

Both theoretical findings and preliminary experimental evidence suggest the possibility of a new mode of channeled propagation in the strong–field regime. A simple picture used to explore the influence of very intense fields on propagation in plasmas led to the conclusion that the Compton intensity establishes the scale for the prominence of very sharply altered conditions of propagation, such as channeling. The analysis presented herein applies generally to situations in which processes resulting in ionization dominate the coupling to the medium, a property of all materials in the strong–field limit being considered.

IV. ACKNOWLEDGEMENTS

The technical assistance of J. Wright, P. Noel and R. Slagle is acknowledged. Support for this research was provided by the Air Force Office of Scientific Research, the Office of Naval Research, the Strategic Defense Initiative Organization, and the Lawrence Livermore National Laboratory.

V. REFERENCES

1. T. S. Luk, A. McPherson, G. Gibson, I. A. McIntyre, X.–J. Pan, K. Boyer and C. K. Rhodes, "Ultrahigh Intensity KrF* Laser System," to be submitted to Opt. Lett.
2. T. S. Luk, U. Johann, H. Egger, H. Pummer, and C. K. Rhodes, "Collision–Free Multiple Photon Ionization of Atoms and Molecules at 193 nm," Phys. Rev. A32, 214 (1985).
3. C. K. Rhodes, "Physical Processes at High Field Strengths," Physica Scripta T17, 193 (1987).
4. T. A. Carlson, C. W. Nestor, Jr., N. Wasserman, and J. D. McDowell, "Calculated Ionization Potentials for Multiply Charged Ions," At. Data 2, 63 (1970).
5. T. S. Luk and C. K. Rhodes, "Multiphoton Dissociative Ionization of Molecular Deuterium," Phys. Rev. A 38, 6180 (1988).
6. A. McPherson, G. Gibson, H. Jara, U. Johann, T. S. Luk, I. A. McIntyre, K. Boyer, and C. K. Rhodes, "Studies of Multiphoton Production of Vacuum Ultraviolet Radiation in the Rare Gases," J. Opt. Soc. Am. B4, 595 (1987).

7. L. A. Jones and E. Källne, "A Study of the VUV Emission from Highly Ionized Krypton in a Theta Pinch Plasma," J. Quant. Spectrosc. Radiat. Transfer 30, 317 (1983).
8. J. Blackburn, P. K. Carrol, J. Costello, and G. O'Sullivan, "Spectra of Xe VII, VIII, and IX in the Extreme Ultraviolet: 4d–np, nf Transitions," J. Opt. Soc. Am. 73, 1325 (1983).
9. S. Bashkin and J. O. Stoner, Jr., Atomic Energy Levels and Grotian Diagram 1, Hydrogen I – Phosphorous XV (North–Holland, Amsterdam, 1975) p. 148.
10. R. L. Kelly and L. J. Palumbo, Atomic and Ionic Emission Lines below 2000 Angstroms, NRL Report 7599 (1973).
11. W. L. Wiese, M. W. Smith, and B. M. Glennon, Atomic Transition Probabilities, Vol. I, Hydrogen through Neon (NSRDS – NBS 4, Washington, D. C., 1966).
12. H. R. Griem, Spectral Line Broadening by Plasmas (Academic Press, New York, 1974).
13. W. B. Mori, C. Joshi, J. M. Dawson, D. W. Forslund, and J. M. Kindel, "Evolution of Self–Focusing of Intense Electromagnetic Waves in Plasma," Phys. Rev. Lett. 60, 1298 (1988).
14. G.–Z. Sun, E. Ott, Y. C. Lee, P. Guzdar, "Self–Focusing of Short Intense Pulses in Plasmas," Phys. Fluids 20, 526 (1987).
15. Ya. L. Bogomolov, S. F. Lirin, V. E. Semenov, and A. M. Sergeev, "Ionization Self–Channeling of Extremely Intense Electromagnetic Waves in a Plasma," Pis'ma Zh. Eksp. Teor. Fiz. 45, 532 (1987) [Engl. transl.: Sov. Phys. – JETP Lett. 45, 680 (1987)].
16. J. C. Solem, T. S. Luk, K. Boyer, and C. K. Rhodes, "X–Ray Amplification with Charge–Displacement Self–Channeling," Phys. Rev. Lett., submitted.
17. P. L. Kelley, "Self–Focusing of Optical Beams," Phys. Rev. Lett. 15, 1005 (1965).
18. E. F. Plechaty, D. E. Cullen, and R. J. Howerton, "Tables and Graphs of Photon–Interaction Cross Sections from 0.1 keV to 100 MeV Derived from the LLL Evaluated–Nuclear–Data Library," UCRL–50400, Vol. 6, Rev. 3, 11 November 1981, p. 31.
19. U. Littmark and J. F. Ziegler, Range Distributions for Energetic Ions in All Elements, Vol. 6 (Pergamon Press, New York, 1980) p. 99.
20. R. D. Evans, The Atomic Nucleus (McGraw–Hill, New York, 1955) p. 624.
21. T. S. Luk, T. Graber, H. Jara, U. Johann and C. K. Rhodes, "Subpicosecond Ultraviolet Multiphoton Electron Spectroscopy of Rare Gases," J. Opt. Soc. Am. B4, 847 (1987).
22. A. J. Alcock and M. C. Richardson, "Creation of a Spark by a Single Subnanosecond Laser Pulse," Phys. Rev. Lett. 21, 667 (1968).
23. M. C. Richardson, private communication.

PART III: Electron Correlations in Radiative Processes

THEORETICAL ASPECTS OF MULTIPHOTON IONIZATION AND HARMONIC GENERATION

Göran Wendin*) and Lars Jönsson*)
Laboratory of Atomic, Molecular and Radiation Physics
Department of Physics, The University of Illinois at Chicago
Chicago, Illinois 60680, USA

ABSTRACT

We consider diagrammatic expansions for p-th order polarization relevant for harmonic generation. We present preliminary results of a numerical calculation of lowest-order multiphoton matrix elements using a complete basis (LDA) of one-electron wave functions for xenon. The present stage is exploratory. Certain aspects (certain paths) of up to 9-th harmonic order have been investigated. We have also investigated partial, infinite order, summations of contributions involving "Rabi-like" coupling of the initial state to the excited states. This takes into account dressing of states through renormalization and effective interactions.

1. INTRODUCTION

Subpicosecond pulsed optical and uv-lasers are now reaching very high intensities, causing massive ionization of atomic gases [1,2] and solids [3] during a single pulse. At very high intensities, the laser radiation not only induces multiphoton transitions but also perturbes the electronic structure. In particular at low laser frequencies and high intensities there is no clear distinction between multiphoton ionization and field ionization via tunneling or escape over the potential barrier.

From a theoretical point of view [4] there are two obvious aspects of this highly non-linear problem, namely (i) the one electron problem with laser-field-perturbed orbitals, and (ii) the many-electron problem. Even the one-electron problem is a major stumbling point: it is very difficult to construct continuum states in the region where the action is (i.e. where photons are being absorbed and emitted and where the photoelectron energy distribution is formed) because the atomic and laser fields have comparable magnitudes and different symmetries.

Concerning the many-electron problem, several points should be noted [4]:

(i) The electronic charge distribution will adjust to the time-dependent laser field (screening, relaxation) and give rise to an effective field, (ii) electron correlation effects may cause multielectron excitations, and (iii) the effective laser field may cause multielectron excitation and ionization.

Strong laser fields will lead to multiple electron ionization. An interesting problem is whether multielectron ionization in practice always is sequential or whether direct ("simultaneous") multiple electron emission should be observable.

For sufficiently strong and short laser pulses one can imagine the distinction between direct and sequential multielectron processes to become unclear: if the interaction energy during the ionization time increases by an amount equal to the Coulomb interaction, electrons could in principle pour out of the atom in a steady current.

In spite of the great challenge involved in the interaction of very intense and short laser pulses with matter, there is a clear necessity for theoretical calculations based on systematic perturbation theory and realistic descriptions of electronic states. The purpose of the present work is therefore to investigate the possibilities and limitations of diagrammatic perturbation theory for calculating the non-linear response of heavy atoms at low-to-moderate intensities.

In this paper we consider diagrammatic expansions for p-th order dipole polarization relevant to high-order harmonic generation, for which there are a number of interesting experimental results [5-7]. Also, recently, there has been considerable progress in calculating high-order multiphoton transition amplitudes [8-10], although so far limited to hydrogenic systems.

Here we present preliminary results of a numerical calculation of multiphoton matrix elements and polarization using a complete basis of one-electron wave functions for xenon in the local density approximation (LDA) (for a discussion of related work on two and three-photon processes, see [11-15]. The present stage is exploratory. Only radial parts of induced densities and perturbed orbitals are considered. Certain aspects (certain paths) of up to 9-th harmonic order has been investigated. We find that only paths involving many s-wave intermediate states seem to pose a real numerical problem.

Finally, we have investigated partial, infinite order, summations of contributions involving "Rabi-like" coupling of the initial state to the excited states. This represents dressing of states through renormalization and effective interactions, and provides an extension of the work in Ref. [4].

2. GENERAL CONSIDERATIONS

The p-th order dipole polarization $P^{(p)}(\omega)$ involved in p-th order harmonic generation can be expressed in terms of a p-th order perturbed orbital $\delta\phi_{li}{}^{(p)}$ according to

$$P^{(p)}(\omega) = \langle i | \hat{\mathbf{e}}\cdot\mathbf{r} | \delta\phi^{(p)} \rangle \tag{1}$$

$$\delta\phi_{li}{}^{(p)}(\mathbf{r};\omega) = \sum_{m} \frac{| ml \rangle \, \tilde{M}_{mi}{}^{(p)}(\omega)}{\tilde{\omega}_{mi}(\omega) - p\omega} \tag{2}$$

$$\widetilde{M}_{ni}^{(p)}(\omega) = \langle n | \mathbf{E}\cdot\mathbf{r} | \delta\phi_{li}^{(p-1)} \rangle \tag{3}$$

In Eq. (2), $\delta\phi_{li}^{(p)}(\mathbf{r};\omega)$ is the p-th order perturbed orbital, $\widetilde{M}_{mi}^{(p)}(\omega)$ is the p-th order dressed multiphoton matrix element, and $\widetilde{\omega}_{mi}(\omega)$ represents dressed excitation energies. The set $\{\phi_m\}$ ($\phi_m \equiv |m\rangle \equiv |ml\rangle$) is a complete basis of free-atom one-electron wave functions. $\hat{\mathbf{e}}$ is the polarization vector of the electric field **E**. We have simplified the notation by omitting complex conjugate quantites. $\widetilde{\omega}_{mi}(\omega)$ contains intensity dependent Stark shifts and $\widetilde{M}_{mi}^{(p)}(\omega)$ includes effects of intensity dependent renormalization.

In the lowest order approximation, we consider absorption of p photons by a free, undressed atom.

$$\delta\phi_{li}^{(p)}(\mathbf{r};\omega) = \sum_m \frac{|ml\rangle\, M_{mi}^{(p)}(\omega)}{\omega_{mi} - p\omega} \tag{4}$$

$$M_{ni}^{(p)}(\omega) = \langle n | \mathbf{E}\cdot\mathbf{r} | \delta\phi_{li}^{(p-1)} \rangle \tag{5}$$

The zeroth order perturbed wave function $\delta\phi_{li}^{(0)}$ is the initial state orbital, $\delta\phi_{li}^{(0)} = \phi_i$.

3. NUMERICAL CALCULATION OF LOWEST-ORDER MULTIPHOTON TRANSITION AMPLITUDES

The goal is to calculate high-order multiphoton matrix elements in the **E·r** gauge by direct construction of intermediate state wave packets (perturbed orbitals, Green functions) from a wave function basis. In this Section we present preliminary results of a study of the development of numerical problems with increasing order.

The present calculation has been pushed to 9-th order along (and beyond) some of the paths possible in Fig. 1.

The basis set consists of numerical one-electron wave functions calculated in the local-density approximation (LDA) without self-interaction corrections [11-15]. For Xe, this potential has no bound excited states, only a *continuum of excited states*. The LDA potential gives a good average description of the Xe excitation spectrum [11] but clearly cannot describe resonance processes involving bound states. In all angular momentum channels there is an excitation gap and then a continuous excitation spectrum.

Roughly speaking, the LDA potential for Xe potential consists of a realistic potential well in the region of occupied states, and at larger distances a constant

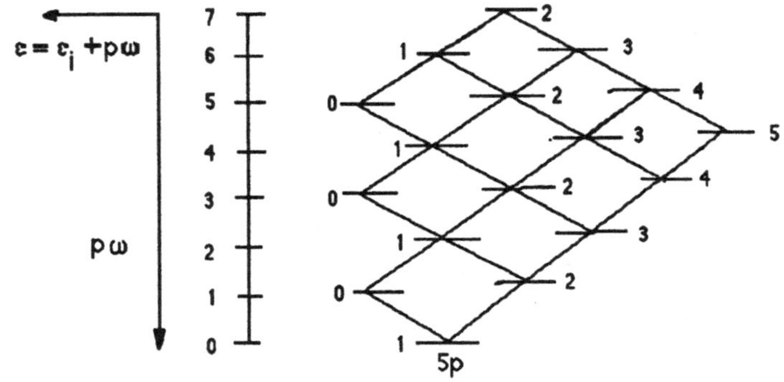

Fig. 1. Level scheme for excitation of the 5p-shell of Xe. The left part shows the order of the multiphoton process plus radiative decay and ionization; the right part shows the possible paths, numbers indicating angular momenta.

potential. As seen in Fig. 2, the outer part of these continuum wave functions are indeed *free-electron* like. The inner parts, however, are *atomic-like*: the potential in this region will serve as a realistic momentum source and sink for absorption and emission of photons.

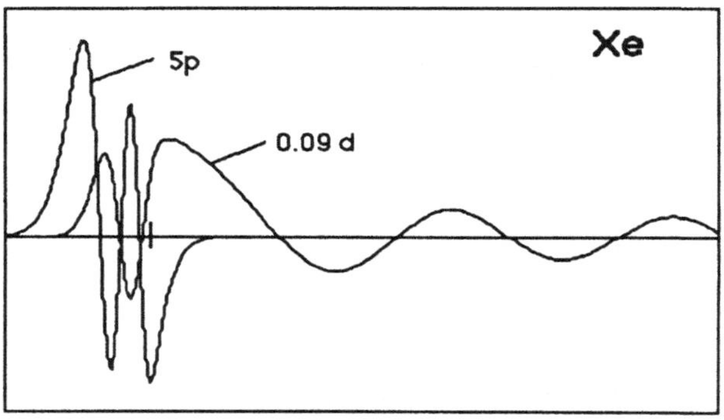

Fig. 2. Xe 5p initial state and εd (ε=0.091 Ry) continuum final state basis wave functions. The horizontal scale covers r=0-56 a.u. The first twenty percent (tickmarked) is *logarithmic* and covers 0-2 a.u. The remaining part (2-56 a.u.) is *linear*. The same is true for Figs. 3-7.

We only consider the *radial* parts of the perturbed orbitals and proceed in an iterative manner

$$M_{ni}^{(p)} = \langle n | r | \delta u_{li}^{(p-1)} \rangle \tag{6}$$

$$\delta u_{li}^{(p)} = \sum_n \frac{|n\rangle M_{ni}^{(p)}}{\omega_{ni} - p\omega} \tag{7}$$

The iteration is started by inserting the Xe 5p ground state orbital for the perturbed orbital, $\delta u_{li}^{(0)} = u_i(r)$ (i=5p) into Eq. (6). The resulting spectrum for the first order transition matrix element is inserted into Eq. (7) for the first order perturbed orbital $\delta u_{li}^{(1)}$. The first order perturbed orbital $\delta u_{li}^{(1)}$ is then reinserted into Eq. (1), and so on.

During the iteration, a combined "filtering" and tail procedure is applied. For $p\omega < |\varepsilon_i|$ (excitation energy less than the gap) the perturbed orbital in Eq. (7) should decay exponentially. However, the high-energy cut off (presently 20 Ry) of the energy sum in Eq. (7) will introduce rapid, and rapidly growing, oscillations in the tail at large distances. This will quickly corrupt the whole scheme since we are working in the **E·r** gauge and are very sensitive to poor behaviour at large distances: the low-energy continuum wave functions will have a large overlap with any "noise" at large radial distances.

To fix this, we apply a smoothing procedure to tail region of the wave packet calculated from Eq. (7), proceed outwards until the behaviour looks exponential and then fit to an exponential tail. In this way, we have been able to iterate Eqs. (6) and (7) up to 9-th order for certain paths in Fig. 1 (notably the 1212121212 path). However, by then the exponential tail of the perturbed-orbital wave packet starts seriously to extend beyond the end point of the radial mesh (presently cut off at 56 a.u.), causing serious oscillations of the matrix elements in the next step of iteration. This happens, according to the present preliminary results, already at 6-th order for the 10101010.... path.

In this paper we shall only present results for *radial* perturbed orbitals corresponding to certain paths in Fig. 1 (we shall use the notation po1212, po1232, etc, which explicitly shows the path in Fig. 1). Angular coefficients will be omitted and the perturbed orbitals will be "arbitrarily" normalized to allow convenient presentation and comparison. Therefore, from the present material there is no information on relative amplitudes for interfering paths. Here we only want to show the radial development of the perturbed orbitals with increasing order.

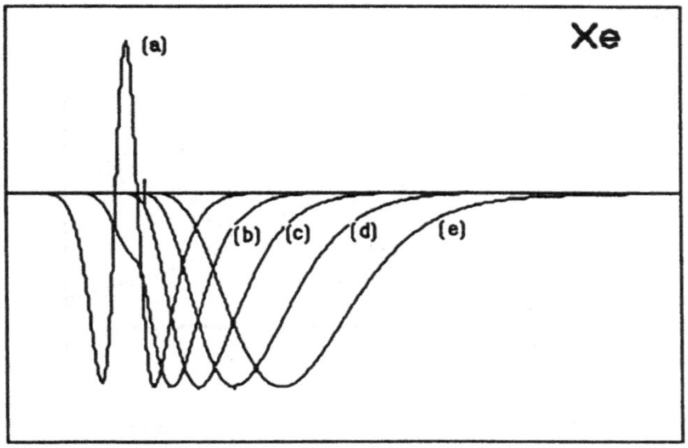

Fig. 3. Perturbed orbitals for the 123456 path (up to 5-th order), ω=0.1 Ry (arbitrary normalization to the same maximum amplitude). (a) po12, (b) po123, (c) po1234, (d) po12345, and (e) po123456.

Fig. 3 shows the results for perturbed orbitals with successively increasing angular momenta, pol2 ($\delta u_{2i}^{(1)}$), pol23 ($\delta u_{3i}^{(2)}$), pol23 ($\delta u4_{1i}^{(3)}$), pol2345 ($\delta u_{5i}^{(4)}$), and pol23456 ($\delta u_{6i}^{(5)}$).

With increasing order, the wave packet expands radially. The reason may be the oscillating character of the matrix element M_{ni} [Eq. (7)] which tends to weight energies closer to the continuum threshold. The expansion is basically not due to the admixture of high angular momenta, because, as shown below, low-angular-momentum paths show the same effect in even more pronounced ways.

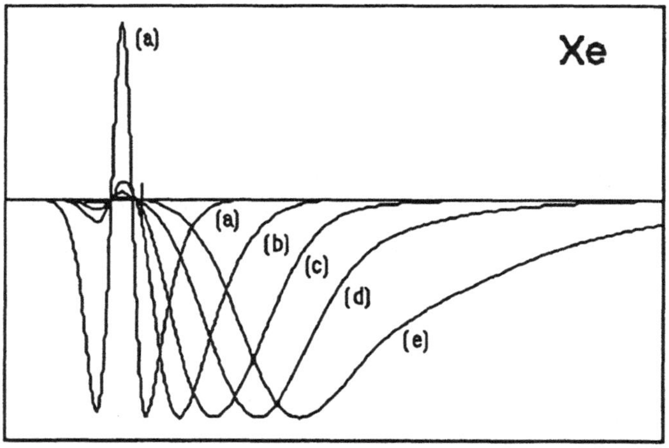

Fig. 4. Perturbed orbitals $\delta u_{2i}^{(p)}$) for ω=0: (a) pol2, (b) pol212, (c) pol21212, (d) pol2121212 and (e) pol212121212.

Figure 4 shows the development of the perturbed orbitals up to 9-th order when the final states are restricted to l=1,2 angular momentum states (see Fig. 1). On this scale the perturbed-orbital wave packets look perfectly regular (apart from irregularities due to the low resolution plot).

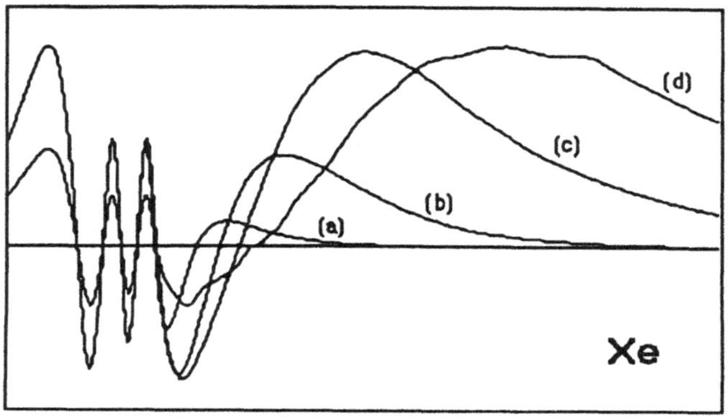

Fig. 5. Perturbed orbitals $\delta u_{0i}^{(p)}$) for ω=0: (a) pol0, (b) pol010, (c) pol01010, and (d) pol0101010 (the last, 7-th order, orbital is totally unreliable).

Figure 5 shows the development of the perturbed orbitals with final states restricted to l=1,0. This is clearly the most difficult case, because already the 5-th order wave packet extends beyond the radial cutoff, which introduces serious oscillations in the 7-th order wave packet. That the 10101010.... path should be the difficult one seems understandable. The matrix elements $M_{ni}^{(p)}$ are large close to threshold (s-wave final states) and the oscillatory character of the matrix element will further weight states close to threshold.

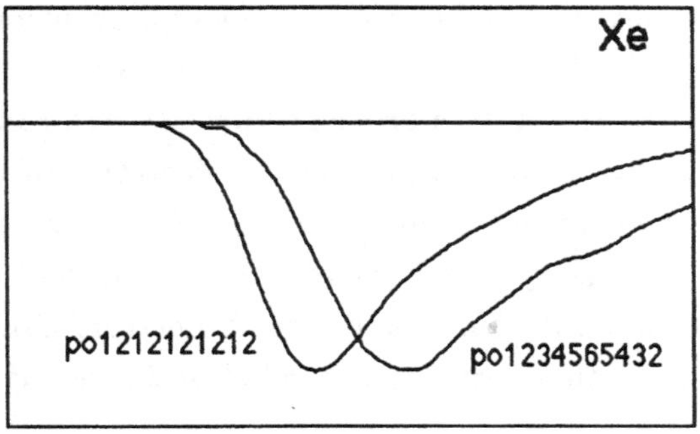

Fig. 6. 9-th order l=2 perturbed orbitals, $\delta u_{2i}^{(9)}$, from two different paths (the po1234565432 orbital is totally unreliable).

Figure 6, finally shows the results of two different paths in 9-th order (the normalization is arbitrary). The result for the 1234565432 path is numerically very poor an is only included to give a qualitative comparison.

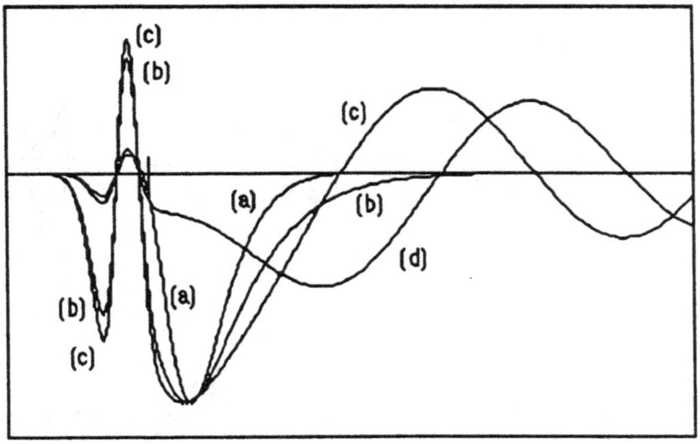

Fig. 7. $\delta u_{2i}^{(3)}$ (po1212) for (a) 3ω=0, (b) 0.60 and (c,d) 0.65 Ry (ω,2ω=0 in the first two steps, for simplicity). The LDA 5p binding energy is 0.617 Ry. (c) is the real part and (d) is the imaginary part of the continuum perturbed orbital for 3ω=0.65.

Figures 4-6 show that, for fixed frequency ω (here ω=0) the wave packets move towards higher r-values with increasing order. Figure 7 illustrates the radial development of the real part of $\delta u_{2i}^{(3)}$ (pol212) for fixed angular momentum when the frequency is increased to just below and just above threshold. For frequency just below threshold, the amplitude to increase in the inner region and the tail begins to extend. For frequency just above threshold, the tail develops a continuum oscillation, and in addition the perturbed orbital develops an imaginary part which is 90 degrees out of phase.

At present, there do not seem to be any fundamental problems that prevent us from going to considerably higher order. The obvious breakdown of the scheme in high order, particularly clear in the 7-th order perturbed orbital in Fig. 5, is mainly caused by the finite radial size (r<56 a.u.) of the one-electron basis. There should be no computational problems involved in extending the radial mesh to 100-200 a.u., and perhaps increasing the energy cutoff somewhat. The present calculations have been done on a MacII with 1 Mb RAM and Absoft Fortran/020 V2.3 in about an hour. This means that calculations with an extended basis should present no problems with a fast computer.

4. PERTURBATION EXPANSIONS AND DRESSED STATES

We first study the linked diagrammatic perturbation expansion for first order (p=1) harmonic generation with inclusion of dressing due to high intensity effects of an adiabatically switched-on time-dependent electric field. Figure 8 shows a few examples of a particular sequence of diagrams exclusively connecting the ground state to the excited states (generalization of ground state Stark shift and Rabi type behaviour to many level systems)

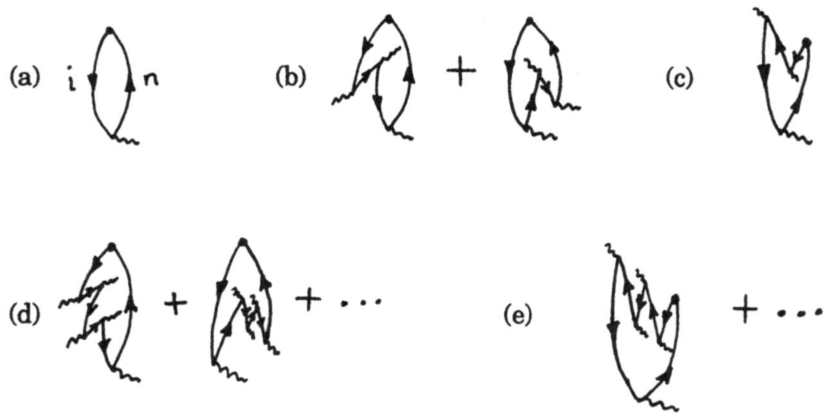

Fig. 8. Ground state Stark shift and renormalization diagrams. (a) is the lowest order polarization contribution $P_0^{(1)}$ to the first harmonic. (b) and (d) represent self-energy corrections (dressed energies), while (c) and (e) give rise to renormalization (dressed wave functions).

Combining all possible time orderings, the result takes a simple form. The third order contribution in Figs. 8b,c split into two parts with different character: (i) $P_0^{(1)}\,\Sigma_i^{(2)}$ from Fig. 8b and (ii) $P_0^{(1)}\,X_i^{(2)}$ fom Fig. 8c, where the zeroth-order polarization $P_0^{(1)}$, the second order self-energy contribution $\Sigma_i^{(2)}$ and the second-order renormalization contribution $X_i^{(2)}$ are given by

$$P_0^{(1)} = -\sum_q \frac{d_{iq}\,V_{qi}}{\omega_{qi}-\omega} \tag{8}$$

$$\Sigma_i^{(2)} = -\sum_q \frac{|V_{qi}|^2}{\omega_{qi}-\omega} = -I\sum_q \frac{|d_{qi}|^2}{\omega_{qi}-\omega} = -\alpha I \tag{9}$$

$$X_i^{(2)} = -\sum_q \left|\frac{V_{qi}}{\omega_{qi}-\omega}\right|^2 = -I\sum_q \left|\frac{d_{qi}}{\omega_{qi}-\omega}\right|^2 = -\gamma I \tag{10}$$

where

$$V_{qi} = \langle q|\mathbf{E}\cdot\mathbf{r}|i\rangle = E\,d_{qi} \tag{11}$$

The fifth order diagrams in Fig. 8d reduce to (i) a fourth order self-energy term $P_0^{(1)}\Sigma_i^{(2)}\,X_i^{(2)}$, and (ii) a fourth order self-energy term $P_0^{(1)}\,[\Sigma_i^{(2)}]^2$, while the fifth order diagrams in Fig. 8e reduce to a single fourth order renormalization term $P_0^{(1)}$ $[X_i^{(2)}]^2$. The remaining fifth order diagrams give rise to cross terms involving self-energy and renormalization contributions.

Extrapolation of the first, third and fifth order results to infinite order suggests that the polarizability can be written on the form

$$P^{(1)} = \sum_n \frac{d_{in}\,V_{ni}\,Z_i}{\omega_{ni}-\omega-\Sigma_i} \tag{12}$$

where

$$\Sigma_i = \Sigma_i^{(2)}\,Y_i \qquad \text{(a)}$$

$$Z_i = 1 + X_i^{(2)} + \ldots \qquad \text{(b)} \quad (13)$$

$$Y_i = 1 + X_i^{(2)} + \ldots \qquad \text{(b)}$$

Σ_i represents the Stark shift of the initial (ground) state energy. Z_i represents *renormalization*: at finite intensity it reduces the weight of the initial state and lowers the polarizability. To first order in the intensity, Z_i and Y_i are identical. However, in higher orders, this does not seem to be true: the series are not geometric, and the behaviour is much more complicated and gives e.g. rise to the Rabi solutions for a two-level system [16].

To the above treatment we must add the processes describing coupling of excited levels by the field, or scattering between different excited states of the unperturbed atom, through absorption and emission of photons [4,11a]. These are the diagrams that relate to the Keldysh approach [17] with a dressed free-electron final state. Here, however, the dressed state is an *atomic* continuum state with a possibility to absorb and emit momentum and involving *real* transitions between ATI ladders [18]. The result may be written as a generalization of Eq. (12) ,

$$P^{(1)} = \sum_n \frac{d_{in} \tilde{V}_{ni} \tilde{Z}_i}{\omega_{ni} - \omega + \tilde{\Sigma}_{ni}} \tag{14}$$

$\tilde{\Sigma}_{ni} = \tilde{\Sigma}_n - \tilde{\Sigma}_i$ where $\tilde{\Sigma}_n$ is the complex Stark shift of the excited state [19], and $\tilde{V}_{ni}$ has become an effective matrix element due to scattering between excited states [4,11a]. Likewise, $\tilde{Z}_i$ and $\tilde{\Sigma}_i$ are modified due to dressed excited states. Moreover, the excited state Stark shift $\tilde{\Sigma}_n$ is modified due to repeated scattering back to the initial state.

Finally, our present understanding is that Eq. (14) can be generalized to p-th order polarization according to

$$P^{(p)} = \sum_{nkqm} \frac{d_{in} \tilde{V}_{nk} \cdots\cdots \tilde{V}_{qm} \tilde{V}_{mi} \tilde{Z}_i}{[\omega_{ni} - p\omega - \tilde{\Sigma}_{ni}] \cdots\cdots [\omega_{qi} - 2\omega - \tilde{\Sigma}_{qi}] [\omega_{mi} - \omega - \tilde{\Sigma}_{mi}]} \tag{15}$$

Equation (15) unifies and generalizes the result derived in Refs. [11a,c] involving field-induced coupling between excited free-atom levels (now via the ground state as well as excited states), and the result in Ref. [4a] involving Stark shift of the initial state. Moreover, the form of Eq. (15) includes many-electron effects, e.g. in the form of screened electron-photon interactions and shifted excitation energies due to relaxation and correlation effects [4,11].

Equation (15) and the discussion leading to it now provides an "explicit" way to construct the *dressed perturbed orbitals* in Eqs. (1)-(3) including finite field effects and many-electron interactions. With the direct method for calculating these perturbed orbitals, it seems relatively straightforward to go beyond perturbation theory. The Dalgarno-Lewis method of differential equations, which so far has been the most common way to calculate multiphoton matrix elements perturbatively (see e.g. Ref. [9]), may possibly be more accurate as a numerical scheme. However, in a strong-field situation one will have to include effective energies and interactions. To our knowledge, such calculations have not yet been performed, and it is difficult to say which will be the most efficient method for calculating high order multiphoton processes in strong-field situations.

5. CONCLUDING REMARKS

Although we have not talked much about it, the many-electron problem [4] is really at the focus of interest for this investigation. In order to get anywhere with the many-electron system, it is essential to be able to perform realistic high-order calculations of absorption and emission of photons. From this work, it seems that reasonably accurate numerical methods can be developed for calculating 10-th to 15-th order multiphoton one-electron matrix elements for heavy atoms. It should therefore be possible to perform perturbative calculations [of the type in Eqs. (14), (15)] of harmonic generation and above threshold ionization on many-electron systems involving dressing of states, relaxation, screening etc., to obtain realistic high-order results for at least moderate laser intensities.

It is important to be able to calculate sufficently high orders at sufficiently high photon energy such that the kinetic energy of the photoelectron is comparable to the electron binding energy $|\varepsilon_i|$. The absorption and emission of real photons will then begin to become inefficient because the potential will vary too slowly at large distances to provide the necessary momentum transfer: there will be a cutoff in the spectrum of harmonics and ATI lines [20]. There should also be a dynamic connection because ionization will tend to quench radiative emission.

ACKNOWLEDGEMENT

We are indebted to C.K. Rhodes and coworkers for hospitality and support during extended stays at UIC (GW, 1987-88; LJ, 1987-89). This work has been supported by the National Science Foundation, the U.S. Air Force Office of Scientific Research, the U.S. Office of Naval Research, the Lawrence Livermore National Laboratory, the Strategic Defense Initiative Organization, and the Swedish National Science Research Council.

REFERENCES

*) Present address: Department of Physics, Chalmers University of Technology, S-412 96 Göteborg, Sweden

1. C.K. Rhodes, Physica Scripta **T17**, 193 (1987)
2. G. Gibson, R. Rosman, T.S. Luk, I.A. McIntyre, A. McPherson, G. Wendin, K. Boyer, and C.K. Rhodes, *OSA Proceedings on Short Wavelength Coherent Radiation: Generation and Application*, 26-29 Sept. 1988, Cape Cod, eds. R.W. Falcone and J. Kirz, in press
3. M.M. Murnane, H.C. Kapteyn, and R.W. Falcone, Phys. Rev. Lett. **62**, 155 (1989)
4. (a) G. Wendin, L. Jönsson, and A. L'Huillier, J. Opt. Soc. Am. B **4**, 833 (1987)
   (b) G. Wendin, L. Jönsson, and A. L'Huillier, Phys. Rev. Lett. **56**,1241 (1986)

5. A. McPherson, G. Gibson, H. Jara, U. Johann, T.S. Luk, I.A. McIntyre, K. Boyer, and C.K. Rhodes, J. Opt. Soc. Am. B **4**, 595 (1987)
6. M. Ferray, A. L'Huillier, X.F. Li, L.A. Lompré, G. Mainfray, and C. Manus, J. Phys. B **21**, L31 (1988)
7. R. Rosman, G. Gibson, K. Boyer, H. Jara, T.S. Luk, I.A. McIntyre, A. McPherson, J. C. Solem, K. Boyer, and C.K. Rhodes, J. Opt. Soc. Am. B **5**, 1237; 1948 (1988)
8. R.M. Potvliege, and R. Shakeshaft, Z. Phys. D **11**, 93 (1989)
9. (a) B. Gao, and A.F. Starace, Phys. Rev. Lett. **61**, 404 (1988);
(b) B. Gao, and A.F. Starace, submitted to Phys. Rev. A
10. Liwen Pan, K.T. Taylor, and C.W. Clark, Phys. Rev. Lett. **61**, 2673 (1988)
11. (a) A. L'Huillier, L. Jönsson, and G. Wendin, Phys. Rev. A **33**, 3938 (1986);
(b) A. L'Huillier, and G. Wendin, J. Phys. B **20**, L37 (1987);
(c) A. L'Huillier, and G. Wendin, Phys. Rev. A **36**, 4747 (1987);
(d) A. L'Huillier, and G. Wendin, Phys. Rev. A **36**, 6532 (1987);
(e) A. L'Huillier, and G. Wendin, J. Phys. B **21**, L247 (1988);
12. A. Zangwill, J. Chem. Phys. **78**, 5926 (1983)
13. K.R. Subbaswamy, and G.D. Mahan, J. Chem. Phys. **84**, 3317 (1986)
14. G. Senatore and K.R. Subbaswamy, Phys. Rev. A **35**, 2440 (1987)
15. A. Liebsch, Phys. Rev. Lett. **61**, 1233 (1988)
16. L. Jönsson, and G. Wendin, these proceedings, and to be published.
17. L.V. Keldysh, Soviet Phys. JETP 20, 1307 (1965)
18. See e.g. A. Giusti-Suzor and P. Zoller, Phys. Rev. A 36, 5178 (1987), for a discussion of the variation of the potential and the range of the ATI spectrum.
19. When intermediate states lie in the continuum, the energy shifts of the continuum electrons are given by $\Sigma_n = \Sigma_k = \ldots = e^2A^2/4mc^2$. This quiver "self-energy" shift is due to the diagonal part of the $\mathbf{A}^2$ operator (the off dagonal parts are zero since **A** is assumed to be constant in space). This shift exists also for the excited bound levels, and also appears in the shift Σ_i of the initial state level. Therefore, *formally*, the ponderomotive shift cancels out in the excitation self-energy shifts Σ_{ni}, which then only depend on **A·p** to second order and higher. *Physically*, however, we think that it is reasonable to include the ponderomotive shifts explicitly, so that a deep atomic orbital will shift very little, highly extended bound states will shift a lot, and continuum states will shift by the full ponderomotive potential. This reflects the fact that a deep orbital will be little distorted by the field, and that we keep the vacuum zero of the free atom as our reference. The ponderomotive shift then corresponds to the energy spent on creating the quiver dressing of a free electron.
20. This is also the result of recent computer experiments by K.C. Kulander, and B.W. Shore, Phys. Rev. Lett. **62**, 524 (1989), and J.H. Eberly, Q. Su, and J. Javanaien, Phys. Rev. Lett. **62**, 881 (1989), who numerically solved time-dependent Schrödinger equations and studied the frequency spectrum of the dipole moment of *a single atom*. However, the harmonic generation experiments are performed in fairly high density target gas for which one must consider effects of propagation [6,7]. X.F. Li, A. L'Huillier, M. Ferray, L.-A. Lompré, and G. Mainfray, Phys. Rev. A, in press, and L.-A. Lompré, A. L'Huillier, and G. Mainfray, these proceedings, invoke the dielectric properties of the target gas to correct the harmonic spectrum for effects of phase matching. The results suggest that there are important, but smooth, corrections to the single-atom picture. (The low-density, single-atom, limit cannot be studied experimentally because of low radiation intensity. Since the ATI ladder spectrum is connected with the collision-free, single-atom limit, this means that the ATI spectrum and the harmonic spectrum cannot be studied simultaneously under the same conditions at present.)

EIGENCHANNEL R-MATRIX CALCULATION OF PHOTOIONISATION PROCESSES IN STRONTIUM

Mireille Aymar

Laboratoire Aimé Cotton, CNRS II, Bat. 505, 91405 Orsay Cedex, France

I. INTRODUCTION

The alkaline-earth atoms have been the subject of numerous recent laser experimental investigations. The major motive for the interest in these spectra is the ability to study the strong interchannel interactions which cause irregularities in bound Rydberg series and autoionising structures. The multichannel quantum defect theory[1–5] (MQDT) has enjoyed remarkable success in interpreting the data obtained in the heavy alkaline earths Ca, Sr and Ba. However most of the previous analyses use empirical MQDT parameters which are adjusted to agree with a particular set of measurements. The complexity of heavy elements is such that empirical MQDT treatment comes against difficulties related to the energy dependence of the parameters, to the occurence of isolated perturbers and to the presence of a too large number of interacting channels[5]. This has made desirable to develop techniques to calculate the energy-dependent short-range MQDT parameters.

Recently, the new eigenchannel formulation of the R-matrix theory developed by Greene[6–8] has proven to be a fast and efficient tool for calculating MQDT parameters in alkaline earths. More precisely the wavefunctions of the pair of electrons outside a frozen core are calculated variationally within a finite reaction volume V, using a variational expression for the normal logarithmic derivative on the surface S of V. By matching on S the R-matrix eigenstates to a channel expansion involving Coulomb functions, MQDT short-range parameters are obtained as explicit functions of the energy.

This approach has been applied with success firstly to light alkaline earths Be[9,10] and Mg[10,11] and lately to heavier elements Ca[12–14] and Sr[15–17] whose description requires somewhat more sophistication. Initial calculations were performed in LS coupling and then the techniques have been extended to include effects of the spin-orbit interaction using a (jj/LS) frame transformation[13,16,17]. Calculations have dealt with either perturbed Rydberg series or photoionisation spectra. For bound levels, energies and oscillator strengths were calculated. Results on photoionisation spectra concern the total and partial photoionisation cross sections, photoelectron angular distributions[13,17] and the alignement of the residual ionic fragments[13]. When the comparison is possible, good agreement was achieved with experiment. Moreover, very recently, the theoretical procedures have been extended for handling open-shell atoms[18].

The present lecture is restricted to alkaline-earths and more precisely to photoionisation spectra of Sr and Ca. The first section explains how to combine the small-scale R-matrix calculation with MQDT and describes the different steps of the computational procedure. The examples presented in section 2 illustrate that the theoretical approach provides a very realistic descripton of Sr and Ca as well as a better understanding of correlation effects in these atoms. Finally, concluding remarks and prospects are given in section 3.

II. METHOD OF CALCULATION

II.1. Eigenchannel R-matrix method in LS coupling

The common point in all R-matrix methods is their solution of the Schrödinger equation within a finite volume V of configuration space. The short-range dynamics of the system is known once the normal logarithmic derivative b defined by

$$\frac{\partial \Psi}{\partial n} + b\Psi = 0 \tag{1}$$

is specified on the surface S enclosing V. In a multichannel problem having several degenerate solutions for a given energy E, the eigenchannel R-matrix selects the eigensolutions Ψ_β with constant b_β across S. These b_β are the reciprocal eigenvalues of the conventional R-matrix[19].

The Schrödinger equation for the two electrons outside a closed-shell core (noted A^{++}) is given (in au) by

$$H\Psi = (-\frac{1}{2}\nabla_1^2 - \frac{1}{2}\nabla_2^2 + U(r_1) + U(r_2) + \frac{1}{r_{12}})\Psi = E\Psi \tag{2}$$

where the interaction of each valence electron with the core is represented by an effective local potential U(r). Heavy alkaline earths are difficult to describe by *ab initio* calculations, partly because of the sensitivity of the ℓ=2 and ℓ=3 orbitals to the e-A^{++} interaction. This difficulty has been circumvented in Sr[15-17] and Ca[12-14] by using empirical potentials U(r) chosen to give an accurate description of the A^+ spectrum.

Using trial functions which are linear combinations of known functions y_k

$$\Psi = \sum_k c_k y_k \tag{3},$$

variational solutions of the Schrödinger equation are determined by the condition that the normal logarithmic derivative b is stationary. The variational principle leads to $\partial b/\partial c_k = 0$, equivalent to an eigenvalue problem for b in the vector space of the coefficients c_k, namely

$$\mathbf{\Gamma}\vec{c} = \mathbf{b\Lambda}\vec{c} \tag{4}$$

where

$$\Gamma_{k\ell} = 2\int_V y_k(E-H)y_\ell dV - \int_S y_k \frac{\partial y_\ell}{\partial n} dS \quad \text{and} \quad \Lambda_{k\ell} = \int_S y_k y_\ell dS. \tag{5}$$

The LS coupled two-electron basis functions y_k are antisymmetrised products of one electron orbitals coupled to form a definite spin and orbital angular momentum,

$$y_{n\ell_1 m\ell_2}(\vec{r}_1,\vec{r}_2) = \mathcal{A}\ \Phi_{n\ell_1}(r_1)\ F_{m\ell_2}(r_2) Y_{\ell_1\ell_2 LM}(\vec{r}_1,\vec{r}_2)\ [\chi_{s_1}(1)\chi_{s_2}(2)]^{SM_S} \tag{6}$$

where the indices n ℓ_1 ℓ_2 define a specific channel converging to the A^+ nℓ_1 threshold, each channel being described by several $y_{n\ell_1 m\ell_2}$ functions characterized by the index m. The $\Phi_{n\ell_1}$ functions describing the A^+ target states are eigenfunctions of the one-electron e-A^{++} Schrödinger equation. It is convenient (but not necessary) to choose the $F_{m\ell_2}$ orbitals to be eigenfunctions of the same one-electron Hamiltonian. Two kinds of orbitals are used which correspond to different boundary conditions on the reaction surface $r=r_0$. A first set of orthogonal 'closed-type' orbitals is obtained by solving the Schrödinger equation with the condition that each orbital vanishes at $r = r_0$. A second set of 'open-type' orbitals is obtained by integrating the Schrödinger equation at several different energies which yield non-zero values of the orbitals and a variety of logarithmic derivatives at $r = r_0$.

The two-electron basis functions are chosen to describe two different groups of channels. A first set contains the open or 'weakly closed' channels (in the sense of quantum defect theory) in which the outermost electron can escape from the reaction volume. These channels require a 'closed-type' orbital to describe the inner electron and several 'open-' and 'closed-type' orbitals ($\sim$ 6-12) for the excited or continuum electron. Let us note that the 'open-type' orbitals are not orthogonal to one another, nor orthogonal to the 'closed-type' orbitals. The non orthogonality adds flexibility to the y_k basis but can lead to numerical inaccuracies when the basis becomes overcomplete and it has been found important to keep 'open-type' orbitals as few as possible ($\sim$ 2-4). For each preselected energy E, the number N of eigensolutions Ψ_β of eq. (4) is equal to the number of open and 'weakly closed' channels. The second set of channels corresponds to the 'strongly closed' channels which contribute at short range but are exponentially small on the surface of the reaction volume. These channels account for correlation effects and for the polarisation-distorsion of the one electron functions induced by the presence of the second electron. The inclusion of such channels also improves the convergence of the variational expansions. These channels are described by 'closed-type' orbitals only.

The choice of the size of the reaction volume requires some considerations. The function $\Phi_{n\ell_1}$ represents a A^+ orbital accurately only when this orbital is contained well within the volume V so that the true function has an exponentially small amplitude at $r=r_0$. Thus the radius r_0 must be large enough to enclose the charge distribution of the A^+ nℓ_1 states involved in open and 'weakly-closed' channels. Accurate descriptions of the Ca[12,13] and Sr[16,17] photoionisation spectra, below the 4p or 5p threshold respectively, have been obtained with $r_0 \sim$ 14-20 au. Predictions for the photabsorption spectrum of Ca up to the 6s threshold were performed[14] using a larger reaction volume ($r_0\sim$ 30-50 au).

Compared to the more standard R-matrix method[19], the eigenchannel version is characterized by a larger flexibility of the basis, a faster convergence of the variational expansion and smaller-scale calculations. The most time consuming part of the calculation, the numerical evaluation of the $1/r_{12}$ matrix elements needs to be performed only once. The eigensystem (4) has to be solved at each energy E. However, Greene and Kim[18] have shown very recently that a transformation of this eigensystem in a new representation permits to speed up considerably this part of the computation; the transformed eigensystem can be solved in a small amount of time on an arbitrary energy mesh.

II.2. LS-coupled reaction matrix

After the N independent $\Psi_\beta(E)$ have been obtained, the solutions on the boundary are matched to linear combinations of energy-normalized regular and irregular Coulomb functions $(f_{\bar\alpha}, g_{\bar\alpha})$[1,20] in each open or 'weakly-closed' channel $\bar\alpha$. These functions are evaluated at the photoelectron energy $\epsilon_{\bar\alpha}$ relative to the average spin-orbit-split A^+ $\bar\alpha$ threshold. It is possible to write

$$< \Phi_{\bar\alpha} \parallel \Psi_\beta >= f_{\bar\alpha}(\epsilon_{\bar\alpha}, r_0) I_{\bar\alpha\beta} - g_{\bar\alpha}(\epsilon_{\bar\alpha}, r_0) J_{\bar\alpha\beta} \tag{7a}$$

$$-b_\beta < \Phi_{\bar\alpha} \parallel \Psi_\beta >= \frac{\partial}{\partial r} f_{\bar\alpha}(\epsilon_{\bar\alpha}, r_0) I_{\bar\alpha\beta} - \frac{\partial}{\partial r} g_{\bar\alpha}(\epsilon_{\bar\alpha}, r_0) J_{\bar\alpha\beta} \tag{7b}$$

where the double bar means surface integration over S leaving behind only the radial degree of freedom of the outermost electron. The LS-coupled reaction matrix of dimension N is simply

$$\mathbf{K}^{LS} = \mathbf{J}\mathbf{I}^{-1}. \tag{8}$$

II.3. jj-coupled reaction matrix

Effects of the spin-orbit interaction on photoionisation spectra of Ca and Sr have been described successfully using a geometric (jj/LS) frame transformation. Indeed in the reaction volume where both electrons are at about equal distance from the core, electrostatic interaction prevails over the spin-orbit coupling which, thus, can be ignored for $r \leq r_0$. In the asymptotic region when the electron is far from the core, it can distinguish the fine-structure levels of the ionic core and the distant electron must be described in jj coupling. The jj-coupled reaction matrix $\mathbf{K}(E)$ is calculated by recoupling the $\mathbf{K}^{(LS)}(E)$ matrices obtained for different LS symmetries using the (jj/LS) geometrical frame transformation (matrix $\mathbf{V}$):

$$\mathbf{K}(E) = \mathbf{V}\mathbf{K}^{(LS)}(E)\mathbf{V}^t. \tag{9}$$

The jj-coupled matrix refers now to the jj-coupled dissociation channels i.

II.4. MQDT analyses

At this point the eigenchannel R-matrix method links naturally with quantum defect theories. The short-range reaction matrix $\mathbf{K}$ referring to open and closed channels does not incorporate any boundary condition at infinity whereby it normally remains a smooth function of energy and does not display resonance effects associated with closed channels. Asymptotic boundary conditions are efficiently imposed by MQDT. Several MQDT formalisms, in particular the eigenchannel method of Lu and Fano[2-5] and the phase-shifted approaches[21-23] can be exploited for calculating the observables on an arbitrarily fine energy mesh. Although algebraic techniques differ according to the formulation, results are independent of it, the alternative sets of energy-dependent parameters being deduced from the same reaction matrix $\mathbf{K}(E)$.

A full exposition of the MQDT procedures can be found in refs. 1-5, 21-24 and only some particular points relevant to the description of ionisation processes are outlined now. In Sr, two distinct one-photon processes are considered corresponding to photoabsorption from either the ground state $5s^2$ or Rydberg states 5sns $^1S_0$. In both cases, the final states correspond to the odd-parity J=1 autoionising spectrum. Calculation of photoionisation cross sections requires the description of initial and final states as well as the determination of dipole matrix elements.

Let us begin with the description of the final state. Within the eigenchannel MQDT formulation, final state channel interaction is described in terms of the the energy-dependent parameters μ_α and $U_{i\alpha}$ which correspond to the eigenvalues and the eigenvectors of the short-range reaction matrix K,

$$K_{ij} = \sum_\alpha U_{i\alpha} \tan\pi\mu_\alpha U^t_{\alpha j}. \tag{10}$$

In the autoionising energy range, for each energy E, the N_O physical solutions (N_O being the number of open channels) are linear combinations of the K-matrix eigenchannels $\Psi_\alpha(E)$ with the coefficients to be determined by large-r boundary conditions. $\Psi(E)$ must decay exponentially in every closed channel ($i \in Q$) while each open channel component ($i \in P$) is required to have a common eigenphase shift $\pi\tau_\rho$. The asymptotic form of the collision eigenchannels Ψ_ρ is

$$\Psi_\rho = \sum_{i\in P} \Phi_i[f_i(r)\cos(\pi\tau_\rho) - g_i(r)\sin(\pi\tau_\rho)]T_{i\rho} + \sum_{i\in Q} \Phi_i P_i(\nu_i, r) Z_{i\rho} \tag{11}$$

where Φ represents the ionic state together with the spin and orbital coupling of the outer electron, ν_i is the effective quantum number in channel i and $P_i(\nu_i, r)$ is an exponentially decreasing Coulomb function[1]. In eq. (11) the coefficients $T_{i\rho}$ and $Z_{i\rho}$ describe the relative admixture of the closed and open dissociation channels respectively. The eigenphase shifts and the mixing coefficients are calculated at each energy in terms of the MQDT parameters.

We turn now to the determination of the dipole matrix elements and begin with photoionisation from ground state. The excitation process takes place entirely within the reaction volume. The D_α matrix elements used by the eigenchannel MQDT are easily deduced from the dipole matrix elements connecting the ground state to the R-matrix eigenfunctions Ψ_β determined for the $^1P^0$ symmetry. The ground state wavefunction is obtained by diagonalising the Hamiltonian (2) using a basis set (6) involving 'closed-type' orbitals only. Finite volume dipole matrix elements are calculated using the length and velocity formulations.

The second ionisation process concerns multistep laser excitation of J=1 odd-parity autoionising levels. Experimental investigations on $5p_{1/2}ns$ and $5p_{3/2}ns$ levels of Sr have been carried out using the isolated core excitation (ICE) technique in which several single electron transitions are driven successively. In the final transition the levels were excited from 5sns $^1S_0$ Rydberg states. Calculation of cross sections requires the knowledge of dipole matrix elements D_i connecting the initial state $\psi_0 \doteq \Phi_0 P(\nu_0, r)$ to the dissociation channels i involved in the final state. The ICE excitation method relies on negligible continuum excitation and on the photoexcitation of the ion core while the Rydberg function of the initial level is only projected on the new Rydberg function of the excited level; thus for $i \in P$ $D_i = 0$ and for $i \in Q$ one has

$$D_i = < \Phi_0 \| D \| \Phi_i > < P_0(\nu_0, r) | P_i(\nu_i, r) > \tag{12}$$

where the explicit form of the overlap integral is

$$< P_0(\nu_0, r) | P_i(\nu_i, r) > = 2 \frac{\sin\pi(\nu_i - \nu_0)}{\pi(\nu_i^2 - \nu_0^2)} \nu_0^{1/2} \nu_i^2 \delta(\ell_0, \ell_i). \tag{13}$$

In the studied case, two D_i quantities, associated to the $5p_{1/2}ns$ and $5p_{3/2}ns$ channels, have non-zero values.

III. RESULTS

III.1. Final state MQDT parameters and channel coupling

Thirteen interacting channels are relevant to the description of the J=1 odd-parity autoionising spectrum of Sr below the $5p_{3/2}$ threshold. The channels involved in LS calculations are:

$$\begin{array}{lllll} 5snp\,^1P, & 4dnp\,^1P, & 4dnf\,^1P, & 5pns\,^1P, & 5pnd\,^1P, \\ 5snp\,^3P, & 4dnp\,^3P, & 4dnf\,^3P, & 5pns\,^3P, & 5pnd\,^3P, \\ 4dnp\,^3D, & 4dnf\,^3D, & 5pnd\,^3D. & & \end{array} \tag{14}$$

The jj-coupled dissociation channels are:

$$\begin{array}{lllll} 5s_{1/2}np_{1/2}, & 5s_{1/2}np_{3/2}, & 4d_{3/2}np_{1/2}, & 4d_{3/2}np_{3/2}, & 4d_{5/2}np_{3/2}, \\ 4d_{3/2}nf_{5/2}, & 4d_{5/2}nf_{5/2}, & 4d_{5/2}nf_{7/2}, & 5p_{1/2}ns_{1/2}, & 5p_{3/2}ns_{1/2}, \\ 5p_{1/2}nd_{3/2}, & 5p_{3/2}nd_{3/2}, & 5p_{3/2}nd_{5/2}. & & \end{array} \tag{15}$$

The channels required to describe the $J=1^o$ spectrum of Ca up to the $4p_{3/2}$ threshold are the same, aside from change in the principal quantum number of the inner electron.

In order to determine the jj-coupled reaction matrix $\mathbf{K}(E)$ as a function of the energy, three independent R-matrix calculations have been done for the $^1P^0$, $^3P^0$ and $^3D^0$ symmetries. Fig. 1 shows the eigenquantum defects μ_α of Sr as functions of the energy. The energy dependence of the μ_α and also of the the $U_{i\alpha}$ parameters is particularly large below the 4d threshold. Moreover the α channels are strongly mixed and, by far, the strongest mixing corresponds to the 4dnf-5pnd one ($\sim$ 55 % between the 4d and 5p thresholds); similar strong mixing ($\sim$ 60 %) was found in Ca[12] between the 3dnf and 4pnd channels. This unusually strong mixing manifests itself in the very large autoionisation widths of the $n_0$pnd (n_0=4 in Ca and n_0=5 in Sr) resonances, as discussed later.

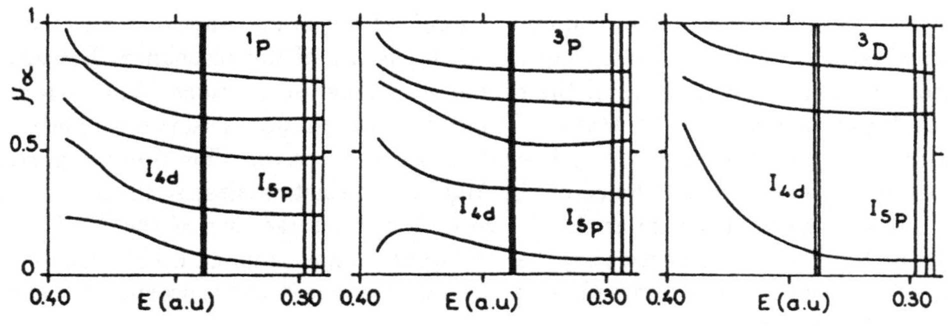

Fig. 1: Eigenquantum defects μ_α for Sr are shown as functions of the energy (from ref. 16)

III.2. Photoabsorption spectra out of the ground state

The Sr photoabsorption spectrum below the $5p_{3/2}$ threshold is dominated by broad $4d_{3/2,5/2}n\ell$ and $5p_{1/2,3/2}n'\ell'$ resonances. The observed features reflect strong interactions amongst the individual doubly-excited series and their coupling with the $5s\epsilon p$ $^1P_1$ and $^3P_1$ continua. Moreover the structures are complicated by a severe breakdown of LS coupling resulting from the spin-orbit interaction. Initial eigenchannel R-matrix calculations[15] in LS coupling were unable to reproduce the experimental features[25–27] whereas, as documented below, good description of the observed spectrum has been achieved using the (jj/LS) frame transformation to include the effects of fine structure[16]. A recent calculation[28] in the same energy range, using the many body perturbation theory, shows inferior agreement between theory and experiment; this is hardly unexpected because spin-orbit effects are completely ignored.

The absorption spectrum of Ca below the 4p threshold is very similar to that of Sr without, however, most of the complications due to the spin-orbit interaction. There also, excellent description of the experimental features were attained by Greene and Kim[12,13] who used the same eigenchannel R-matrix approach. In contrast with Sr, the dominant experimental features can be described in Ca ignoring the spin-orbit interaction[12]; however improved results were obtained when fine-structure effects are accounted for[13]. By far, poorer description of the Ca experiment was achieved using the conventional R-matrix method[29] or the many body perturbation theory[30].

Detailed comparison between calculations and experiments can be found in ref. 16 for Sr and in refs. 12,13 for Ca; here I will discuss some particular points only.

Let us begin with the Sr absorption spectrum below the $4d_{3/2}$ threshold. The autoionising structures correspond mainly to the 4dnp and 4dnf resonances. In addition resonances associated with the 5p6s levels have been observed in the low energy range while the 5p5d levels, expected to lie in the same energy range, are surprisingly unknown. Our calculation has provided an accurate description of most features of the observed spectrum, despite extremely strong correlations between the outer electrons and a severe breakdown of LS coupling. Moreover our study was not restricted to the determination of the positions, shapes and intensities of the resonances but also provided a deeper understanding of the effects of channel interactions. The introduction of 5pnℓ' channels were found crucial to reproduce the shapes of 4dnℓ resonances whose widths and intensities evolve very irregularly as n increases. This trend is partly due to the strong energy dependence of the MQDT parameters already illustrated on fig. 1. However the main irregularities derive from the interactions of the 4dnℓ channels with the 5pnℓ' ones. In particular, the 4dnf-5pnd channel mixing is so strong that the 5p5d levels are completely embedded in several resonances and consequently cannot be ascribed to particular structures.

Below the 4d threshold the MQDT parameters calculated with the eigenchannel R-matrix method depend strongly on energy; moreover the resonances supported by the thirteen interacting channels often interfer and overlap. Thus, it is difficult to envisage how the MQDT parameters could be obtained empirically from experiment. Finally, note that the conclusions obtained for Sr essentially apply to Ca.

We turn now to the Sr photoabsorption spectrum between the $4d_{5/2}$ and $5p_{3/2}$ thresholds whose structure is due to transitions to the five 5pns,nd channels. Simultaneously we consider the homologous spectrum of Ca.

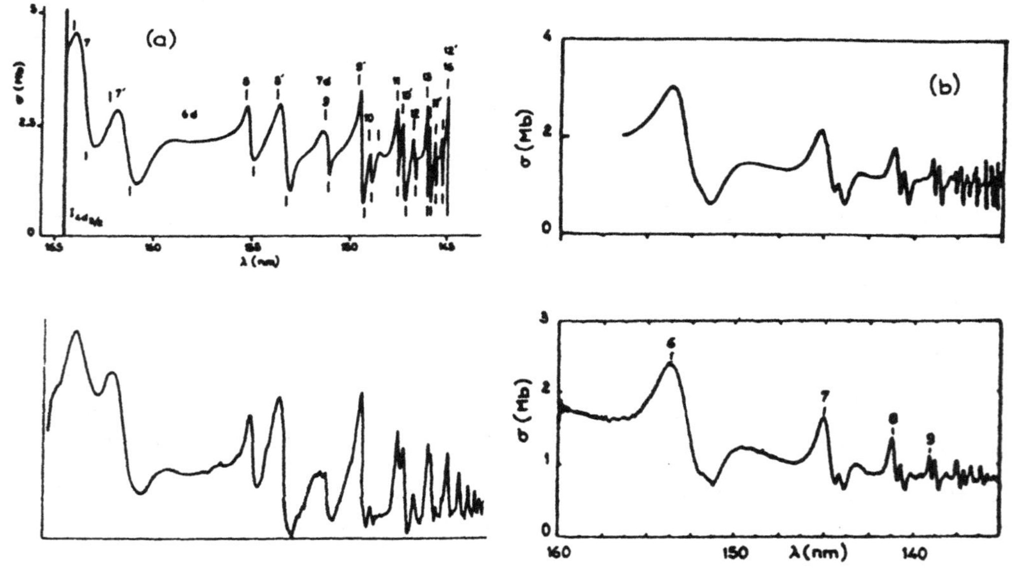

Fig. 2: Absorption spectra of Sr (a) and Ca (b) between the $(n_0$-1)d and n_0p thresholds. The upper curves dipsplay the eigenchannel R-matrix calculations of Aymar[16] in Sr and Kim and Greene[13] in Ca; the lower curves give the experimental spectra of Connerade et al[26] in Sr and Griesmann et al[31] in Ca(n stands for $n_0p_{1/2}$ns and n' for $n_0p_{3/2}$n's).

Low-resolution investigation of the photoabsorption spectra of Ca and Sr was performed, in this energy range, by Connerade et al[26]. Fig. 2a shows that our calculation in Sr reproduces nicely these experiment data (note that in the lower wavelength range the experimental resolution is too low for a precise comparison). Very recently new measurements have been done in Ca[31] using synchroton radiation. As shown on fig. 2b, the eigenchannel R-matrix calculation of Kim and Greene[13] perfectly reproduces these recent measurements; when compared with older experimental data[26] agreement was inferior. The only remaining discrepancy concerns the absolute scale of the cross sections. The comparison of figs. 2a and 2b clearly illustrates the similarity between Ca and Sr. For both spectra, the sharpest resonances correspond to ns levels while the broad structures are associated with the nd resonances. The large autoionisation widths of $n_o$pnd levels result from their strong coupling with the $(n_o\text{-}1)\epsilon$f continua. The increasing role of the spin-orbit interaction when going from Ca to Sr is reflected by the higher intensity of $5p_{3/2}$ns resonances in Sr compared to the $4p_{3/2}$ns resonances in Ca.

Fig. 3 compares the photoionisation cross sections calculated near the $5p_{1/2}$ threshold of Sr with the experimental results of Brown et al[27].

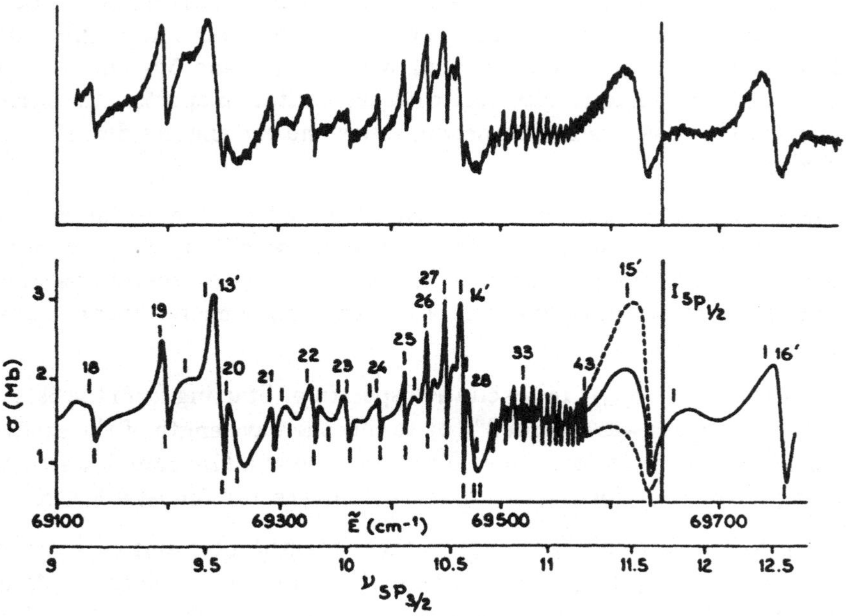

Fig. 3: Sr photoionisation cross sections near the the $5p_{1/2}$ threshold. (a) experimental results of Brown et al[27]; (b) eigenchannel R-matrix calculation of Aymar[16].

The overall agreement between theory and experiment is excellent in spite of the complexity of the stucture. The various characteristics of experimental resonances (position, width, intensity) are accurately reproduced. The $5p_{1/2}ns$ and $5p_{3/2}ns$ levels correspond to the sharpest features while the shoulders or broad resonances between the 5pns peaks are associated to $5p_{1/2}nd$ or $5p_{3/2}nd$ levels. The intensity of the resonances varies periodically with $\nu_{5p_{3/2}}$, with maximum intensity for the $5p_{3/2}ns$ peaks; the average value of the rapid oscillations below the $5p_{1/2}$ threshold joints smoothly to the cross section just above the threshold. As a matter of fact, although new continua are available for decay, autoionisation mainly proceeds to the continua built on the $4d_{3/2,5/2}$ Sr^+ levels.

Another interesting feature came out from the calculation. Near the $5p_{1/2}$ threshold the dominant partial cross sections are the ones producing Sr^+ $4d_{3/2,5/2}$ ions resulting in a remarkable population inversion of Sr^+ ions. This implies severe breakdown of the independent-particle model which should predict that only ground state Sr^+ 5s is populated by the photoionisation process from the $5s^2$ state of Sr. Similar population inversion deriving from strong electron correlations was found for Ca[13].

III.3. Photoionisation from Rydberg states

A complete set of measurements have been reported recently by Xu et al[32] and Zhu et al[33] for the $5p_{1/2}ns$ and $5p_{3/2}ns$ J=1 autoionising states of Sr. These levels were excited through the $5sn_0s$ $^1S_0$ Rydberg levels using the experimental ICE method. Isolated $5p_{1/2}ns$ and $5p_{3/2}ns$ resonances having approximately Lorentzian line profiles as well as structured $5p_{3/2}ns$ states strongly interacting with $5p_{1/2}ns$ and $5p_{1/2}nd$ levels have been investigated. Experimental results deal with the excitation spectra, the positions and widths of the levels as well as the branching ratios and the angular distributions of the photoelectrons.

All the above-mentioned observables have been calculated and the overall agreement with experiment is satisfactory, accounting for the extreme sensitivity of the observables to small changes in the MQDT parameters. Detailed accounts on the comparison between theory and experiment can be found in ref. 17 and only some results are discussed here.

Fig. 4 deals with the 5s14s-$5p_{3/2}$ 14s excitation spectrum. Starting from the 5s14s $^1S_0$ Rydberg level, this spectrum was recorded[32] by tuning the wavelength of the excitation laser in the vicinity of the 5s-5p ionic line. The spectrum exhibits complex structures due to the mixing of the $5p_{3/2}14s$ level with nearby $5p_{1/2}ns$ and $5p_{1/2}nd$ levels.

It is interesting to compare this spectrum to that obtained after single-photon excitation from the ground state (fig. 3). In both cases, the same five autoionising Rydberg series interact wich each other. The complexity in fig. 3 arises because the excitation amplitudes to the discrete autoionising states and to the continua are comparable leading to Beutler-Fano interferences. The spectrum of fig. 4 is simpler because, using the ICE method, there is virtually no continuum excitation and the observed resonances are free of the Beutler-Fano interferences; moreover only a specific discrete character of an autoionising resonance is accessible, here the $5p_{3/2}ns$.

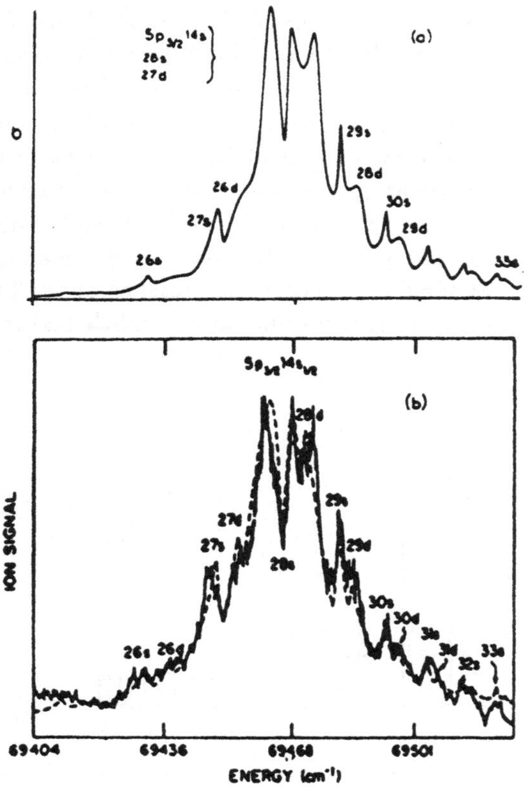

Fig. 4: The 5s14s- $5p_{3/2}14s$ excitation spectrum. (a) eigenchannel R-matrix calculation of Aymar and Lecomte[17]; (b) observed ion signal and MQDT simulation (- - -) from Xu et al[32].

The calculated total photoionisation cross sections very well reproduce the recorded spectrum, not only the central features involving three strongly interacting levels but also the positions and shapes of the satellite $5p_{1/2}ns$ and $5p_{1/2}nd$ resonances. The shape of the spectrum is mainly related to the variation of the $Z_{i\rho}$ admixture coefficients describing the closed part of the Ψ_ρ wavefunctions (eq. 11); indeed the spectrum is within the central lobe of the overlap integral (eq. 13). Although only one D_i dipole matrix element associated with the i= $5p_{3/2}ns$ channel (eq. 12,13) has a non negligible amplitude, the $5p_{1/2}ns$ and $5p_{1/2}nd$ resonances show up in the spectra because of their coupling with the $5p_{3/2}14s$ level.

In fig. 5 we show a different spectrum recorded in Sr[32] by populating the 5s20s $^1S_0$ level and scanning the excitation laser to the blue of the 5s-$5p_{1/2}$ ionic line; the spectrum corresponds to the shakeup satellite spectrum of the 5s20s- $5p_{1/2}20s$ transition observed in the perturbed region near the $5p_{3/2}14s$ level which includes $5p_{1/2}ns$ the levels with n = 23 - 36.

The precise form of the satellite peaks depends now on $Z_{i\rho}$ coefficients and on the variation of the overlap integrals involved in two dipole matrix elements D_i. The squared overlap integral is maximum for $\nu_i = \nu_0$, equals to zero for $\nu = \nu_0 + k$ where k is a non zero integer and has subsidiary maxima for $\nu_i = \nu_0 + k + 1/2$. The minima between each peak correspond to the zeros of the overlap integral associated to $i = 5p_{1/2}ns$ while the maxima correspond approximately to those of the $Z_{i\rho}$ coefficients for particular ρ eigenchannels. The strong asymmetry of the peaks results from the distorsion due to the overlap integral whose zeros are very close to the maxima of the admixture coefficients. The irregular behaviour near n=27-30 reflects the presence of the $5p_{3/2}14s$ perturber. The calculated cross sections perfectly reproduce all the subtle features occurring in the recorded spectrum.

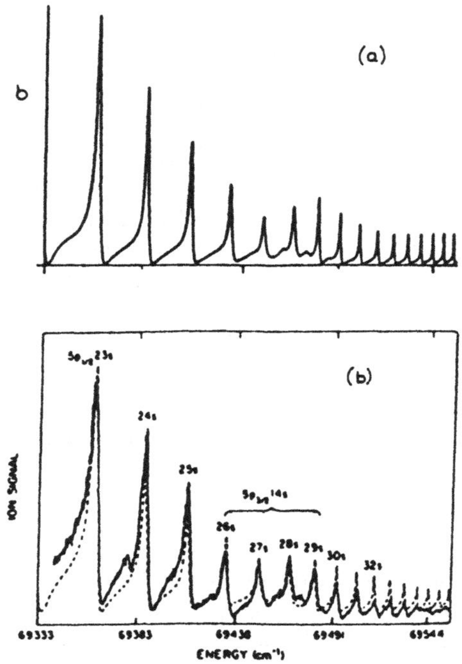

Fig. 5: Shakeup spectrum from the 5s20s to the $5p_{1/2}ns$ levels. (a) eigenchannel R-matrix calculation of Aymar and Lecomte[17]; (b) observed ion signal and MQDT simulation (- - -) from Xu et al[32].

We consider now the results on the energy and angular distributions of electron ejected from the $5p_{1/2}ns$ and $5p_{3/2}ns$ levels[33]. Below the $5p_{1/2}$ limit, experimental data concern autoionisation to the $5s_{1/2}$ and $4d_j$ ionic states, the two photoelectron groups corresponding to decay into the $4d_{3/2}$ and $4d_{5/2}$ ionic states being not resolved. Above the $5p_{1/2}$ limit, data also concern decay to the $5p_{1/2}$ ionic level. Our calculated branching ratios for the autoionisation to the different available Sr^+ ionic levels are in good agreement with experiment. Our calculation has confirmed that about 60% -80% of the electrons decay to the $4d_{3/2,5/2}$ ions resulting in a population inversion of the Sr^+ ions, similar to that observed after single-photon excitation from ground state. Moreover it has shown that the formation of the 4d ions corresponds to ejection of mainly p electrons and that above the $5p_{1/2}$ limit, $5p_{1/2}$ ions are formed with ejection of about two times more d electrons than s electrons. These latter informations cannot be measured directly.

Turning to the β asymmetry parameters which characterize the angular distributions for the ejected electron, the calculated β_{4d} and $\beta_{5p_{1/2}}$ (above the $5p_{1/2}$ threshold) compare well with experiment while some discrepancy was found for the asymmetry parameters β_{5s}. Deviation from $\beta_{5s}=2$ is related to the parity unfavoured term[24] present because of the spin orbit interaction. The calculated β_{5s} values are all smaller than experimental ones, mainly for the $5p_{1/2}$ns levels, indicating that our theoretical treatment overestimates the weight of the unfavoured process. The main approximation introduced in the calculation consists in neglecting spin-orbit effects in the reaction volume, i.e. fine structure effects of the Sr^+ ion are included whereas spin-orbit effects related to the outer excited and continuum electrons are neglected. Such an assumption is very likely to be no longer completely justified when concerned with the description of the extremely sensitive β_{5s} quantities.

Very recently, similar laser measurements, however more precise than in Sr, have been performed by Lange et al[34] for $4p_{1/2}$ns and $4p_{3/2}$ns autoionising states in Ca. Results were compared with calculations done with the MQDT parameters calculated by Kim and Greene[13]. Conclusions obtained for Sr essentially hold for Ca. There also, the asymmetry parameter β_{4s} revealed some discrepancy, however less pronounced than in Sr.

Finally, I would mention another interesting conclusion attained in our study on laser excited autoionising levels of Sr. Experimental data were previously analysed[32,33] using a simplified 6-channel MQDT model depending on parameters determined empirically from the experiment. Such MQDT simulation can correctly reproduce experimental data on total photoionisation cross sections (see fig. 3 and 4). In contrast, it is almost impossible to obtain empirically a complete set of constant parameters allowing to reproduce other observables such as the branching ratios and the photoelectron angular distributions which are much more sensitive to the detailed form of the wave functions. This stresses the great interest presented by calculations of the full short-range reaction matrices referring to the whole set of interacting channels.

IV. DISCUSSION

The studies of ref. 12-17 have demonstrated (as illustrated by the examples discussed in this lecture) that a very realistic description of Ca and Sr can be attained by combining the eigenchannel R-matrix and MQDT methods. Impressive agreement between theory and experiment can be obtained not only for the photoabsorption spectra but also for other observables which are much more sensitive to channel interactions. Calculation of MQDT parameters permits to bypass the difficulties encountered in empirical treatment of complex atoms[5] and furthermore is of great practical use for predicting or interpreting other experiments.

The studies performed in Sr and Ca have provided a clearer understanding of channel mixing in these atoms and pointed out many similarities between Ca and Sr. It will be particularly interesting to extend the comparison to Ba where a wealth of data is available[5,35,36]. In Ba, the effect of the spin-orbit terms in the Hamiltonian is so strong they must be included explicitely within the short-range calculations instead of by the (jj/LS) frame transformation. This was accomplished[37] using the conventional formulation of R-matrix theory; however this previous study, restricted to a limited number of resonances, did not give direct access to the short-range reaction matrix or MQDT parameters required to calculate any desirable observables.

Another field, still little explored by theoreticians, concerns the high-lying doubly-excited states of alkaline earths where experimental data become to be available[38,39]. Some predictions on the photoabsorption spectrum of Ca up to the 6s threshold has been provided by Kim and Greene[14] who used the eigenchannel R-matrix method. Moreover, higher-lying Wannier states $n\ell n\ell'$ ($n \leq 12$) have been investigated in alkaline earths[14,40]. However, the predictions are restricted to the level positions and theoretical investigation on the excitation processes and the autoionisation properties of these high-lying states is worthwhile.

A further topic of future research should be to extend the theoretical approach to the calculation of multiphoton ionisation processes. MQDT has been already used to handle the resonance structures in the final state involved in mutiphoton ionisation processes in rare gases[41]. Several experimental and theoretical works have been devoted to multiphoton ionisation processes in alkaline earths (see ref. 42 and references therein). Calculation of multiphoton ionisation cross sections necessitates an accurate description of the resonances due to intermediate Rydberg states and final autoionising states which could be provided by eigenchannel R-matrix calculations. In addition it requires the determination of multiphoton dipole matrix elements, which involve infinite summations. An MQDT parametrisation for perturbative two-photon amplitudes of many electron atoms has been recently proposed[43]; in addition to the K matrices characterizing the channel mixing in intermediate and final states, it introduces a set of dipole matrix elements which are smooth functions of energy. This technique has been applied to two-photon ionisation of atomic hydrogen only; the difficulties involved in applying this approach to more complex situations remain to be assessed.

REFERENCES

1. Seaton M J, Rep. Prog. Phys. **46** 167 (1983)
2. Lu K T, Phys. Rev. **A4** 579 (1971)
3. Lee C M and Lu K T, Phys. Rev. **A 8** 1241 (1973)
4. Fano U and Rau A R P Atomic Collisions and Spectra, Academic press, INC, Orlando (1986)
5. Aymar M, Phys. Rep. **110** 163 (1984)

6. Greene C H, Phys. Rev. A **28** 2209 (1883)
7. Greene C H, Phys. Rev. A **32** 1880 (1985)
8. Greene C H, in Fundamental Process of Atomic Dynamics, Eds. Briggs J, Kleinpoppen and Lutz H (Plenum) (1988)
9. O'Mahony P F and Greene C H, Phys. Rev. A **31** 250 (1985)
10. O'Mahony P F Watanabe S, J. Phys. B **18** L239 (1985)
11. O'Mahony P F, Phys. Rev. A **32** 908 (1985)
12. Greene C H and Kim L, Phys. Rev. A **36** 2706 (1987)
13. Kim L and Greene C H, Phys. Rev. A **36** 4272 (1987)
14. Kim L and Greene C H, Phys. Rev. A **38** 2361 (1988)
15. Aymar M, Luc-Koenig E and Watanabe S, J. Phys. B **20** 4325 (1987)
16. Aymar M, J. Phys. B **20** 6507 (1987)
17. Aymar M and Lecomte J M, J. Phys. B **22** 223 (1989)
18. Greene C H and Kim L, Phys. Rev. A **38** 5953 (1988)
19. Burke P G and Robb D, Adv. At. Mol. Phys. **11** 143 (1975)
20. Greene C H, Fano U and Strinati G, Phys. Rev. A **19** 1485 (1979)
21. Giusti-Suzor A and Fano U, J. Phys. B **17** 215 (1984)
22. Cooke W E and Cromer C L, Phys. Rev. A **32** 2725 (1985)
23. Lecomte J M, J. Phys. B **20** 3645 (1987)
24. Dill D, Phys. Rev. A **7** 1976 (1973)
25. Hudson R D, Carter V L and Young P A, Phys. Rev. **180** 77(1969)
26. Connerade J P, Baig M A, Garton W R S and Newsom G H, Proc. R. Soc. Lond. A **371** 295 (1980)
27. Brown C M, Longmire M S and Ginter M L, J. Opt. Soc. Am. **73** 985 (1983)
28. Frye D and Kelly H P, J. Phys. B **20** L677 (1987)
29. Scott P, Kingston A E and Hibbert A, J. Phys. B **16** 3945 (1983)
30. Altun Z, Carter S L and Kelly H P, Phys. Rev. A **27** 1943 (1983)
31. Griesmann U, Shen N, Connerade J P, Sommer K and Hormes J, J. Phys. B L **21** 83 (1988)
32. Xu E Y, Zhu Y, Mullins O P and Gallagher T F, Phys. Rev. A **33** 2401 (1986)
33. Zhu Y, Xu E Y and Gallagher T F, Phys. Rev. A **36** 3751 (1987)
34. Lange V, Eichmann U and Sandner W, to be published (1989)
35. Kachru R, Tran N H, Pillet P and Gallagher T F, Phys. Rev. A **31** 218 (1985)
36. Sandner W, Eichmann U, Lange V and Võlkel M, J. Phys. B **19** 51 (1986)
37. Bartschat K, Rudge M R H and Scott P, J. Phys. B **19** 2469 (1986)
38. Boulmer J, Camus P, Lecomte J M and Pillet P, J. Opt. Soc. Am. B **5** 2199 (1988)
39. Morita N, Susuki T and Kato K, Phys. Rev A **38** 551 (1988)
40. Aymar M, J. Phys. B. to be published (1989)
41. Gangopadhyay P, Lambropoulos P and Shakeshaft R, Phys. Rev. A **34** 2998 (1986)
42. Lambropoulos P, Tang X, Agostini P, Petite G and L'Huillier A, Phys. Rev. A **38** 6165 (1988)
43. Fink M G F and Zoller P, Phys. Rev. A to be published (1989)

RESONANT MULTIPHOTON EXCITATION OF AUTOIONIZING RYDBERG STATES AND PLANETARY ATOMS

Wolfgang Sandner
Fakultät für Physik, Universität Freiburg
Hermann Herder Str. 3, D7800 Freiburg, W. Germany

1. INTRODUCTION

The interaction of laser light with atoms still appears as a process with many facets. Reduction of the problem to a small number of fundamental phenomena suffers partially from the fact that we know very little about atomic systems that have absorbed a certain amount of light energy, but distributed it among several electrons. These multi-electron excited atoms, their internal dynamics and their interaction with laser light will be the topic of the present overview.

Many experimental results on laser-atom interactions (even if obtained from heavy atoms) are today expressed and analyzed in terms of hydrogenic systems, i.e. as the interaction of light with a *two body Coulomb system*. Deviations from hydrogen are obviously observed, but often taken into account in a phenomenological way, as by adjustment of energy levels, or by the introduction of quantum defects in the electron's wave function. In this sense the laser interaction with hydrogenic systems is certainly "fundamental", and it appears to be well understood over a wide range of the parameter space that characterizes today s laser light sources. Considering that most atoms are non-hydrogenic, laser interaction with multi-electron atoms is certainly equally fundamental, but is considerably more complex. This is true not only in case of extremely strong laser fields (resulting in spectacularly high degrees of ionization and/or high order nonlinear frequency conversion processes), but even in the low-intensity regime, where the light field is only a small perturbation for the initial state of the atom. The complexity arises, of course, from multi-electron excited states of atoms. They need to be considered whenever the photon energy is less than the energy for double (or multiple) ionization of the atom, and when the light energy is not absorbed purely sequentially, i.e. by excitation and ionization of one electron at a time.

The present overview will be restricted to the discussion of *highly excited states* of a small number of atomic electrons, and their interaction with the laser field. Any such discussion must start with the intrinsic properties of these states in the absence of external electromagnetic fields; questions of basic interest are the energy spectrum and the stability against radiative and autoionization decay. This is where the interest in the fundamentals of laser-atom interaction is met by the interest in the internal dynamics of an n-body Coulomb system, consisting of an ion and n-1 excited electrons.

2. AUTOIONIZING RYDBERG STATES

2.1 Classification and Theory:

Following the general characteristics of the well known "Rydberg states" of atoms [1], "Autoionizing Rydberg States" of an N-electron atom may be defined through the requirement that the configuration space for one highly excited electron can be separated into a "core" and an outer region [2]. Here the core comprises, in general, an ion with N-1 electrons and is governed by many-particle interactions which need not be specified; the outer region is assumed to be fully determined by the Coulomb interaction between the Rydberg electron and the core. Furthermore, we must require the core to be in an excited state (with an excitation energy larger than the binding energy of the Rydberg electron), otherwise no autoionization would occur if the Rydberg electron exchanges energy with the core. Speaking in terms of periodic scattering of the Rydberg electron from the core, *inelastic scattering* into continuum states may occur, in addition to elastic scattering as in regular bound Rydberg states. Various examples of autoionizing Rydberg states exist, depending on the particular core configuration. Within the present context we shall focus on autoionizing Rydberg states of the alkaline earth metals. There, one valence electron is highly excited and the other is in a low excitation state such that is still part of the spatially confined core, but may exchange energy with the Rydberg electron.

The theoretical description of autoionizing Rydberg states follows the spirit of scattering theory, with the allowance for elastic and inelastic (i.e. multi channel) scattering. The relevant theory is the Multi Channel Quantum Defect Theory MQDT [3]. A "channel" in the MQDT sense is defined through a complete set $\{i\}$ of quantum numbers describing all (N-1) electrons inside the core together with the angular and spin quantum numbers of the Rydberg electron, leaving only the energy ε of the latter as free variable. Associated with each channel is the ionization limit $I^{(i)}$, whereby the arrangement of ionization limits reflects the energy levels of the core. A channel is said to be "closed" in the energy region below its ionization limit, $\varepsilon < I^{(i)}$, while above $I^{(i)}$ it is "open"; at any given energy the total number M of channels may be divided into M_o open and M_c closed channels. In principle, all channels belonging to a given value of total angular momentum J and parity Π can interact with each other, whether they are closed or open. The interaction must occur inside the core at $r < r_0$ (therefore also called the "reaction zone"), since the only long range interaction considered is the Coulomb interaction which preserves the quantum numbers $\{i\}$ of a channel. A very compact description of the channel interactions is given by the *short range scattering matrix* $\boldsymbol{\chi}$, an weakly energy dependent (MXM) matrix with the properties $\boldsymbol{\chi} = \boldsymbol{\chi}^T$ and $\boldsymbol{\chi}\boldsymbol{\chi}^* = 1$. The matrix $\boldsymbol{\chi}$ differs from the usual scattering matrix **S** in that it connects *all* channels, not just the open ones; reduction from $\boldsymbol{\chi}$ to the familiar ($M_o \times M_o$) matrix **S** depends on the total energy of the system and the arrangement of ionization limits [3].

Once the short range scattering matrix $\boldsymbol{\chi}$ and the arrangement of ionization limits is known, the MQDT wave function $\Psi(E,\rho)$ for the outer electron of an autoionizing Rydberg state is easily obtained (here ρ is a scaled radial variable to allow for arbitrary charge Z of the core [3]). It is derived from a complete set of M linear independent solutions of the M-channel problem, commonly represented by a (MXM) function matrix $\mathbf{F}(\boldsymbol{\chi},\rho)$

$$\mathbf{F}(\boldsymbol{\chi},\rho) = \varphi^-(\rho) - \varphi^+(\rho)\,\boldsymbol{\chi} \qquad (1)$$

where $\varphi^-(\rho)$ and $\varphi^+(\rho)$ are diagonal matrices of asymptotically incoming and outgoing Coulomb wave functions [3] (alternatively, **F** may be represented in *any* basis of Coulomb functions, defining different channel interaction matrices which may all be transformed into one another). A unique solution $\Psi(E,\rho)$, *the* MQDT wave function for a given problem, is obtained by regarding the specific boundary conditions of the experimental situation, and the proper partitioning of **F** into closed and open channels. In general, $\Psi(E,\rho)$ is a vector with M components describing the channel composition; Ψ is constructed as a linear combination of the columns of the function matrix **F**. Observable quantities (like photoionisation cross sections, autoionization electron angular distributions etc.) are obtained by substituting $\Psi(E,\rho)$ into the relevant transition matrix elements.

2.2 Experiments on laser excited autoionizing Rydberg states

The experimental investigation of two-electron excited states with lasers has concentrated on the alkaline earth atoms [2]. As a typical example we may consider the calcium level diagram (fig. 1), which is, in fact, representative for the heavy alkaline earth atoms. We recall that these atoms all have an ns^2 ground state configuration outside closed shells, with n = 4,5, and 6 for Ca, Sr, and Ba. The second ionization limit lies about 1 eV above the first limit and belongs to a (n-1)d configuration of the ion; the np configuration lies about 2.5 eV above the first limit. Both the (n-1)d and the np configurations exhibit some fine structure splitting. Autoionizing Rydberg states are found just below each of these limits, when one electron is highly excited and the second is in a low np or (n-1)d state.

Cooke and Gallagher [4-6] have first proposed the following excitation scheme for autoionizing Rydberg states, which will be explained by the Ca level diagram of fig. 1, but holds, *mutatis mutandis,* for Sr and Ba as well: in a first step a bound $4sn\ell$ Rydberg state is excited by one- or two-photon absorption out of the ground state, eventually through the intermediate 4s4p state which lies roughly half way between the ground state and the ionization limit. The Rydberg electron is then in a s,p, or d state, with some selectable principal quantum number n; for the sake of simplicity we shall assume a 4sns ($^1S_0$) state. Then, in a second step, a further laser drives the transition to the autoionizing Rydberg states, say, of the $4p_{3/2}m\ell'$ series below the $4p_{3/2}$ ionization limit. This laser

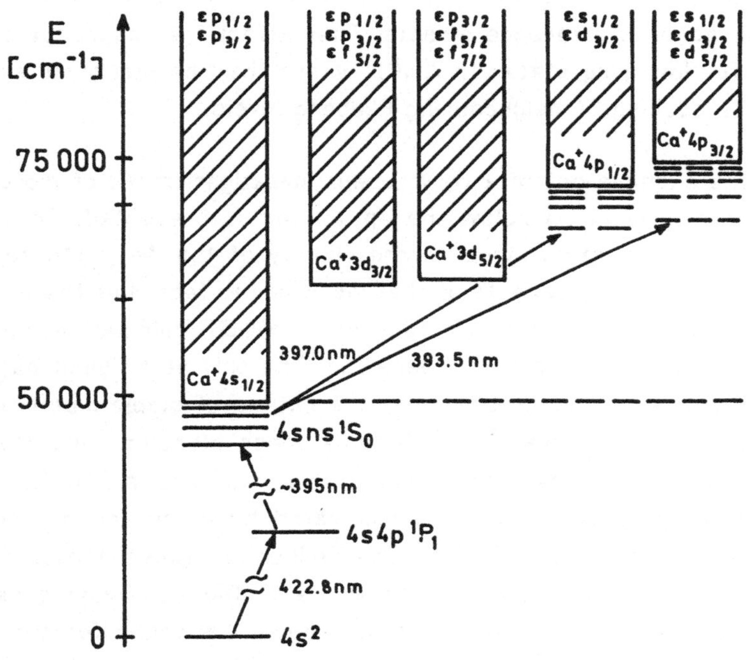

Fig.1: "Isolated core excitation" of autoionizing Rydberg states in Ca

is scanned in the immediate vicinity of the $4s_{1/2}$- $4p_{3/2}$ ion resonance, thus driving the inner electron transition ("isolated core excitation"). Near the ion resonance frequency, the oscillator strength for the transition is close to 1. If the last laser is detuned, the Rydberg electron adjusts its energy but remains approximately in its quantum state ns, provided that the laser detuning from the ion resonance is less than the energetical spacing between successive Rydberg levels ("near resonant core excitation"). The oscillator strength drops rapidly on either side of the ion resonance [5], but has a useful range of more than half the spacing to the neighboring resonance. By changing the *initial* Rydberg state ns the whole autoionisation region can be covered in overlapping steps (note, however, that the signals from adjacent initial states ns,(n$\pm$1)s are generally different in the overlap regions).

The great advantage of this excitation scheme is its *channel selectivity,* i. e. in most cases only *one* component of the final state's MQDT wave function $\Psi(E,\rho)$ is excited. As a consequence, complications due to interfering excitation amplitudes do in general not occur (in contrast to UV excitation from the ground state [7], where in the present example, 13 complex dipole amplitudes, which one is usually not interested in, would interfere). In calcium, due to the small (222.9 cm^{-1}) fine structure splitting of the $4p_{1/2}$-$4p_{3/2}$ inner electron transition, two dipole amplitudes to the two fine structure channels may contribute to the signal and may exhibit some interference effects. They affect the resonance structure predominantly for low lying states ($n \lesssim 10$), where the fine structure splitting is smaller than the spacing of neighbouring Rydberg levels.

On fig. 2 one example out of a very recent investigation [8] of more than 15 autoionizing resonances in Ca below the $4p_{1/2}$ limit is presented. Shown is the near resonant core excitation signal in the vicinity of the $4p_{3/2}14s$ resonance. One of the most interesting aspects of that work is the fact that the experiment can be directly compared with a full 13 channel MQDT calculation based on a interaction matrix given by Kim and Greene [9]. The only other input parameters were the experimental Ca ionisation limits, the ratio ρ of dipole matrix elements $4s\rightarrow4p_{1/2}$ and $4s\rightarrow4p_{3/2}$ ($\rho = 2^{-1/2}$) for the inner electron, and the energy dependence of the near resonant core excitation cross section [5]. The electron angular distribution parameters β_l were calculated from the angular momentum transfer theory of Fano and Dill [10]. Comparison of experimental data with theory shows that the position, width and shape of the autoionising resonance (fig. 2a) are very well reproduced. Some discrepancy is found for the electron branching ratio r_d into the $Ca^+(3d)$ continuum (fig. 2b), where the experimental Ca^+ population inversion is slightly higher than predicted (especially around the

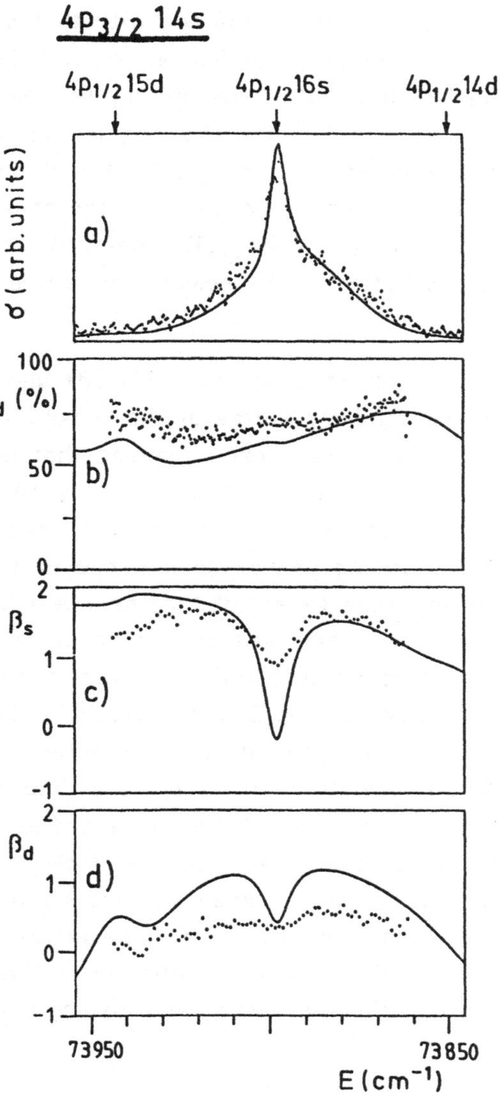

Fig. 2: Experimental (dots) and theoretical (solid lines) results on the $4p_{3/2}14s$ autoionizing Rydberg state in Ca. The positions of perturbing $4p_{1/2}ns,nd$ states are shown on top of the figure.

a) Isolated core excitation cross section (arb. units).

b) Branching ratio r_d (percentage fraction of autoionization electrons emitted into the [unresolved] $Ca^+(3d_{3/2,5/2})$ continua)

c) Angular distribution parameter β_s for electron emission into $Ca^+(4s)$

d) Angular distribution parameter β_d for electron emission into $Ca^+(3d)$

$4p_{1/2}16s$ perturber, whose position - together with that of other perturbers - is shown on top of fig. 2). The residual figures 2c and 2d present the energy dependence of the angular distribution asymmetry parameters β_s and β_d, respectively. These parameters determine the angular distribution $\sigma(\Theta)$ of the autoionisation electrons emitted into the $Ca^+(4s)$ and $Ca^+(3d)$ continua, respectively. A noticeable discrepancy is observed at the characteristic central dip in the β_s parameter caused by the $4p_{1/2}16s$ perturber. In contrast, the β_d parameter (fig. 2d) agrees better in the resonance centre than in the wings of the resonance.

Summarizing the information from this [8] and similar [11-13] very recent work on autoionizing Rydberg states of the heavy alkaline earths, we may identify several interesting aspects. First, it seems that both the general form of the MQDT interaction matrix and its energy dependence is rather well represented by theory in Ca [9] and Sr [11]. This overall agreement is particularly interesting in view of the strong energy dependence of the ns-(n-1)d channel interactions, which are generally expected for the heavy alkaline earth atoms, but may be restricted to lower energies than covered by the present experiments [9]. Second, it was found that the most sensitive observables like the angular distribution parameters β_s and β_d still revealed some discrepancies between theory and experiment. Thereby, the β_s parameter allows for a particular transparent interpretation: deviations from $\beta_s = 2$ (most pronounced at the $np_{1/2}ms$ resonances of Ca (n=4) and Sr (n=5)) are due to parity unfavoured transitions which occur as a consequence of spin orbit interaction; when going from Ca to Sr it appears as if these transitions are increasingly *overestimated* by theory with increasing atomic number. The β_d parameter has a more complicated dependence on the transition moments, and may additionally be affected by anisotropic long range interactions between the escaping electron and the aligned $Ca^+(3d)$ ion.

In general, further information is required to relate the observed discrepancies to particular approximations of the theory, like the derivation of a jj-coupled interaction matrix from an LS-coupled one through simple angular momentum algebra [9], or the neglect of anisotropic long range interactions in the MQDT approach. It will certainly be interesting to follow the comparison between theory and experiment through the sequence of heavier alkaline atoms up to Ba. Experimental data on the energy dependence of angular distributions and branching ratios in Ba exist [14] and have already revealed some qualitative differences to both Sr and Ca, but can, at present, not yet be compared with theory on a sufficiently fine energy mesh. However, we may already conclude that this particular class of doubly excited states is essentially understood and, with minor restrictions, well represented by an effective one-electron theory such as MQDT. Furthermore, most phenomena can even be qunatitatively reproduced by model calculations. We may therefore proceed to more complicated situations where the independent particle model is expected to break down.

3. PLANETARY ATOMS

The term "Planetary Atom" has been introduced by Percival [15] in his semiclassical treatment of atoms with two highly excited electrons. Without restriction to the semiclassical theory we shall use this notation to describe a situation where, in contrast to autoionizing Rydberg states, two (or even more) electrons are highly excited. We note that sometimes the term "Double Rydberg state" is also used, indicating that none of the two electrons is part of the core any more. We can imagine planetary atoms to evolve from autoionizing Rydberg states through excitation of the inner electron, with the outer electron being held fixed in a given quantum state. Speaking in terms of MQDT, the size of the reaction zone (core) is increased at the expense of the outer electron's Coulomb zone. However, it is expected that MQDT will soon come to its limits, certainly then when the inner electron is so highly excited that the identification of an "outer region" looses its meaning. Even long before that one might expect the inner electron's wave function to loose its "stiffness" against non-central electron-electron interactions. In the simplest picture, the inner wave function will be adiabatically polarized by the slowly moving outer electron. These effects should be most pronounced for high angular momentum states of the inner electron, which do not penetrate the core and are therefore extremely polarizable.

Conceptually, the transition from autoionizing Rydberg states to planetary atoms appears as straightforward; nevertheless, it is experimentally difficult. Most experiments to date have again used isolated core excitation. As can be seen from the Ba level diagram of fig. 3, transition to planetary states (located in the uppermost part of the level diagram, just below each of the Ba^+ ion states) requires between 5eV and 10eV excitation energy for the inner electron, if one starts from abound Rydberg state $6sn\ell$. *Resonant* multiphoton laser excitation of the inner electron was only rarely used as it principally suffers from the rapid autoionization decay of the intermediate states. In order to avoid these losses, non-resonant [16,17] or ingenious near-resonant [18] excitation schemes have been most extensively employed. In almost all cases a two-photon excitation of

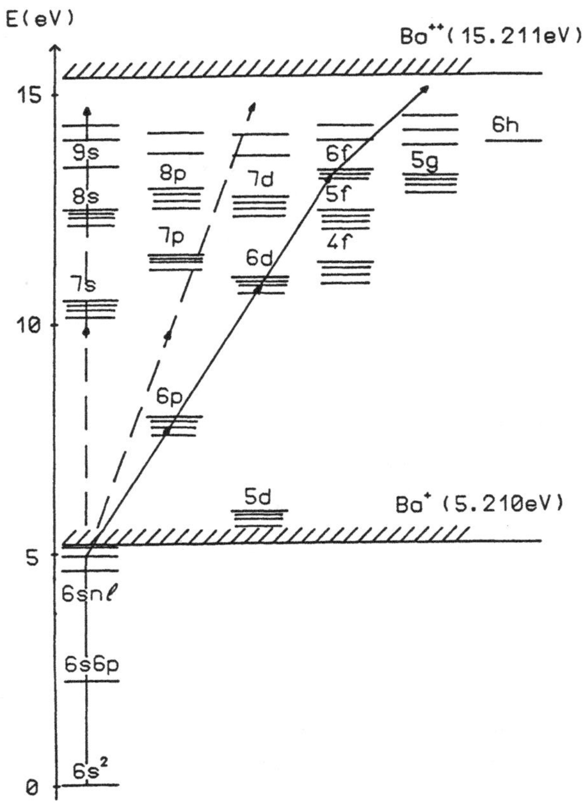

Fig. 3: Non-resonant 4-photon (dashed line) and resonant 6-photon (solid lines) excitation of planetary states in Ba

the inner electron was employed (dashed arrows on fig.3), limiting the accessible two-electron configurations to mdnℓ or msnℓ planetary states.

Much of the experimental work on planetary atoms has concentrated on laser excited barium. In terms of approximate independent electron quantum numbers mℓ' nℓ the earlier experiments [16-19] investigated mainly msnd states with $7 \leq n \leq 11$ and $13 \leq m \leq 85$. The observed resonances were typically interpreted in the spirit of MQDT, i.e. quantum defects and widths were related to the systematic variation of *size* of the reaction zone [20], which reflects only globally the influence of the inner electron. The same approach has been used to interprete the systematics of the transition moments: a dipole transition of the inner electron is typically assumed, accompanied by a simultaneous "shake" transition of the outer one to account for the change in quantum defects [20]. Even though many observations agree qualitatively with the MQDT picture, a quantitative comparison with theory is still missing in most cases.

Very recently [21,22] the studies have been extended to states with two very highly excited electrons, i.e. m around 30 and n around 45. Even there the MQDT approach has been applied; however, a breakdown of this model became finally obvious in two places: first, the observed resonance structure disappeared when the principal quantum numbers of the two electrons came close to each other (i.e. their ratio differed by less than roughly 25% from unity [21]). Second, extra structures appeared [22] at positions corresponding to different angular momentum states of the inner electron. Similar results have been obtained in doubly excited Ca [23].

The latter observations, although not yet quantitatively explained, are certainly the most spectacular manifestations of *positional correlations* between the two excited electrons that have been observed in laser excited planetary atoms. Nevertheless, it is useful to recall that they became only obvious at rather high excitation states of the two electrons, and even there only as a gradual breakdown of the independent particle model. This is to be contrasted with the experiment on even the lowest doubly excited states in helium, obtained 25 years ago by Madden and Codling [24] with VUV synchrotron light: there, the correlation effects were so dramatic that only one out of three expected series was clearly visible, the second was extremely weak and the third was completely missing. That experiment triggered an avalanche of theoretical work on correlated states of an electron pair [25]. Experimentally, it is still one of the outstanding manifestations of two-electron correlations in photon excited planetary states, despite of the enormous efforts and progress with laser excited systems during the last years.

When we ask about the reasons for this drastic difference between doubly excited barium (in high quantum states) and doubly excited helium (even in low quantum states), then one immediate answer is the ℓ-degeneracy of the inner electron's wave function in He. This ℓ-degeneracy is responsible for the extremely high polarizability of the inner electron's wave function in the field of the oute electron. As a consequence, ℓ looses its meaning and is to be replaced by new quantum numbers like "+" or "-" in case of the $He^+(n=2)$ core [25]. The correlated wave function differs from angular momentum eigenstates, leading to the unexpected photoabsorbtion cross section in doubly excited He. In contrast, we may read from the level diagram of fig. 3 that the Ba^+ core states are far from degenerate, at least for angular momenta $\ell < 4$ (up to f states). Hence, they remain approximate angular momentum eigenstates even in the presence of correlation interactions, which explains the qualitative difference between Ba and He planetary states.

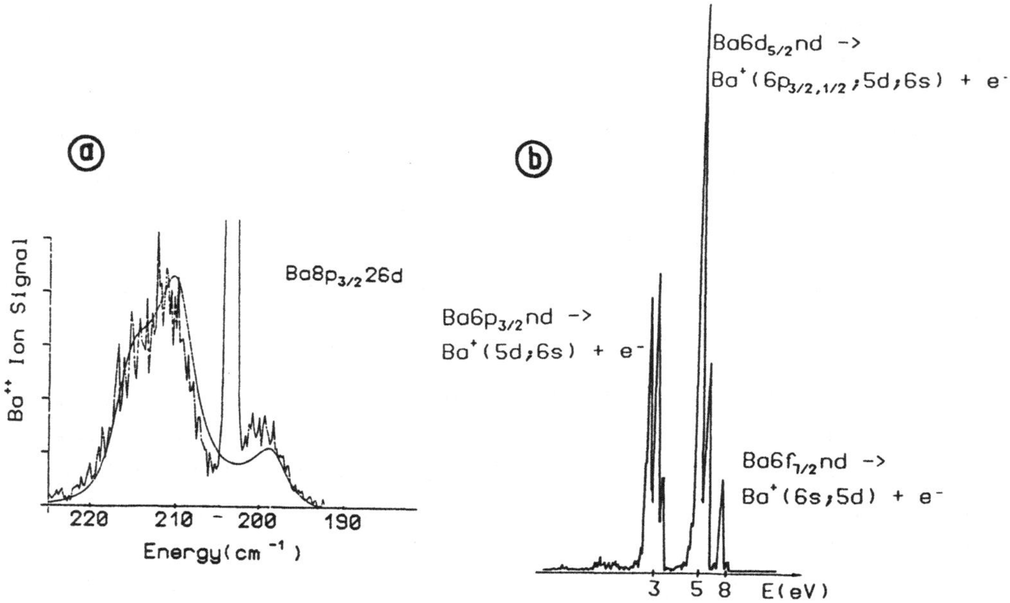

Fig. 4: a) Photoabsorption signal from the Ba $8p_{3/2}26d_{(J=2,3)}$ state
b) Electron spectrum from Ba $6f_{7/2}nd$ states (n>30)

In the light of such considerations, we have recently applied a resonant multiphoton excitation scheme [26] for planetary states in barium, which should ultimately lead to He-like, non-core penetrating states for the inner electron. It is shown on fig. 3 (solid arrows) and uses four different light frequencies (four lasers) to excite the inner electron through resonant intermediate states up to g-states. Together with two lasers for the outer electron, this is, in fact, a 6-laser excitation mechanism for planetary states. First results after five laser excitation to 8pnd and 6fnd states are shown on fig. 4a and 4b, respectively; six laser excitation is in progress. Fig. 4a displays the total photoabsorption signal of the $8p_{3/2}26d_{(J=2,3)}$ states, together with a restricted (three channel) MQDT analysis (solid line; note that the strong line near $205cm^{-1}$ does *not* belong to the neutral Ba spectrum). The MQDT analysis of a whole series of these states yielded first information on the quantum defects and total decay widths of these planetary states, together with a revised value for the Ba^+(8p) ionization limit. Fig. 4b shows first results on the electron spectroscopy of planetary states, in particular, of a $6f_{7/2}nd$ (n>30) state in Ba after resonant five laser excitation. Apart from strong autoionization signals from intermediate resonances (around 3eV and 5eV, respectively), the signal from electrons emitted from the $6f_{7/2}nd$ state into the Ba^+(6s) continuum, occuring at about 8eV, is clearly seen.

This experimental result, although rather qualitativ, is highly interesting for at least two reasons: first, it shows that 6fnd planetary states in Ba are still strongly "core penetrating" states, otherwise they would have negligible overlap with the final $Ba^+(6s)$ state and would not autoionize into this particular continuum. Second, considering that the initial Rydberg state was a 6snd state, the electron signal around 8eV may be interpreted as the *three photon above threshold ionisation (ATI) signal* in barium (starting from a well defined Rydberg state). The remarkable fact is, that this signal was obtained with laser intensities in the 10^5 W/cm^2-regime, while conventional ATI-spectroscopy requires laser intensities beyond 10^{10} W/cm^2. Of course, the experimental situation differed from usual ATI experiments in that each of the "above threshold photons" had a different frequency; only this way the strong resonant enhancement of the process could be achieved. On the other hand, it raises the interesting question about the possible role of doubly excited states in conventional, single frequency multi-photon absorption of atoms by high intensity laser light.

4. LASER INTERACTION WITH MUTLIPLY EXCITED ATOMIC STATES

The question raised at the end of the preceeding section has recently been taken up by a collaboration between the University of Virginia, USA, and the Laboratoire Aimė Cotton, Orsay, France, with participation from the Universität Freiburg, West Germany [27-29].

The experimental method is easily explained using the level diagram of fig. 5: On fig. 5, two possible ionization paths are shown, 1 and 2, which lead to the production of doubly ionized barium. Path 1 is a direct eight-photon double-ionization through broad autoionization resonances of neutral Ba (i.e. through autoionizing Rydberg states and planetary states, schematically shown as broad intermediate structures in fig 5). If the double ionization proceeds that way, the two electrons are obviously ionized *simultaneously* by the laser light. One expects resonant enhancement of the Ba^{++} production, whenever one of the intermediate resonances lies at or near a multiple integer of the photon energy $h\nu$. Path 2, on the other hand, describes a *sequential* double ionization process: First, Ba^+ is created in a well defined one electron configuration through multi-photon single ionization by a number of photons (three in the present example). Starting from this configuration, Ba^{++} is created through a second multi-photon ionization process. The latter (and thus the total Ba^{++} production) may also be resonantly enhanced through excited ionic states. However, these one-electron resonances

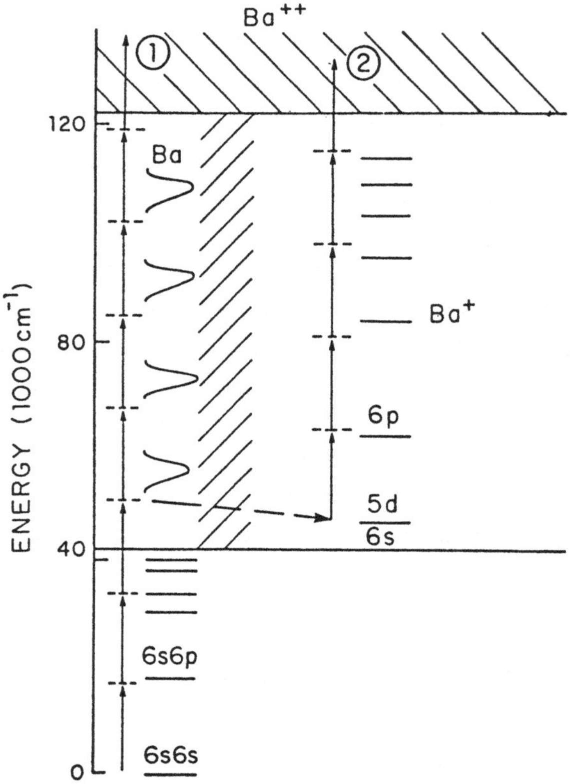

Fig. 5: Simultaneous (path 1) and sequential (path 2) double ionization in Ba

are encountered at different photon frequencies compared with the corresponding autoionization resonances of path 1, since in path 2 some energy is lost in the kinetic energy of the first ionized electron (dashed arrow in fig. 5).

The experiment has been performed such that laser light from pulsed, *tunable* dye lasers was used in the wavelength region from 280 to 700 nm; the laser intensity was between 10^9W/cm^2 and 10^{10}W/cm^2, pulse duration was 5 nsec. Observed was the Ba^{++} production as a function of laser wavelength. Using the above consideration it was then decided by the energetical position of the resonances whether the double ionization occured simultaneously (path 1) or sequentially (path 2). More than sixty resonances have been found in the investigated wavelength region [28], and have been assigned to possible intermediate states. As an example, we may arbitrarily pick the photon energy region 19600 cm^{-1} $\leq$ $h\nu \leq$ 20000cm^{-1}, shown on fig. 6. A total of 8 resonances of varying widths and intensities has been catalogued in this region, almost all of which could be identified with intermediate resonances in path 2, starting from either the $Ba^{+}5d_{3/2,5/2}$ or $Ba^{+}6p_{1/2,3/2}$ initial state. A similar situation was found over the

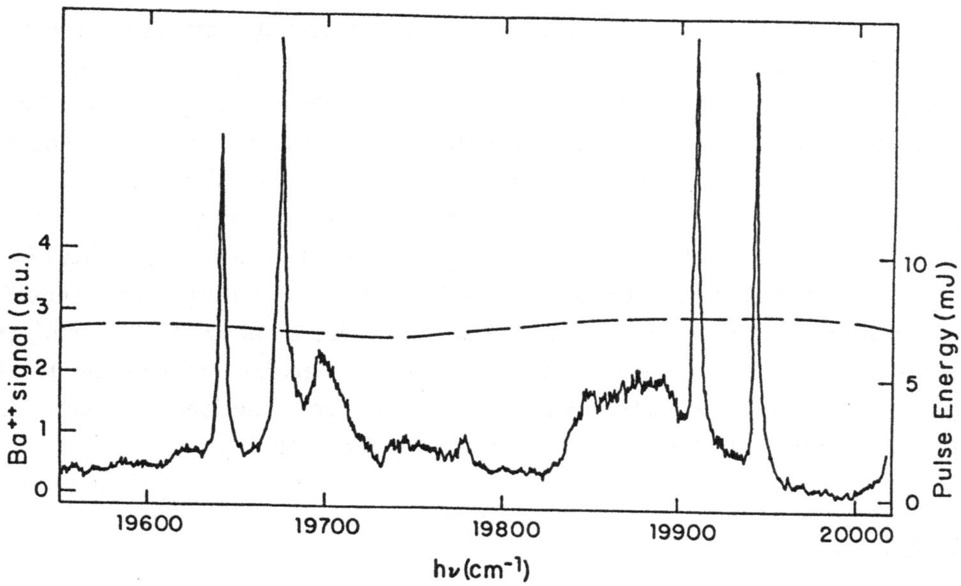

Fig. 6: Ba^{++} production with tunable laser light ($19600cm \leq h\nu \leq 20000cm^{-1}$)

whole photon energy region under investigation. While some small features remained unexplained, there was *no positive identification of any two-electron excited resonance* in the production of Ba^{++} under the conditions of the present experiment [27-29].

The somewhat surprising conclusion from this experiment is that autoionizing Rydberg states and planetary atoms do not play any substantial role in the resonant enhancement of double ionization of barium, i.e. double ionization occurs in a sequential manner. At present, it is unclear how these results transform into observations at considerably higher laser intensities and/or shorter pulse durations (apart from the trivial limiting case of extremely strong and short laser pulses where any multiple ionization will necessarily become "simultaneous"). The basic question is whether two electrons in the Coulomb field of an ion can simultaneously absorb photons from a laser field while going through a sequence of *correlated three-body Coulomb states*. Further investigations along this line, using different atoms and lasers, are certainly highly desirable.

5. CONCLUSION

Laser-atom interaction beyond the hydrogenic (two-body Coulomb) approximation for the atom is still at its beginning, but has already produced some very promising results. Various very efficient laser excitation schemes have been developed to prepare doubly excited electronic states of atoms, so called autoionizing Rydberg states and planetary atoms. They require only very moderate laser powers, thus allowing the investigation of these states in a quasi field-free environment. This is of eminent importance as these states are experimental realizations of the quantum mechanical n-body Coulomb problem, one of the last fundamental unsolved problems in atomic physics. On the other hand, is has been shown in barium that these states play, at best, a minor role as resonances in the one-frequency laser-atom interaction at intensities of about 10^{10} W/cm^2 and pulse durations of 5nsec. Further experiments are required to investigate the general role of simultaneouas ("collective") electronic motions in the multiphoton multiple ionization process of atoms by laser light.

REFERENCES

[1] R.F. Stebbings and F.B.Dunning (eds.),"*Rydberg States of Atoms and Molecules*" Cambridge University Press (1983)

[2] W. Sandner, Comm. At. Mol. Phys. **20** (1987) 171

[3] M.J. Seaton, Rep. Progr. Phys. **46** (1983) 167

[4] W.E. Cooke and T.F. Gallagher, Phys. Rev. Let. **41** (1978) 1648-1652

[5] S.A. Bhatti, C.L. Cromer and W.E. Cooke, Phys. Rev. A **24** (1981) 161-165

[6] N.H. Tran, P.Pillet, R. Kachru and T.F. Gallagher, Phys.Rev A **29** (1984) 2640

[7] J.P. Connerade, A.M. Lane and M.A.Baig, J.Phys.B:At.Mol.Phys. **18** (1985) 3507-3527

[8] V. Lange, U. Eichmann and W. Sandner, J.Phys B:At.Mol.Opt.Phys., in print

[9] C.H. Greene and L. Kim, Phys. Rev. A **36** (1987) 2706-2717
L. Kim and C.H. Greene, Phys. Rev. A **36** (1987) 4272-4279

[10] U. Fano and D. Dill, Phys. Rev. A **6** (1972) 185-192

[11] M. Aymar and J.M. Lecomte, J. Phys. B: At. Mol. Opt. Phys. **22** (1989) 223-245

[12] E. Xu, Y. Zhu, O. Mullins and T.F.Gallagher, Phys. Rev. A **33** (1986) 2401-2409

[13] Y. Zhu, E. Xu and T.F. Gallagher; Phys. Rev. A **36** (1987) 3751-3767

[14] V. Lange, U. Eichmann and W. Sandner (unpublished)

[15] I.C. Percival, Proc. R.Soc. Lond. A **353** (1977) 189

[16] R.M.Jopson, R.R.Freeman, W.E.Cooke and J.Bokor, Phys.Rev.A **29** (1984) 3154

[17] J.Boulmer, P.Camus, J.M.Gagnė and P.Pillet, J.Phys.B:At.Mol.Phys. **20** (1987) L143

[18] T.F.Gallagher,R.Kachru,N.H.Tran, H.B.v.Lindenvd.Heuvel,Phys.Rev.Lett. **51** (1983) 1753

[19] T.F. Gallagher, K.A. Safinya and W.E.Cooke, Phys. Rev. A**24** (1981) 601

[20] L.A.Bloomfield, R.R.Freeman,W.E.Cooke and J.Bokor, Phys.Rev.Lett. **53** (1984) 2234

[21] J.Boulmer, P.Camus and P.Pillet, in "*Electronic and Atomic Collisions*", H.B.Gilbody, W.R.Newell, F.H.Read and A.C.H.Smith (eds.), Elsevier (1988) 759

[22] J.Boulmer, P. Camus, J.M.Lecomte and P.Pillet, Proceedings of 11.ICAP (in print)

[23] N. Morita and T. Suzuki, J.Phys. B:At.Mol.Opt.Phys. **21** (1988) L439-L444;
N. Morita, T.Suzuki and K. Sato, Phys.Rev A **38** (1988) 551-554

[24] R.P.Madden and K.Codling, Phys.Rev.Lett.**10** (1963)516; Astrophys. J.**141** (1965) 364

[25] U. Fano, Rep.Prog.Phys. **46** (1983) 97

[26] U. Eichmann, P. Brockmann, V.Lange and W. Sandner (unpublished)

[27] U. Eichmann, Y. Zhu and T.F. Gallagher, J.Phys.B **20** (1987)4461-4471

[28] Y.Zhu, R.R.Jones, W.Sandner,T.F.Gallagher, P. Camus, P.Pillet, and J.Boulmer, J.Phys.B: At.Mol.Opt.Phys. (1989) in print

[29] R.R.Jones,Y.Zhu, U.Eichmann and T.F.Gallagher, IEEETrans.PlasmaSc.**16** (1988) 534

Electron Correlations in Time and Space

W.E. Cooke, J.G. Story, and X. Wang
Physics Dept.
University of Southern California
Los Angeles, CA 90089-0484

Introduction

Two-electron atomic wavefunctions are normally constructed from products of one-electron wavefunctions which are solutions to spherically-symmetric potentials. Thus, there is a natural inclination to view correlations in a time independent model, with solutions classified according to their various symmetries. For example, the wavefunctions are usually written as stationary linear combinations of products which have "good" values of parity and total angular momentum, J, and "almost good" values of spin, S, and orbital angular momentum, L. The rationale for this is that stationary states have well defined energies and that the electrostatic interaction preserves S, L, J, and parity. However, this view can obscure both the temporal behavior (since no temporal variation can occur in a stationary state) and the spatial correlation (which couples the individual angular momenta, l_1 and l_2, to make L). Multichannel Quantum Defect Theory[1–5] (MQDT) quite naturally incorporates such time-dependent correlations through its wavefunction normalization. The spatial correlations, however, are more problematical. Some recent attempts to construct two electron wavefunctions have used new representations which explicitly emphasize the spatial correlations; however, they still categorize solutions by their overall symmetries.[6–8] In our measurements of the $6pnf$ states of barium, we have found that the multiplets appear to have little regard for these basic symmetries.

In all the experiments reported here, we have used the Isolated Core Excitation[9,10] (ICE) method to excite barium atoms from their ground state, $6s^2\ ^1S_0$, to an autoionizing state. This method begins by exciting one $6s$ electron to a high Rydberg state (it is subsequently called the Rydberg electron), and then excites the second $6s$ electron to the valence $6p$ state (it is subsequently called the core electron). The excitation moment for the second step is overwhelmingly dominated by a one-electron picture of a core excitation, while the Rydberg electron merely shakes into a final state which has the same l value and is always very close in binding energy to its initial state. (Because the Rydberg electron sees different potentials with and without the core electron excitation, it will have different sets of wavefunctions before and after the core excitation, so that *some* change will generally occur. The ICE excitation is strongest for those states which minimize the energy change.) By using the ICE method to excite a specific target configuration, we accomplish two simplifications: (1) only the part of the spectrum which involves the target state is present; and (2) the transition strength is a direct measure of the admixture of the target configuration. These are particularly useful simplifications for analyzing configuration interactions.

Time Dependent Correlations

Using the ICE technique, one is naturally inclined to think in terms of temporal evolution. The power of the technique arises from the temporal separation of the excitation steps since *first* one excites the Rydberg electron, and *then* one excites the core electron *before* the Rydberg electron has time to change its wavefunction. Consequently, it is only natural to think of any other processes in terms of the temporal development of the excited core state which was initially prepared. For example, if we excite a $6p_{3/2}11d$ state of barium using the ICE technique, then initially, it will have *only* the $6p_{3/2}11d$ configuration. After a short time, the state will evolve into a continuum configuration (usually, the $Ba^+(6p_{1/2})+e^-$ continuum), as the atom autoionizes.[11] Of course, autoionization times are usually so short that current technology cannot easily measure such sub-picosecond time development.

However, the atom itself can be used to time this development, by choosing a state which "autoionizes" into a second, bound configuration. The temporal evolution of such a state will develop in two steps, with only the second step leading to an electron-ion pair. Under the right circumstances, this configuration can be long-lived, and thus "trap" the atom, so that the internal conversion process can be detected. In the first experiment to be described here, this second configuration will then fluoresce (with a decay time of 5 ns) as evidence to show the time scale of the net decay processes.

The $6p_{3/2}11d$ state of barium lies just above the $Ba^+(6p_{1/2})$ ionization limit, so that the lower energy tail of the state interacts with an infinite number of Rydberg states converging onto that ionization limit. A normal Configuration Interaction (CI) analysis of this state would be difficult because of this large number of configurations, and because of the large spatial extent of the wavefunctions. However, MQDT easily treats this case by including an entire Rydberg series as a single channel, since it enforces the boundary conditions which quantize the channel only *after* formally solving for the admixtures of various wavefunctions. Since MQDT solves for wavefunctions before quantizing, it is also easier to see temporal correlations in the MQDT formalism.

In Fig. 1, we show the energy diagram for exciting the $6p_{3/2}11d$ state using the ICE technique. First, two dye lasers are used to excite the $6s11d, {}^1D_2$ bound state using the resonant $6s6p, {}^1P_1$ intermediate state. Next, a third laser is used to drive the "core" electron transition, so that the atom ends up in the $6p_{3/2}11d$ autoionizing state. This doubly excited state will then produce either ions, or fluorescence or both.

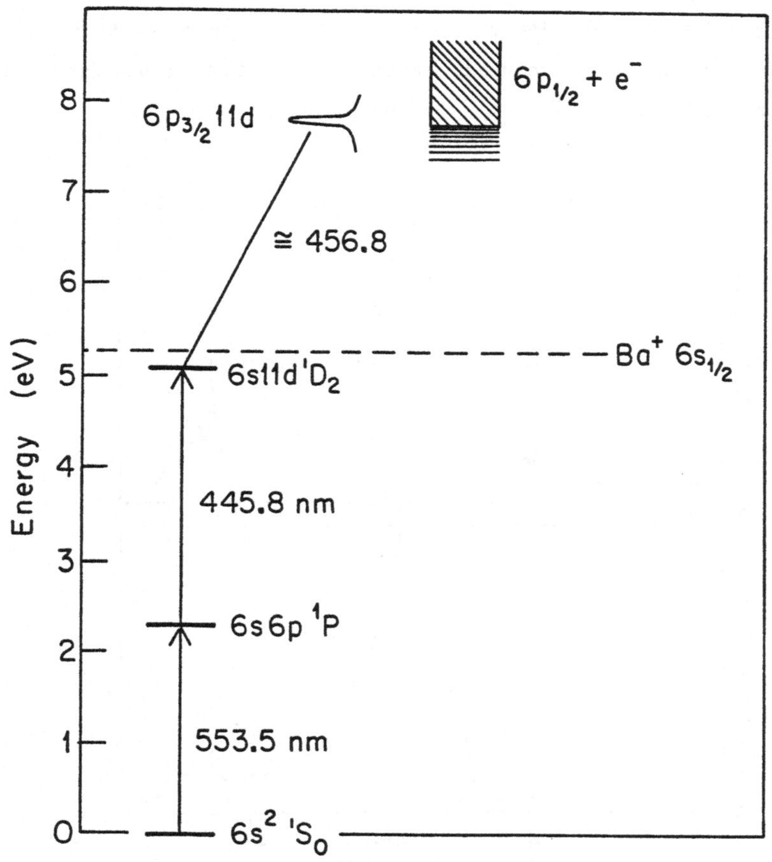

Figure 1: The energy level diagram for the excitation of the $6p_{3/2}11d$ state of barium.

The three lasers are used to excite barium atoms from one of two atomic beams. In one case, the lasers cross the barium beam between two parallel capacitor plates. The bottom plate has a small, positive voltage bias, and the top plate has a hole in it, so the final ions are forced through the hole, and into a microchannel-plate particle multiplier. In the second case, the lasers cross the atomic beam in a region which is imaged onto a monochromator, so that any fluorescence may be detected. A spectrum is obtained by setting the first two lasers to excite the $6s11d$ Rydberg state, and then scanning the third laser while monitoring the resultant ions or fluorescence.

Figure 2 shows typical spectra. The top spectrum is the total ion signal, recorded at a relatively high power, so that the dominant feature is broadened when all of the available bound states are ionized near the line center. The lower spectrum shows the fluorescence produced on the $6p_{1/2} \rightarrow 6s$ ionic transition. The dashed vertical line shows the location of the $Ba^+(6p_{1/2})$ ionization limit. Below this limit, the fluorescence signal falls off rapidly, since no $Ba^+(6p_{1/2})$ ions can be produced. Nevertheless, the fluorescence signal *does* extend somewhat below this limit.

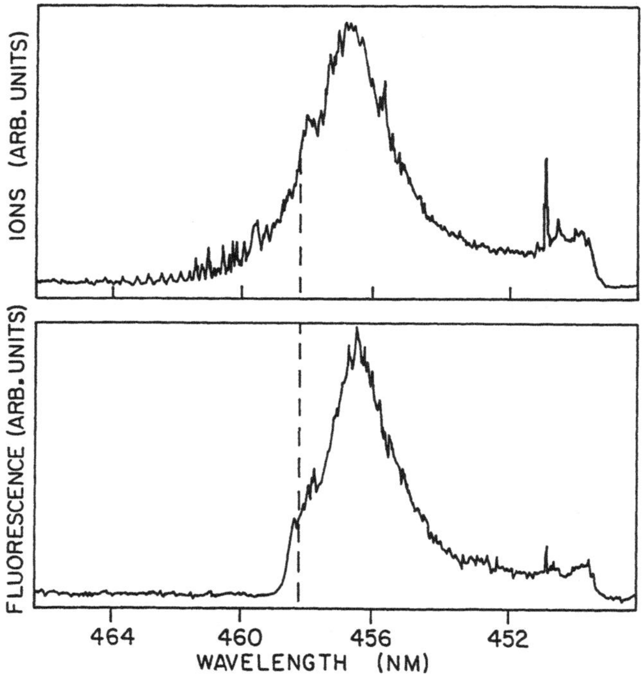

Figure 2: Ion and fluorescence yield from the $6p_{3/2}11d$ state of barium.

One expects the signal to be continuous across the ionization limit, of course, since there is no difference between the small r behavior of wavefunctions for a free electron or for a bound one, in the limit of very high principle quantum numbers. In Fig. 3, we show an expanded view of the region near the ionization limit, with insets to show greater expansion for selected regions. In the expanded regions, a clear Rydberg sequence can be seen, which we have identified as the $6p_{1/2}ng$ series from the states' quantum defects, widths, and symmetries. This represents doubly excited states *fluorescing* to produce *bound* $6sng$ final states. In the regions very near the ionization limit, the Rydberg series is obscured or destroyed, presumably by stray electric fields. (A field of only 3 V/cm will ionize an $n=100$ state.)

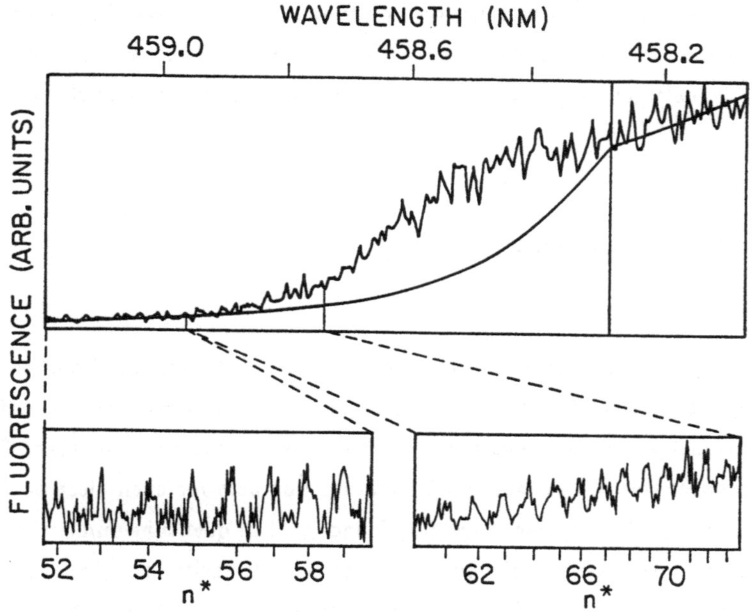

Figure 3: Expanded view of the fluorescence yield from the $6p_{3/2}11d$ state of barium.

This Rydberg sequence is a natural clock to time the development of the continuum wave character of the doubly excited state. The $6p_{1/2}ng$ states will fluoresce at the ion transition, but the natural decay time for this process is of the order of 5 ns. Figure 1 shows that *some ions* were still produced, even at the energies where fluorescence occurred. Consequently these ions must have developed on a time scale of 5 ns, even though the lifetime of the initial $6p_{3/2}11d$ state is less than a picosecond. This strongly suggests that the initial $6p_{3/2}11d$ configuration *first* converts internally into a $6p_{1/2}ng$ configuration on a time scale of picoseconds, and that then this configuration traps the atom until it fluoresces. In such a case, the fluorescence yield will only depend on the ratio of the radiative rate to the autoionization rate for each $6p_{1/2}ng$ state. Since the autoionization rate of the $6p_{1/2}ng$ states increases rapidly, as n^{-3}, with decreasing n, the fluorescence yield rapidly decreases. The solid line in Fig. 3 represents this model.

It may appear that the atom violates time reversibility by converting quickly into the $6p_{1/2}ng$ configuration, and then decaying slowly from that configuration; however, a classical picture can clarify this. Time reversibility implies that the electron has as much chance to go from the $6p_{1/2}ng$ configuration *to* the $6p_{3/2}11d$ configuration *per orbit* as it does going the other way. However, the *orbit times* of the two configurations are vastly different. This is what "traps" the electron in the high n configuration. MQDT naturally includes this same effect by using wavefunctions that are normalized per unit energy for continuum and bound states. This normalization scales the wavefunction by the square root of the classical orbit period so that all interactions are calculated *per Rydberg orbit.* This is the only reasonable normalization for calculating quantities that are localized near the core of the atom.

Space Dependent Correlations

The ICE technique also encourages one to think in terms of the symmetries of each electron's wavefunctions, rather than the total symmetry of the two-electron system. After all, one begins with a model where each electron is excited independently of the other. Thus, in the spectra of the $6p_jnl$ states, the individual quantum numbers for each electron are specified rather than the composite angular momenta, L and J. The quantum number J can often be identified by using polarization techniques which selectively excite one or another J state, but L is usually completely scrambled by the large fine structure of the core electron. In this section, we will describe our measurements of the $6p_j19f$ states, which show very little dependence on J, and appear to be best described in terms of the spatial characteristics of the two electrons' wavefunctions. The dominant characteristic of these states is their very fast decay rate - possibly too fast for the angular momentum couplings to be established. These states decay with a lifetime that is less than a classical Rydberg orbital period.

These short lifetimes create overlapping spectra, since the lifetime broadening is considerably larger than the splittings which may separate the various components of the multiplet. Consequently, it becomes very important to isolate the J states to the greatest extent possible using polarization techniques. We excite these $6pnf$ states with four linearly polarized lasers, using the $6s5d$ intermediate two photon resonance and the $6snf$ three photon intermediate resonance, as shown in Fig. 4. The first three lasers all have parallel polarizations to insure that the $6snf$ bound state is prepared in a pure $J=3$, $m=0$ state. The final core excitation laser, which drives the transition to the $6p_jnf$ state, can be polarized either parallel or perpendicular to the first two lasers. If it is parallel polarized, only $m=0$, $J=2$ or 4, states will be excited; whereas, if it is perpendicularly polarized, all three J states will be excited. The differences between the various J states then appear as different structure in the two excitation spectra.

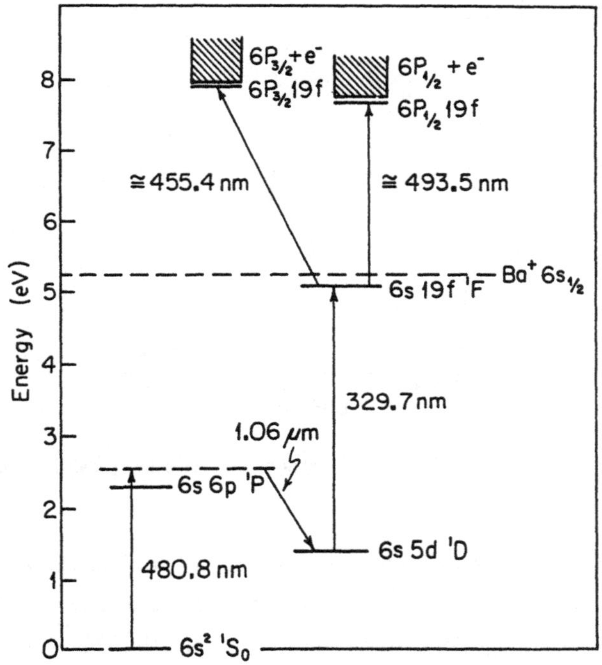

Figure 4: Energy level diagram for the excitation of the $6p_j19f$ states of barium.

In Fig. 5, we show the two spectra for exciting the $6p_{1/2}19f$ state, using either parallel or perpendicular polarizations. These two spectra, should include the various J states in the proportions 6:0:8, and 2:7:5 for J values of 2, 3, and 4 respectively. From these figures, and our least squares fits to these spectra, we can find no evidence of more than one spectral component. Consequently, the splittings between the J states must be small when compared to the linewidths of the J states, and *more surprisingly* the linewidths of the components must also be nearly identical.

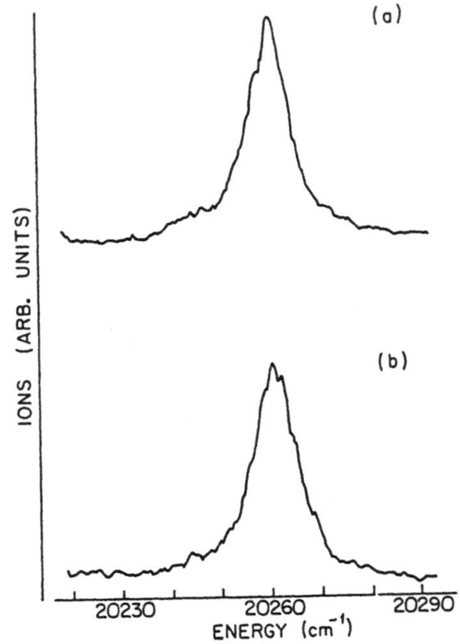

Figure 5: Excitation spectra for exciting the $6p_{1/2}nf$ states of barium using parallel (a) and perpendicular (b) core laser polarizations.

Since this spectra appears to be only one broad state, rather than the expected four states (two $J=3$ and one each of $J=2$ and $J=4$), it seems reasonable to assume that some interactions do not have enough time to develop. Equivalently, this would mean that some interactions are just much smaller than the states' linewidths, and can therefore be ignored. If we begin by ignoring the coupling of the outer electron's spin, then the four states separate into two distinct pairs, depending on whether the f is coupled to the $p_{1/2}$ to make an intermediate angular momentum of 7/2 or 5/2. Even these two pairs are only separated by $\frac{2}{15}F^2$, where F^2 is the Slater integral for the electric quadrupole interaction. Since the linewidths are so large, even this may just be too small to create structure.

It is not possible, however, to use any of these approximations to explain the similarity of the linewidths of the different components of the multiplet. The linewidths of these states are determined by autoionization, which intrinsically depends on the intial configuration *and* on the final configuration. The final configurations are strongly sensitive to the angular momentum, J, and to the spin, S. For example, the $6p_{1/2}nf$, $J=2$ states can decay to the $6s\epsilon d$ ion-electron state, but the $J=4$ state cannot. Our calculations suggest that this rate will be one of the largest, so that the $J=4$ state should be significantly narrower. Since the *area* of this resonance should dominate in the case of parallel polarization, it should easily be the largest feature, and therefore visible.

As a second example, the $5d\epsilon d$ decay channel has a strong dependence on the electron spin, S. For small values of the radial electron coordinate, r, the singlet states have far more electron density than the triplet states. This is because the $5d$ and ϵd electronic wavefunctions are nearly identical in this region, so antisymmetrization eliminates overlap in the case of triplet states. Again, our calculations suggest that this channel to be the second most important one; nevertheless, we can see no evidence of more than one feature in our spectra.

The $6p_{1/2}nf$ spectra thus appear to be completely independent of the core laser polarization. This surprising fact does reflect, however, the spatial isotropy of the $Ba^{+}(6p_{1/2})$ core wavefunction, suggesting that the spatial characteristics of the core dominate this rapidly decaying state. Data for the $6p_{3/2}nf$ states also show too few states; however, in that case the spectrum does change with core laser polarization. A broader feature appears when all the lasers have parallel polarization. Again, this seems to represent the spatial anisotropy of the $6p_{3/2}$ core wavefunction, more than it represents the overall symmetries.

Conclusion

Using the ICE method, we have shown that the autoionizing spectra can be explained in terms of temporal evolution and spatial correlations, rather than in terms of the energy eigenstates or the overall symmetries. In this sense, the doubly excited atom may be approaching closer to classical physics than single electron Rydberg states. New theoretical methods are now being developed for the high doubly excited atoms, *i.e.* planetary atoms. This work suggests that even lower lying core-excited states require new theoretical methods to describe their decay rates accurately.

Acknowledgements

This work was supported by the National Science Foundation under grant PHY85-00885.

References

1. M.J. Seaton, Rep. Prog. Phys. 46, 167 (1983).
2. U. Fano, Phys. Rev. A 2, 353 (1970).
3. W.E. Cooke and C. L. Cromer, Phys. Rev. A 32, 2725 (1985).
4. M. Aymar, J. Phys. B 20, 6507 (1987).
5. A. Giusti-Suzor and H. Lefebvre-Brion, Phys. Rev. A 30, 3057 (1984).
6. A.R.P. Rau, J. Phys. B 16, 699 (1983).
7. J.M. Feagin and J. Macek, J. Phys. B 17, 245 (1984).
8. H. Klar, Phys. Rev. Lett. 59, 66 (1986).
9. W.E. Cooke, T.F. Gallagher, S.A. Edelstein, and R.M. Hill, Phys. Rev. Lett. 41, 178 (1978).
10. S.A. Bhatti and W.E. Cooke, Phys. Rev. A 28, 756 (1983).
11. J.G. Story, L.D. Van Woerkom, and W.E. Cooke, Phys. Rev. A 34, 4508 (1986).

CORRELATIONS IN MULTIPLY EXCITED ATOMS

Hubert Klar
Fakultät für Physik, Universität Freiburg
Hermann-Herder-Straße 3, D-7800 Freiburg, FRG

Introduction. In this article we consider not too heavy atoms such that a non-relativistic description within the frame of the Schrödinger equation is meaningful. Stationary atomic states may be excited e. g. by a weak radiation field in the usual sense of perturbation theory. Modifications of the atomic spectrum by the field itself are not the subject of this article. Often a light sourse of suitable frequency to excite a doubly excited state or a double continuum state is not yet available experimentally. Due to this lack we consider here also inelastic electron scattering because e. g. the scattering of fast electrons at small momentum transfer yields cross sections directly related to photoexcitation and photoionization cross sections.

It is usually believed that neither the Schrödinger equation nor the classical equations of motions for a N-body Coulomb system ($N \geq 3$) are separable in any coordinate system. Moreover, these mechanical systems are nonintegrable in the sense that the number of conserved quantities is smaller than the number of degrees of freedom. In principle correlation is therefore always present. Nevertheless it is known since a long time that single-particle approximations like e. g. the Hartree-Fock approximation represent often suprisingly good descriptions of the reality. This is true in particular (i) if only the lower part of the atomic spectrum is considered, (ii) if only physical quantities insensitive to correlations (e. g. energy levels) are calculated, and (iii) if closed shells are considered because coorelation effects are at least partly washed out in this case. There are however exceptions. In negative ions e. g. correlation is always strong as shows the H^- ion for which the Hartree-Fock modell fails to predict the weakly bound ground state. On the other hand the same modell predicts a reasonable value for the He ground state binding energy. Its correponding wave function, however, yields much less accurate values for photoabsorption matrix elements.

In contrast to the situations mentioned above we focuss in this paper on situations where correlation is strong or even dominant. Typical examples are doubly and multiply excited states as well as multi-electron continuum states, situations where a few (at least two) weakly bound or even unbound electrons move in open shells. Two-electron continuum states are of particular interest. They may be created either in a single ionization by electron impact or in a double photoionization. These states have the advantage that the two escaping electrons can be detected in coincidence, i. e. one can measure energy and momentum of each electron. Such an experiment performed at fully determined kinematics yields all information about the correlation between the electrons. Experiments of this kind have been performed with increasing accuracy since 20 years for electron impact, for photon impact only integrated cross sections have been reported until now due to the smallness of multiply differential cross sections and due to the lack of suitable light sources. Doubly excited bound states (resonances) are of course also of great importance. Their complete analysis analogous to the above case of two-electron continuum states is however a practically impossible experiment, it would require to measure the energy position, all partial widths, and the angular distribution of autoionization electrons for each available ion state. We will see below that lower and intermediate doubly excited states are essentially understood, but for high double excitation there is neither experimental data nor a convincing theory except for fragments.

Doubly excited states. The breakdown of independent particle models for a many-body system is not equivalent to its strict nonseparability. It means only that the system is not separable within single particle coordinates. A natural question is therefore: Are there other, collective coordinate frames which are better suited for the problem under consideration, or, is there a 'best' coordinate frame? As a measure for the quality of a coordinate frame we have here in mind the amount of coupling between eigenmodes along generalized coordinate lines. Consider e. g. a two-dimensional problem to make clear this point. Let us ask for standing waves inside a nearly circular domain. We will then use plane polar coordinates, and we expect weakly coupled radial and angular eigenmodes. This coupling will decrease for decreasing departures from the circle. It is however obvious that these coordinates are not nessessarily the best ones, if the geometry of the domain is e. g. closer to an ellipsis than to a circle because the Helmholtz equation is exactly separable also in elliptic coordinates. To my knowledge it has never been tried to find systematically particularly useful coordinate frames in the above explained sense. A straightforward minimization of the channel coupling is an extremely difficult mathematical problem, and its eventual solution would yield presumably a hopelessly complicated result. The only realistic way to introduce better than single particle coordinates is presently intuitition.

Two sets of collective coordinates have been used particularly for two-electron atoms: Hyperspherical coordinates and more recently 'molecular' coordinates. We mention the hyperspherical coordinates here only very briefly, and refer the reader to Fano's review article /1/. A brief description of the method with applications to photo-processes was given elsewhere /2/. The molecular approach transfers the traditional Born-Oppenheimer treatment of the H_2^+ molecule to helium-like atoms /3,4/.

The key quantity in the hyperspherical approach is the hyperradius of the two electrons given by

$$R = \sqrt{r_1^2 + r_2^2} \tag{1}$$

where $\vec{r}_1$ and $\vec{r}_2$ are the electron positions. It is essential that for finite values of R the size of the atom is finite. The remaining five coordinates may therefore be choosen as angles. The time-independent Schödinger equation for a helium-like atom with nuclear charge Z

$$\left\{-\frac{1}{2}(\Delta_1 + \Delta_2) - Z\left(\frac{1}{r_1} + \frac{1}{r_2}\right) + \frac{1}{|\vec{r}_1 - \vec{r}_2|}\right\}\Psi = E\Psi \tag{2}$$

reads in these coordinates

$$\left\{-\frac{1}{2}R^{-5}\frac{\partial}{\partial R}R^{5}\frac{\partial}{\partial R}+\frac{\Lambda^{2}}{2R^{2}}+\frac{1}{R}C(\omega)\right\}\Psi=E\Psi \tag{3}$$

where the quantity Λ^2 is the quadratic Casimir operator for the six-dimensional rotation group, and acts only in the angles ω. There are infinitely many possibilities to parametrize the hypershere. Two of these parametrizations are particularly useful for three-body problems /5/, a special choice, however, is not relevant for the following. The Schrödinger equation is then conveniently solved in two steps /6/. In a first step one determines eigenvalues and eigenfunctions at constant values of R, i. e. one disregards the derivatives with respect to R,

$$\left\{\frac{\Lambda^{2}}{2R^{2}}+\frac{1}{R}C(\omega)\right\}\phi_{\mu}(\omega;R)=U_{\mu}(R)\,\phi_{\mu}(\omega;R). \tag{4}$$

These eigenvalues and eigenfunctions depend parametercally on the hyperradius R. In a second step one expands the exact wave function into a series of the above determined eigenfunctions with coefficients depending on R,

$$\Psi=\sum_{\mu}F_{\mu}(R)\,\phi_{\mu}(\omega;R). \tag{5}$$

Substitution of this ansatz into the Schrödinger equation yields a set of coupled ordinary differential equations,

$$\left\{-\frac{1}{2}\frac{d^{2}}{dR^{2}}-\frac{5}{2R}\frac{d}{dR}+U_{\mu}-E\right\}F_{\mu}=\sum_{\nu}\left\{P_{\mu\nu}\left(\frac{d}{dR}+\frac{5}{2R}\right)+\frac{1}{2}Q_{\mu\nu}\right\}F_{\nu} \tag{6}$$

where the coupling is provided by the matrix elements given by

$$P_{\mu\nu}(R)=\int d\omega\;\phi_{\mu}^{*}\frac{\partial}{\partial R}\phi_{\nu} \tag{7}$$

and

$$Q_{\mu\nu}(R)=\int d\omega\;\phi_{\mu}^{*}\frac{\partial^{2}}{\partial R^{2}}\phi_{\nu}. \tag{8}$$

This adiabatic expansion is formally analogous to the Born-Oppenheimer method of molecular physics. Small coupling matrix elements suggest then the adiabatic approximation, i. e. one disregards all off-diagonal coupling terms. This is not quite the same as the usual Born-Oppenheimer approximation which neglects any channel coupling. This difference results from the coupling matrix Q which has both diagonal and off-diagonal terms whereas the matrix P is purely antisymmetric and has therefore only off-diagonal elements. Many numerical examples, see /1/, have shown that the channel coupling is infact small at least for lower excitation. An adiabatic approximation is then a relatively simple and surprisingly good one. On this physical basis many doubly excited states of He, and a few states of H^- and of the system 'proton + electron + positron' have been accurately calculated /7,8/. The validity of this adiabatic approximation confirms at least in the spectral range of not too high double excitation the quasiseparability of the two-electron atom. The strong correlation between the electrons is then perfectly described by the channel function whose pattern of node lines does not change structurally as a function of R. The eigenvalues $U_{\mu}(R)$ may then be regarded as potentials that control the motion of the electron pair as a whole along the collective coordinate R. We stress that the adiabatic approximation fails in two situations. There may be at some value of R a near-degeneracy between two adiabatic potentials. The cooresponding coupling matrix element becomes then large, and the adiabatic approximation fails locally. Figure 1 shows a typical example for this situation in the $^1P^o$ symmetry of H^-, a similar effect was observed in the system 'H + positron' /8/. The socalled diabatic approximation allows then to circumvent this difficulty. One assumes that the potential curves really cross and one disregards any coupling for the crossing channels. The second kind of breakdown of adiabaticity is more serious and occurs in all three-body Coulomb systems, and presumably in all many-body Coulomb systems. This has to do with the long range Coulomb force which implies an accumulation of adiabatic potentials at the three-body break-up threshold, i. e. at the threshold for double ionization. This situation differs from the preceeding one in two aspects. The channel coupling is then everywhere, and not only locally, large. And second, there are many channels strongly coupled in contrast to two coupled channels above. There is no simple procedure available like the diabatic approximation in the two-channel case which repairs this unpleasent situation. In principle one had to solve numerically extremely large sets of strongly coupled systems of differential equations.

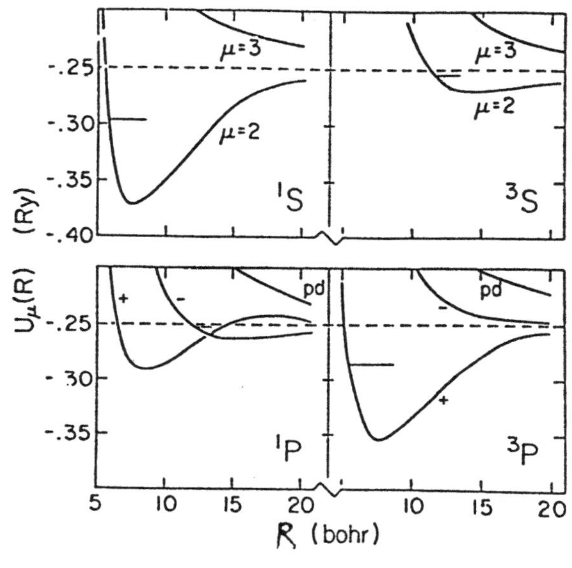

Figure 1: Hypersperical potential curves of H^- converging to the n=2 level of the hydrogen atom (from /6/)

The molecular approach for the two-electron atom is not too far away from the hyperspherical one. Also this approach becomes particularly useful within an adiabatic expansion technique, the abiabatic coordinate is here the electron-electron separation $\vec{R}$, not to be confused with the above hyperradius, see Figure 2. The position of the two-electron center-of-mass relative to the nucleus is $\vec{r}$. As in molecular calculations it is convenient to consider the six-dimensional $\vec{r},\vec{R}$ space in terms of three Euler angles which rotate from a space-fixed to a body-fixed frame with z-axis along $\vec{R}$, the length R itself and the coordinates of $\vec{r}$ in the body-fixed frame. A wave function of the atom is then represented in the form

$$\Psi_{LM}(\vec{r},\vec{R}) = \sum_K \sum_i D^L_{MK}(\varphi,\theta,0) R^{-1} f^L_{ik}(R) \varphi_{ik}(\vec{r},R) \tag{9}$$

where $D^L_{MK}(\varphi,\theta,0)$ are rigid top wave functions describing the rotation of the three-body system with total orbital angular momentum L. They are eigenfunctions of $\vec{L}^2$ with eigenvalue L(L+1) and eigenfunction of L_z with eigenvalues M and K in the space-fixed and body-fixed frames, respectively. The functions φ_{ik} are the MO wave functions describing the motion of $\vec{r}$ for given values of R. They are analogous to the channel functions in the hyperspherical approach. The corresponding eigenvalues play again the role of potentials for the correlated motion of the electron pair. The coefficient functions $R^{-1} f^L_{ik}(R)$ describe the vibration of the coordinate R between the two electrons.

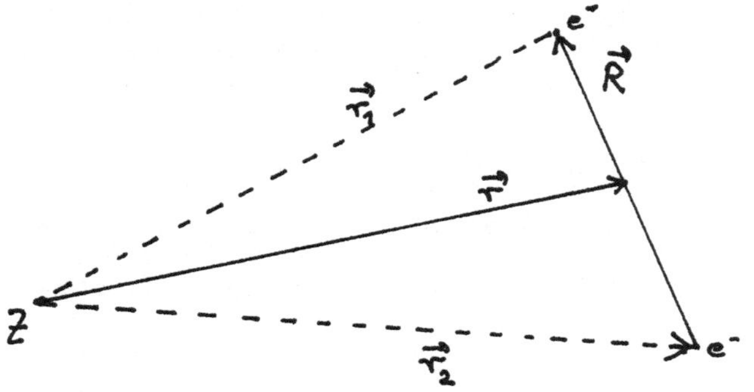

Figure 2: Molecular coordinates for a helium-like atom.

The treatment is formally analogous to the Born-Oppenheimer approximation of the H_2^+ molecule where the adiabatic coordinate is of course the proton-proton separation. It is important to stress that traditional molecular physics justifies the Born-Oppenheimer approximation with the small electron/nucleus mass ratio. Here in the two-electron atom, however, one keeps fixed the distance between the light particles. The quasiseparability results here therefore directly from the three-body dynamics rather than from a kinematical factor. Exact symmetries of the atom like exchange symmetry P_{12} and parity P are straightforward to incorporate. The eigenvalue of the product $P \cdot P_{12}$ is +1 (-1) for gerade (ungerade) states. The MO decomposition gives further approximate quantum numbers. One of them is the K quantum number, the projection of the orbital angular momentum onto the interelectron distance. An essential feature of the molecular model is the goodness of this quantum number, i. e. the absence of external rotational coupling. Two further quantum numbers result from the separability of the adiabatic Hamiltonian, i. e. the Hamiltonian for a fixed value of the interelectronic distance, in confocal elliptic coordinates

$$\lambda = (r_1 + r_2)/R \tag{10}$$

$$\mu = (r_1 - r_2)/R \tag{11}$$

The λ solution of the separated equation has elliptical nodal surfaces with the two electrons as foci. The solution in μ has corresponding hyperbolic nodal surfaces. The quantum numbers n_λ, and n_μ of these nodal surfaces are conserved for all values of R and allows one to connect separated atom and united atom limits. In the united atom limit (R=0) the coordinates become spherical ones. Then $n_\lambda = n - \ell - 1$, the number of radial spherical nodes, and $n_\mu = \ell - |m|$, the number of planar nodes in θ. In the separated atom limit the coordinates become the Stark-state parabolic ones. The two parabolic quantum numbers are $n_1 = n_\lambda$ and $n_2 = n_\mu/2$ if n_μ is even and $(n_\mu-1)/2$ if n_μ is odd. There remains a further constant of the motion in the MO model. This arises from the separability of the adiabatic Hamiltonian in elliptic coordinates, indeed the separation constant itself, which depends on R. This additional constant discussed by several authors /9/ is for our case of two electrons and one nucleus of charge Z conveniently expressed as the operator

$$\Omega = (\vec{\ell}_1 + \vec{\ell}_2)^2/4 - 3(\vec{\ell}_1 - \vec{\ell}_2)^2/4 - \vec{R}\cdot(\vec{a}_1 - \vec{a}_2)/2 \qquad (12)$$

where $\vec{\ell}_i$ are the individual orbital angular momenta of the electrons, and $\vec{a}_i$, i=1,2, are the correponding Runge-Lenz vectors. A typical set of potential curves for H^- converging to H(n=2) is shown in Fig. 3.

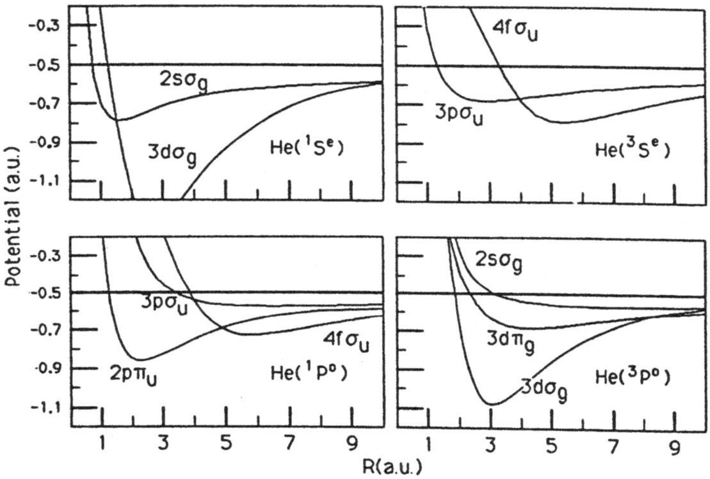

Figure 3: Molecular potential curves for H^- converging to H(n=2) (from /4/)

The potential curves obtained in the molecular approach show a similar topology as those obtained with hyperspherical coordinates as may be seen e. g. comparing Figure 1 and 3. There are, however, differences in their interpretation. For example, in the hyperspherical approach the two lower curves of the $^1P^o$ symmetry do not really cross, but the distance between the two adiabatic eigenvalues is very small at that particular value of R. The local breakdown of the adiabatic approximation requires a diabatic interpretation of the channels as discribed above in order to identify the plus and minus states. In the molecular approach, however, the cooresponding curves do cross, these are two channels of different symmetry, the difference consists of different values of the angular momentum projection onto the interelectronic axis (quantum number K). The channels which carry the plus/minus states are here adiabatic rather than diabatic ones. There is also a fundamental mathematical difference between the two methods. For hyperspherical coordinates the channel spectrum is discrete in any event because for a constant, finite value of the hyperradius one does a quantization within a finite volume. The number of channels, of course, is infinite. This method represents therefore also single and double continuum states within a discrete basis. For a free particle plane wave state this would be an angular momentum expansion. In the case of molecular coordinates the channel spectrum for constant values of the electron-electron separation consists of both a discrete and a continous part. This makes the set of coupled equations non-numerable and therefore very complicated.

We conclude this subsection with a more general remark. We have seen that the adiabatic approximation describes successfully doubly excited states within at least two rather different coordinate frames. The essential element linking the two methods is the existence of one adiabatic coordinate which has to do with the geometrical size of the three-body complex. We conclude therefore that for not too high double excitation the three-body Coulomb system is quasiseparable. The choise of coordinates is more or less a question of secondary importance. For high double excitation, on the other hand, the adiabatic method becomes unsuitable for both and presumably for all other coordinate frames.

High double excitation. The breakdown of approximations based on the assumption of a quasiseparability at high double excitation like the adiabatic approximations described above indicate substantially different and new modes of motion as compared to lower double excitation. In this spectral range there are still many open questions, presently we have only a fragmentary physical understanding which I try to summerize in this section. Basically there are two theoretical routes which make reasonable predictions for this situation. One of them employs classical mechanics, and the other is based on a diabatic approximation which couples approximately an infinite series of adiabatic channels.

We start here with a description of the classical methods. The key work within this frame was done long ago by Wannier /10/ who investigated the strongly correlated motion of two electrons in the field of a point nucleus at zero total energy, i. e. at the threshold for double escape. Without going into mathematical details his main conclusions are as follows:

(i) At lowest, even zero energy the long range parts of the Coulomb interractions are most important. The gradients of the multidimensional potential surface are there small, increasingly so for increasing interparticle distances. The quantum mechanic should therefore converge to its classical limes.

(ii) Trajectories describing a pair of electrons far away from the nucleus are statistically extremely unlikely (of measure zero) with respect to those trajectories whose time evolution leads to one electron far away from the nucleus whereas the other remains closer to the nucleus. This observation is made on the energy surface in the phase space, i. e. one compares sets of trajectories with the same total energy (equal to zero here). This means simply that each doubly excited state, bound or not, is degenerate with an infinity of singly excited states. Therefore, for statistical reasons the probability to excite a doubly excited state should be vanishingly small, for the same reason e. g. also the probability for double photoionization should be vanishingly small as compared to single photoionization plus excitation. There is immense experimental data showing that this statistical consideration leads to a heavily wrong result. The conclusion is therefore that trajectories leading to two electrons far away from the nucleus need a dynamical stabilization provided by an equilibrium configuration of the electron pair.

(iii) Such equilibrium configurations must be unstable in order to describe the competition between single and double excitation. The amount of instability measured by local Liapunov stability indices determines then the energy dependence of cross sections near threshold.

We consider here two recent experiments, double and triple photoionizations, as examples. In the case of double photoionization of helium /11/

$$h\nu + He \longrightarrow He^{++} + 2e^-$$

two continuum electrons move slowly in the field of a nucleus. The equilibrium trajectory is here a pair of hyperbolas which may be seen as follows. We inspect the classical equations of motion which read in obvious notation

$$\ddot{\vec{r}}_1 = -\frac{Z\vec{r}_1}{r_1^3} + \frac{\vec{r}_1 - \vec{r}_2}{|\vec{r}_1 - \vec{r}_2|^3} \tag{13a}$$

$$\ddot{\vec{r}}_2 = -\frac{Z\vec{r}_2}{r_2^3} + \frac{\vec{r}_2 - \vec{r}_1}{|\vec{r}_1 - \vec{r}_2|^3} . \tag{13b}$$

The static equilibrium configuration is the collinear configuration, the nucleus in the middle and the electrons at equal distances at the ends,

$$\vec{r}_1 = - \vec{r}_2 .$$

Substituting $\vec{r}_1 = \vec{\rho}$, $\vec{r}_2 = -\vec{\rho}$ into eqs. (13a,b) we find one equation for

$$\ddot{\vec{\rho}} = -(Z - 1/4)\, \vec{\rho}\, \rho^{-3} . \tag{14}$$

This is the equation for a Kepler motion, the trajectory itself is for positive energy a hyperbola. The two electrons move therefore on two hyperbolas as shown in Fig. 4 such that they are always on opposite directions but at equal distances from the nucleus. The stability index μ obtained from a linearization of the equations of motion determines with some other assumptions, see /10/ and below, the cross section near threshold,

$$\sigma(E) \propto E^{\mu - 1/4} \tag{15}$$

with μ given by

$$\mu = \frac{1}{4}\left(\frac{100Z - 9}{4Z - 1}\right)^{1/2}. \tag{16}$$

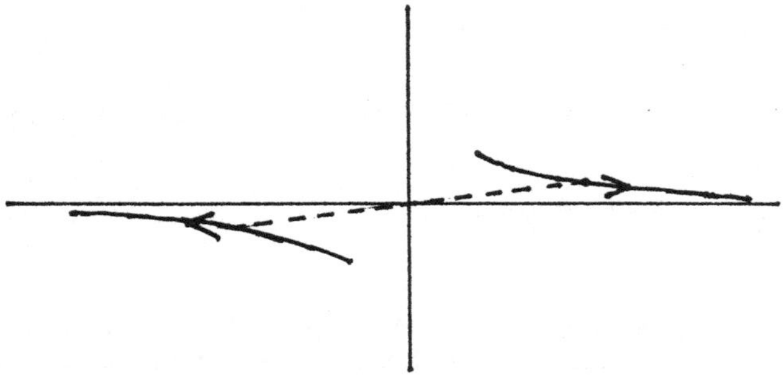

Figure 4: Two electrons escaping on equilibrium orbits, see text, (schematically)

In the above case we have Z=2 which leads to an exponent of 1.056 in eq. (15). This compares favourably with the experimental value of 1.05 found by ref. /11/. The energy range of validity of this threshold law is about 1 eV. We mention also, that this Wannier law was already tested experimentally in other situations, see e. g. /12/.

We come now to the triple photoionization. The escape of three electrons in the field of an ion was theoretically studied by Klar and Schlecht /13/. The equations of motion are now

$$\ddot{\vec{r}}_1 = \frac{Z\vec{r}_1}{r_1^3} - \frac{\vec{r}_1 - \vec{r}_2}{|\vec{r}_1 - \vec{r}_2|^3} - \frac{\vec{r}_1 - \vec{r}_3}{|\vec{r}_1 - \vec{r}_3|^3} \tag{17a}$$

$$\ddot{\vec{r}}_2 = \frac{Z\vec{r}_2}{r_2^3} - \frac{\vec{r}_2 - \vec{r}_3}{|\vec{r}_2 - \vec{r}_3|^3} - \frac{\vec{r}_2 - \vec{r}_1}{|\vec{r}_2 - \vec{r}_1|^3} \tag{17b}$$

$$\ddot{\vec{r}}_3 = \frac{Z\vec{r}_3}{r_3^3} - \frac{\vec{r}_3 - \vec{r}_2}{|\vec{r}_3 - \vec{r}_2|^3} - \frac{\vec{r}_3 - \vec{r}_1}{|\vec{r}_3 - \vec{r}_1|^3} \tag{17c}$$

The equilibrium configuration, as shown in /13/, is now an equilateral triangle, the electrons at the corners and the ion in the center. To find the classical equilibrium trajectories it is convenient to parametrize the electron position vectors as follows,

$$\vec{r}_1 = \vec{\varrho} \tag{18a}$$

$$\vec{r}_2 = \underline{M}\,\vec{\varrho} \tag{18b}$$

$$\vec{r}_3 = \underline{M}^2\,\vec{\varrho} \tag{18c}$$

where the matrix $\underline{M}$ performs a rotation of 120° in the xy-plane,

$$M = \begin{vmatrix} \cos 120° & -\sin 120° & 0 \\ \sin 120° & \cos 120° & 0 \\ 0 & 0 & 1 \end{vmatrix} \tag{19}$$

Substitution of eqs. (18) and (19) into (17) yields one equation for $\vec{\varrho}$,

$$\ddot{\vec{\varrho}} = \left(Z - 3^{-1/2}\right) \vec{\varrho}\, \varrho^{-3} \tag{20}$$

which is the analog of eq. (14). The equilibrium trajectory of each electron is therefore again a hyperbola, the correlated motion of the three electrons is shown in Fig. 5. The stability analysis is here more complicated and leads to the analytic result /13/ for the cross section

$$\sigma(E) \propto E^{\frac{1}{2}(\mu-1)} \tag{21}$$

where the index μ is now given by

$$\mu = \left(\frac{13 Z\sqrt{3} - 11 + 2(1 + 108 Z^2)^{1/2}}{Z\sqrt{3} - 1}\right)^{1/2}. \tag{22}$$

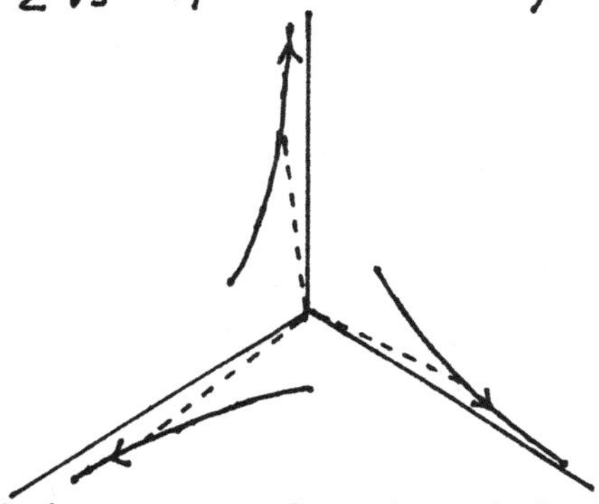

Figure 5: Triple equilibrium escape along hyperbolas (schematically)

Samson et al /14/ have performed the photoionization experiments

$$h\nu + Ne \;\text{-------}\!\!\rightarrow Ne^{+++} + 3e^-$$

and

$$h\nu + O \;\text{--------}\!\!\rightarrow O^{+++} + 3e^-.$$

For Ne and O they found the exponents 2.170 and 2.176, respectively, see Fig. 6; the theoretical value universal for all atoms is acording to eqs.(21) and (22) 2.162. This is for the first time that predictions for triple photoionization /13/ have been tested experimentally.

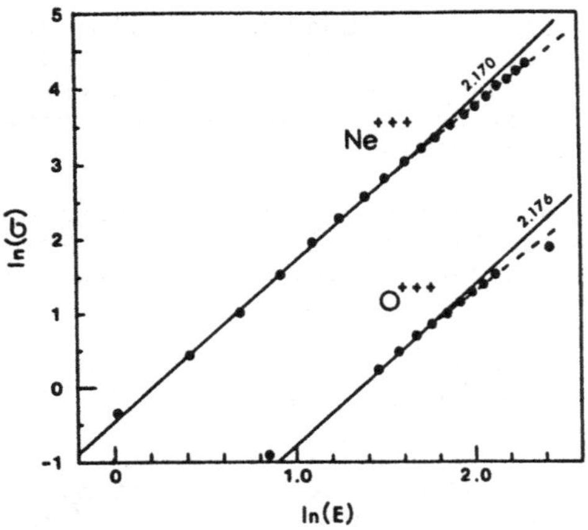

Figure 6: Double logarithmic plot of triple photoionization cross sections vs the excess energy for neon and oxygen, the slopes m of the lines are indicated for each atom (from /14/). The theoretical value for m is universally for all atoms m=2.162, see /13/.

These experiments confirm doubtlessly the main ingredient of the Wannier theory, namely that the strongly correlated electrons perform motions close to classical equilibrium orbits, hyperbolas in the above examples at energies slightly above threshold for double or triple escape. Below this threshold these classical orbits are of course ellipses, again strongly correlated as shown e. g. for two electrons in Fig.7a. These orbits are now periodic and one expects that they correspond to resonances. It is however presently not fully clear how these orbits manifest themselves say e. g. in a photoelectron energy spectrum. In an integrable mechanical system there is a simple one-to-one correspondence between one periodic orbit and one resonance peak in a measured spectrum. The position of the resonance can be obtained from semi-classical quantization (Bohr-Sommerfeld). In a non-integrable system, however, this is not neccessarily true. Nevertheless, there is some evidence that at least very close to the threshold of double escape a Bohr-Sommerfeld quantization of the ellipses constitutes a reasonable approximation. The continuation of the Wannier theory through the thrshold down to high resonances and the problem of the approximate quantization has been investigated theoretically by Macek et al /15/, their result for these Wannier ridge resonances is

$$E_n = -\frac{4(Z - 1/4 - \sigma/4)^2}{(n-d)^2} \qquad (23)$$

where the screening constant depends on the energy itself,

$$\sigma \sim \sqrt{|E|}\,\ln|E| . \qquad (24)$$

The pair of ellipses shown in Fig. 7a is not the only equilibrium orbit for a two-electron atom. Two other orbits which may be derived analytically /16-19/, a rotor and a top, are shown in Fig. 7b and 7c. Their Bohr-Sommerfeld quantization yields in the limes of very high excitation a Rydberg-like level spectrum similar to the above ridge resonances. Until now neither Wannier ridge resonances nor rotating rigid body resonances have been observed in photoabsorption. The analysis of inelestic electron helium scattering, however, indicates the existence of Wannier ridge resonances.

$Ns^2\ ^1S$ states of two-electron atoms have been obtained recently also from a diabatic molecular description /20/. The authors employ the molecular coordinates $\vec{R}=\vec{r}_1-\vec{r}_2$ and $\vec{r}=(\vec{r}_1+\vec{r}_2)/2$. They write the wave function in the form

$$\psi(\vec{r}_1,\vec{r}_2) = \frac{f(R)}{R}\,\phi(\vec{r},\vec{R}) \tag{25}$$

and use for the function $\phi(\vec{r},\hat{R})$ a product of hydrogenic functions normalized to one with respect to the coordinate $\vec{r}$ only. The function f(R) satisfies then the equation

$$\left\{-\frac{d^2}{dR^2}+\frac{1}{R}+U(R)\right\}f(R) = E\,f(R) \tag{26}$$

with the potential U(R) given by

$$U(R) = \left\langle \phi \left| -\nabla_R^2 - \frac{1}{4}\nabla_r^2 - \frac{Z}{|\vec{r}-\vec{R}/2|} - \frac{Z}{|\vec{r}+\vec{R}/2|} \right| \phi \right\rangle . \tag{27}$$

The numerical calculation of ref. /20/ has used a simple 1s function with one adjustable parameter. The potential (27) converges then to the double ionization threshold, and carries a series of bound states surprisingly close to the positions of $Ns^2\ ^1S$ resonances. The curve (27) can be viewed as passing diabatically through the adiabatic curves separating to single-ionization limits.

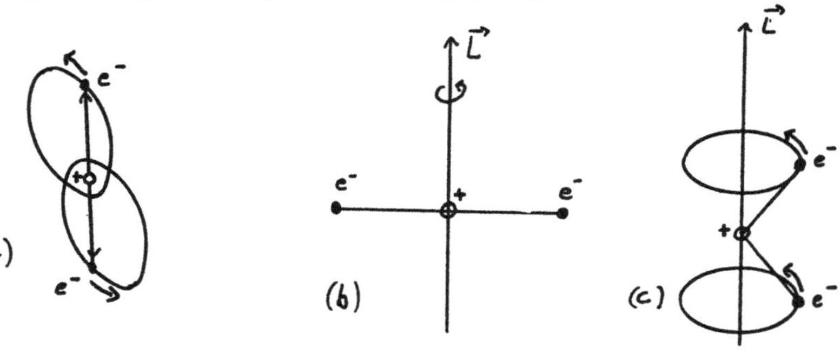

Figure 7: Periodic equilibrium orbits for two-electron atoms, (a) ellipses, (b) rigid rotor, (c) top.

<u>Correlation between</u> <u>fast electrons</u>. The preceeding pages have shown that the importance of correlation in a free atom increases with increasing energy. In particular we have shown that correlation becomes even dominant near thresholds of double or multiple ionization. On the other hand, for still much higher energy the importance of correlation must decrease again because in the limes of infinite energy the whole interaction becomes negligigle and the electrons behave like free particles. Moreover it has been believed

until recently that e. g. two electrons with very different velocities, say a fast one and a slow one, are more or less uncorrelated with respect to each other. The slow one, of course, is correlated to the ion core. A typical example for this situation occurs in the electron impact ionization of H(1s),

$$e(E_0, \vec{k}_0) + H \longrightarrow p + e(E_a, \vec{k}_a) + e(E_b, \vec{k}_b)$$

where E_0, E_a, E_b and $\vec{k}_0$, $\vec{k}_a$, $\vec{k}_b$ are the energies and the momenta of the incoming, the fast scattered and the slow ejected electron, respectively. The calculation of triply differential cross sections which have been measured experimentally, see /21/ for a review, led, however, to serious difficulties. In a typical situation of E_0 =250 eV and E_b=5 eV corresponding to E_a =231.4 eV the first Born approximation fails. An approximate second order Born approximation /22/, a formal pseudostate expansion /23/, a classical post collision interaction (PCI) correction to a Born or Glauber approximation /24-25/, and an effective charge approach /26/ lead to results in agreement with the experiment. The physical pictures of these treatments, however, are very different from each other. References /22/ and /23/ assume a double collision between projectile electron and target atom, i. e. a short range correlation, to explain the departures from the first Born approximation, whereas references /24/ and /26/ attribute these departures to long range correlations. The theoretical background for impact ionization by charged particles has been reconsidered recently /27,28/ without using perturbation theory or semiclassical methods. The essential point is that the identification of a two-electron continuum wave function as solution of an elliptic partial differential equation (Schrödinger equation) needs appropriate boundary conditions on a large hypersphere. These boundary conditions are non-trivial because the Coulomb interaction forbids free-particle states at infinite separations. It can be shown, see /27,29/, that for a two-electron atom the plane wave asymptotic reads

$$\Psi \propto \exp\left[i\left(\vec{k}_a\cdot\vec{r}_a + \vec{k}_b\cdot\vec{r}_b + \gamma(\vec{k}_a,\vec{r}_a) + \gamma(\vec{k}_b,\vec{r}_b)\right)\right] \times \exp\left[-i\gamma(\vec{k}_a-\vec{k}_b, \vec{r}_a-\vec{r}_b)\right] \qquad (28)$$

where the function γ is given by

$$\gamma(\vec{k},\vec{r}) = \frac{1}{k}\ln(kr+\vec{k}\cdot\vec{r}) . \qquad (29)$$

This result was first reported by Redmond, see /30/, its fundamental importance for the exact description of the PCI in many-body Coulomb systems, however, seems not to have been recognized. We remark that the phase $\gamma(\vec{k},\vec{r})$ is just the well known logarithmic phase shift of the familiar two-body Kepler continuum wave function $\psi(\vec{k},\vec{r})$, see e. g. /31/. The structure of the above asymptotic form suggests an approximate two-electron wave function written as product of three Kepler wave functions, one for each relative two-body motion, except for a plane wave factor for the relative electron-electron motion, see /28/ for further justification,

$$\Psi^- \propto \Psi(\vec{k}_a, \vec{r}_a)\,\Psi(\vec{k}_b, \vec{r}_b)\,\Psi(\vec{k}_a - \vec{k}_b, \vec{r}_a - \vec{r}_b) \times \exp\left[-i(\vec{k}_a - \vec{k}_b)\cdot(\vec{r}_a - \vec{r}_b)\right]. \tag{30}$$

With this wave function normalized to a six-dimensional delta function ref. /27/ has calculated the triply differential cross section given by

$$\frac{d^3\sigma}{d\Omega_a\, d\Omega_b\, dE_b} = (2\pi)^4 \frac{k_a k_b}{k_0} |T|^2 \tag{31}$$

with the T-matrix element

$$T = \left\langle \Psi^- \left| -\frac{1}{r_a} + \frac{1}{r_{ab}} \right| \varphi_{1s}(r_b),\ (2\pi)^{-3/2} e^{i\vec{k}_0\cdot\vec{r}_a} \right\rangle \tag{32}$$

where the initial state consists of the 1s bound state and a plane wave with momentum $\vec{k}_0$. Results of this calculation at incident energies of 250 and 150 eV as measured experimentally /32/ are shown in Fig. 8 and 9. The experimental data have been normalized to the calculation in one point. At increasing energy (a few keV) this

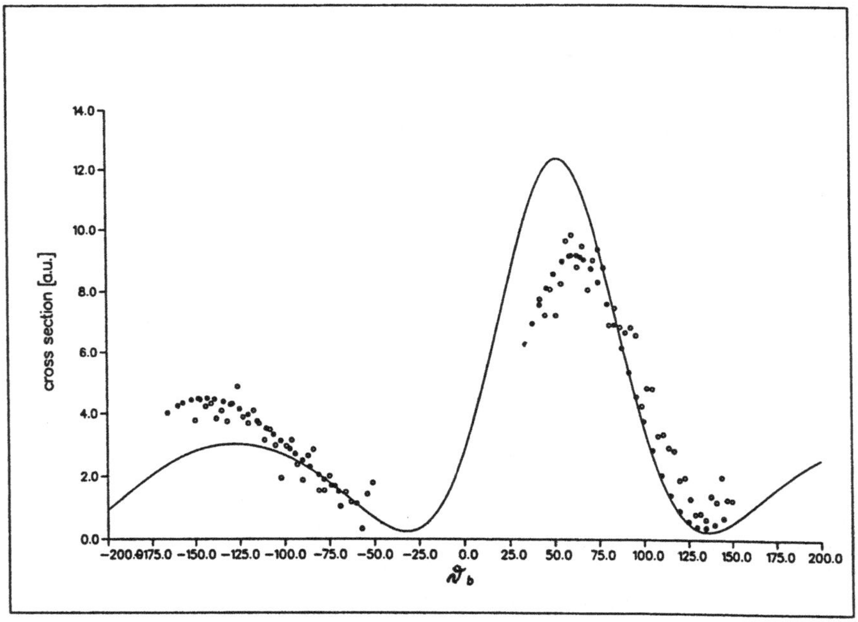

Figure 8: Coplanar triply differential cross section for electron impact ionization of hydrogen at incident energy E_0=250 eV, ejectred electron energy E_b=5 eV, scattering angle=3°. Solid line = Born approximation, open points ref. /32/, full points ref. /27/.

method converges (although slowly) to the first Born approximation. The wave function given by (30) satisfies boundary conditions for two correlated outgoing electrons. It describes therefore also the final state of a double photoionization. Triply differential cross sections for this case, not yet been considered neither theoretically nor experimentally, are expected to depend very sensitively on the correlation of the photo-excited atomic complex.

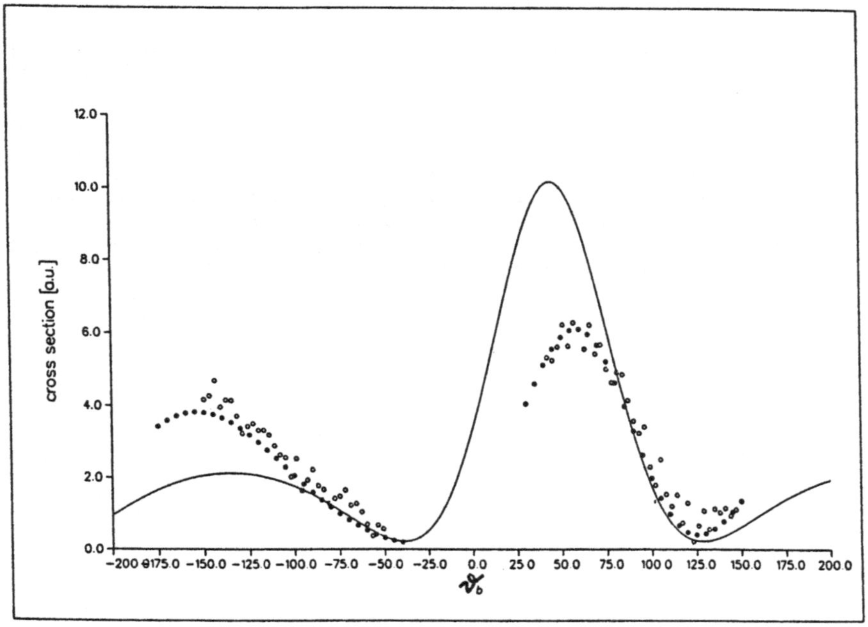

Figure 9: Same as Figure 8, E_a=150 eV, E_b=5 eV, scattering angle=4°

References

/1/ U. Fano, Rep. Prog. Phys. 46,97(1983)
/2/ H. Klar, Lect. Not. Phys. 229,149(1985)
/3/ J. M. Feagin and J. S. Briggs, Phys. Rev. Lett. 57,984(1986)
/4/ J. M. Feagin and J. S. Briggs, Phys. Rev. A37,4599(1988)
/5/ H. Klar, J. Math. Phys. 26,1621(1985)
/6/ J. Macek, J. Phys. B1,831(1968)
/7/ H. Klar and M. Klar, J Phys. B13,1057(1980)
/8/ E. Pelikan and H. Klar, Z. Phys. A310,153(1983)
/9/ C. A. Coulson and A. Joseph, Int. J. Quantum Chem. 1,337(1967)
/10/ G. H. Wannier, Phys. Rev. 90,817(1953)
/11/ H. Kossmann, V. Schmidt and T. Andersen, Phys. Rev. Lett. 60,266(1988)
/12/ H. Klar, Comm. At. Mol. Phys. D19,171(1987)
/13/ H. Klar and W. Schlecht, J. Phys. B9,1699(1976)
/14/ J. A. R. Samson and G. C. Angel, Phys. Rev. Lett. 61,1584(1988)
/15/ J. H. Macek and J. M. Feagin, J. Phys. B18,2161(1985)
/16/ H. Klar, Phys. Rev. Lett. 57,66(1986)
/17/ H. Klar, Z. Phys. D3,353(1986)
/18/ H. Klar, Z. Phys. D6,107(1987)
/19/ H. Klar, J. Opt. Soc. Am. B4,788(1987)

/20/ J. M. Rost and J. S. Briggs, J. Phys. B21,L233(1988)
/21/ H. Ehrhardt, K. Jung, G. Knoth and P. Schlemmer, Z. Phys. D1,3(1986)
/22/ H. Ehrhardt, M. Fischer, K. Jung, F. W. Byron, C. J. Joachain and B. Pireaux, Phys. Rev. Lett. 48,1807(1982)
/23/ E. P. Curran and H. R. J. Walters, J. Phys. B20,337(1987)
/24/ H. Klar, A. Franz and H. Tenhagen, Z. Phys. D1,373(1986)
/25/ H. Klar, A. C. Roy P. Schlemmer, K. Jung and H. Ehrhardt, J. Phys. B20,821(1987)
/26/ S. Jetzke, J. Zaremba and F. H. M. Faisal, Z. Phys. D (in press)
/27/ M. Brauner, J. S. Briggs and H. Klar, Z. Phys. D (in press)
/28/ M. Brauner, J. S. Briggs and H. Klar, Z. Phys. D (to be published)
/29/ H. Klar, to be published
/30/ L. Rosenberg, Phys. Rev. D8,1833(1973)
/31/ H. A. Bethe and E. E. Salpeter, Quantum mechanics of one- and two-electron atoms, Plenum, New York (1977)
/32/ H. Ehrhardt, G. Knoth, P. Schlemmer and K. Jung, Phys. Lett. 110A,92(1985) and privat communication

PART IV: Rydberg Atoms in External Fields

STRONG FIELD COHERENCE EFFECTS IN THE PROCESSES OF PHOTOIONIZATION FROM RYDBERG LEVELS

M.V. Fedorov
General Physics Institute of the USSR Academy of Sciences, 38 Vavilov Street, 117942 Moscow, USSR

I. INTRODUCTION

Rydberg atom is the atom at highly excited (Rydberg) discrete levels $E_n=-1/2\,n^2$, $n \gg 1$, n is the principal quantum number, atomic system of units is used throughout this paper. In the hydrogen atom there is no quantum defect and all the levels E_n are degenerated with respect to orbital momentum (ℓ) and its z-projection (m). Rydberg states have many unusual features making them interesting. The length of Rydberg orbit is large ($\sim n^2 \gg 1$), the period of classical Kepler motion is long ($\sim n^3 \gg 1$), the electron binding energy is small $|E_n| \ll 1$, the levels E_n are almost equidistant: in a vicinity of some central level E_{n_0}

$$E_n \simeq E_{n_0} + n_0^{-3}(n-n_0) \tag{1}$$

if $|n - n_0| \ll n_0$. The space between neighbouring Rydberg levels is equal to the classical Kepler frequency n^{-3}.

The region of Rydberg levels $n \gg 1$ is the region where the Schrödinger equation can be solved in semiclassical approximation [1]. Semiclassical approach is applicable also in a region of continuous spectrum close to the ionization threshold $0 < E \ll 1$. In semiclassical approach matrix elements of radiative transitions have rather simple form [2,3]

$$V_{EE'} \sim \mathcal{E}_0 |E-E'|^{-5/3}, \quad V_{nE} \sim \mathcal{E}_0 n^{-3/2} |E_n - E|^{-5/3}, \quad V_{nn'} \sim \mathcal{E}_0 (nn')^{-3/2} |E_n - E_{n'}|^{-5/3} \tag{2}$$

Where $\mathcal{E}_0$ is the electric field strength amplitude, $V_{ij} = \langle \varphi_i | \frac{1}{2} \vec{d}\vec{\mathcal{E}}_0 | \varphi_j \rangle$, $\vec{d}$ is the dipole moment. If ω is the frequency of the field, inducing transitions, and if the initial and final states i, j obey the energy conservation rule $|E_i - E_j| = \omega$ all the matrix elements (2) are determined by the only constant V:

$$V \equiv \varepsilon_0 \omega^{-5/3}, \; V_{EE'} \sim V, \; V_{nE} \sim V n^{-3/2}, \; V_{nn'} \sim V(nn')^{-3/2} \tag{3}$$

One of the questions which appear in connection with Rydberg states is: how close are they or their superpositions to a picture of localized classical particle in Coulomb field? It is clear that even semiclassical wave functions of stationary states $\varphi_n(r)$ [I] are rather far from classical. $\varphi_n(r)$ is smeared out more or less homogeneously over a large distance ($\sim n^2$) and hence $\varphi_n(r)$ does not describe a localized state. Localization appears if $\varphi_n(r)$ is replaced by a coherence superposition of $\varphi_n(r)$ [4-6]

$$\psi(r,t) = \sum_n C_n \, \varphi_n(r) \exp(-iE_n t) \tag{4}$$

In the case of Gaussian distribution

$$C_n = \pi^{-1/4} (\Delta n)^{-1/2} \exp\left(-\frac{(n-n_0)^2}{2\Delta n^2}\right), \quad n_0 \gg \Delta n \gg 1 \tag{5}$$

the time dependent wave function (4) can be reduced to the form [6]:

$$\psi(r,t) = C(2ir)^{-1/2} (p_{n_0}(r))^{-1/2} \left\{ \exp\left[i\theta_{n_0}(r) - \frac{\Delta n^2 (t - \tau(r) - 2\pi k n_0^3)^2}{2n_0^6} \right] - \right.$$

$$\left. - \exp\left[-i\theta_{n_0}(r) - \frac{\Delta n^2 (t + \tau(r) - 2\pi(k+1) n_0^3)^2}{2n_0^6} \right] \right\} \tag{6}$$

where $C = const.$, $2\pi k n_0^3 \le t \le 2\pi(k+1) n_0^3$, $n = 0, 1, 2, \ldots$

$$p_{n_0}(r) = [2(E_{n_0} - U(r))]^{1/2}, \quad \theta(r) = \int_a^r p_{n_0}(r)\,dr, \quad \tau(r) = \int_a^r p_{n_0}^{-1}(r)\,dr \tag{7}$$

"a" is the left border of classically achievable region, $U(a) = E_{n_0}$, $U(r)$ is the potential energy including centrifugal term, $\tau(r)$ is the classical time of motion between the points "a" and r. $\tau(r)$ ($\le \pi n_0^3$) determines a classical trajectory of electron motion via expressions $t = \tau(r) + 2\pi k n_0^3$ or $t = 2\pi(k+1) n_0^3 - \tau(r)$ where K = 0,I,2,... and $2\pi k n_0^3 \le t \le (2k+1)\pi n_0^3$ or $(2k+1)\pi n_0^3 \le t \le 2\pi(k+1) n_0^3$, respectively. The wave function $\psi(r, t)$ (6) is localised near r(t), the length of localization is $\delta r \sim v n_0^3 (\Delta n)^{-1} \sim n_0^2 (\Delta n)^{-1} \ll n_0^2$, where $v \sim n_0^{-1}$ is the average electron velocity at the Kepler orbit.

Wave packets (4), (5) can be formed in the process of excitation from the ground level by short laser pulses [4,5]. Photoionization of such states shows both their classical and quantum features [5]. If the ionizing field has a frequency $\omega > (2n_0^2)^{-1}$ and pulse duration τ, and if the pulse has its maximum at $t = t_0 \neq 0$, the probability w for the atom to be ionized during the pulse duration in its dependence on t_0 has periodical "revivals" when $t_0 = 2\pi K n_0^3$, $K = 1, 2, \ldots$ (Fig.I).

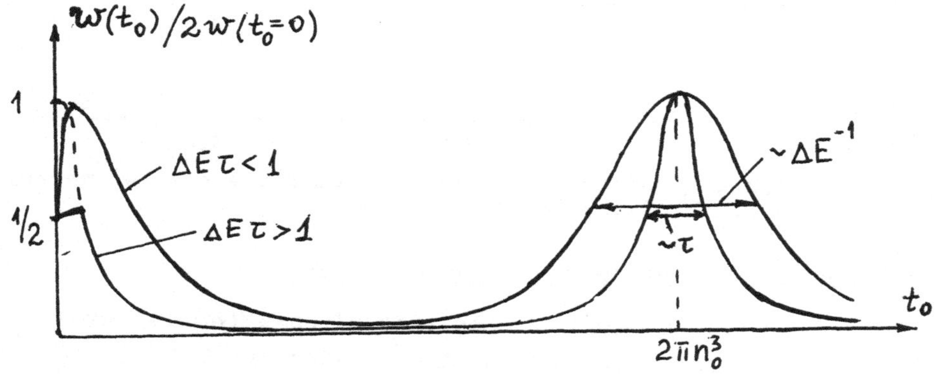

Figure I: Probability of photoionization from the state (4) versus the delay time of ionizing pulse t_0: a) $\Delta E\tau < 1$ b) $\Delta E\tau > 1$, $\Delta E = \Delta n \times n_0^{-3}$

This result has a very simple classical explanation. Localized excited electron (6) returns to the atomic nucleus after each Kepler period. These moments are most suitable for ionization resulting in enhanced probability $w(t_0)$ at $t_0 = 2\pi K n_0^3$ (Fig.I).

On the other hand, the pulse duration dependence of the photoionization probability $w(\tau)$ reflects quantum mechanical features of the initial (unperturbed) state (4) [5]. In its dependence on τ, $w(\tau)$ at first has a fast linearly growing part $w \propto \tau V^2 \Delta n \times n_0^{-3}$ which occurs as long as $\tau < n_0^3 (\Delta n)^{-1}$ and then, $w(\tau)$ becomes constant, i.e. it has a region of "plateau" $w = \text{const} \sim V^2 \ll 1$. When τ becomes larger than n_0^3 linear dependence of $w(\tau)$ appears again but now it is much slower than at the initial part: $w \propto \tau V^2 n_0^{-3}$ (Fig.2). Both initial fast linear growth and subsequent "plateau" $w = \text{const}$ are explained by interference of transitions to the continuum from sublevels of the wave packet (4) [5]. Thus, these interference effects occur already in the weak field limit ($V \ll 1$) if there is an initial coherent population of Rydberg levels. Below another group of interference effects will be described. The ionizing field will be supposed to be strong ($V \gtrsim 1$). Now initial coherent population is not necessary. The effects occur even if a single Rydberg level E_{n_0} is initially occupied

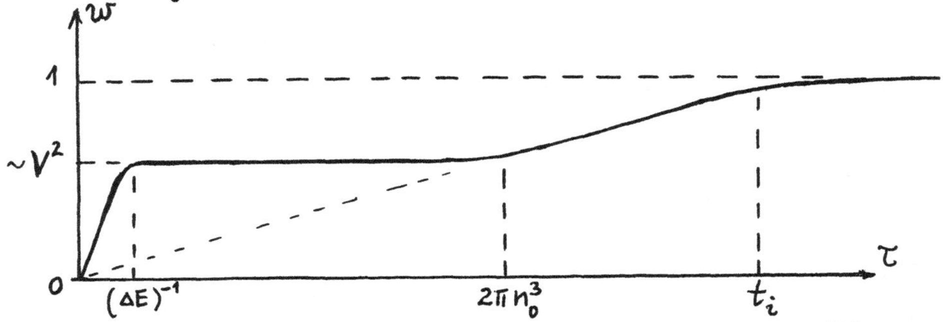

Figure 2. Probability of photoionization from the state (4) versus duration τ of the ionizing pulse.

2. INTERFERENCE PHENOMENA IN A STRONG FIELD

So let the initial atomic state be that of some Rydberg level E_{n_0} φ_{n_0}, $n_0 \gg 1$. Let interaction with the ionizing field $\mathcal{E}_0 \cos\omega t$ ($\omega > (2n_0^2)^{-1}$) be switched on suddenly at t = 0. Let the field be arbitrary strong or weak $V \gtrless 1$. To begin with it is worth to give a definition of a rather simple model (though not the simplest one, see section 7) in which all the main results have been predicted [7,8] . Later, the model can be done more complicated to become more realistic , or it can be simplified to become more transparent. But at the present stage the model is formulated as follows. All the above threshold transitions will be ignored (they will be taken into account later, section 6). All the Rydberg levels and the levels in the continuum will be considered as nondegenerated and connected each with other by the dipole matrix elements (2), (3). At this stage only the main (trivial) part of the a.c.Stark shift will be supposed to exist. This shift is equal to the average free electron vibration energy $\mathcal{E}_0^2/4\omega^2$ which does not depend on E or E_n and does not affect the dynamics of photoionization. This general shift will be supposed to be included in energies E_n, E. Corrections to this main part of the a.c.Stark matrix will be discussed separately in Section 5. No transitions from E_{n_0} to discrete levels with smaller energy ($\sim E_{n_0} - \omega$) will be considered at all. In a strong field these transitions can be shown to be much less important than transitions to the continuum [9] . Thus, the only transitions taken into account at present are the multiple transitions between Rydberg levels E_n and the continuum. Under the formulated assumptions atomic wave function ψ has a form

$$\psi = \sum_n e^{-i\omega t} A_n(t)\varphi_n + \int dE\, A_E(t)\,\varphi_E \qquad (8)$$

with initial conditions $A_n(o) = \delta_{n,n_o}$, $A_E(o) = 0$.

The time-dependent amplitudes $A_n(t)$, $A_E(t)$ obey equations following from the Shrödinger equation in the rotating wave approximation. Solution of these equations is straight forward [8] . It is not reproduced here. The only comment which must be done here concerns with the so-called "pole approximation" [IO] . Laplace transformed aplitudes $A_n(t)$, $\tilde{A}_n(p)$ contain factors $(ip - E_n - \omega)^{-I}$ with $p=-iz+\delta$, $\delta \to 0$ and real z. Such factors are replaced by $-i\pi\delta(z - E_n - \omega)$, the principal value of the integral is dropped out. There are some numerical calculations [II] , as well as qualitative arguments [I2] showing that this approximation is rather good, at any rate in the range of microwave fields $\omega \sim IO^{IO} - IO^{II} s^{-I}$ and $E \sim \omega$. Under the formulated assumptions the following physical results have been derived [7,8] .

I). Long time limit of the photoelectron energetic distribution is given by

$$w(E) = |A_E(t)|^2\Big|_{t\to\infty} = \frac{|V_{En_0}|^2}{(E-E_{n_0}-\omega)^2\left[1+\pi^2\left(\sum_n |V_{nE}|^2 (E-E_n-\omega)^{-1}\right)^2\right]} \qquad (9)$$

Although rather simple this formula predicts a rather interesting and unusual result. In the weak field limit $V \ll I$ equation (9) describes a curve close to the Lorenzian having its maximum at $E = E_{n_o} + \omega$ with the width $\Gamma = V^2 n_o^{-3}$. Quite a different structure of $w(E)$ (9) occurs when the field becomes strong $V \gtrsim I$. The resulting distribution is described at Fig.3.

The peaks of the curve at Fig.3 are localised near $E^{(n)} + \omega$ where

$$E^{(n)} = \frac{1}{2}(E_n + E_{n+1}) \simeq E_{n_0} + n_0^{-3}\left(n - n_0 + \frac{1}{2}\right) \qquad (IO)$$

are the quasienergies of the atom in a strong field (in other words $E^{(n)}$ are the energies of a "dressed atom"). Structure of each peak at Fig.3 is given by

$$w^{(n)}(E) = \frac{1}{\pi^6 V^2 n_0^3 (n - n_0 + \frac{1}{2})^2 \left[(E - E^{(n)} - \omega)^2 + (\pi^3 V^2 n_0^3)^{-2}\right]} \qquad (II)$$

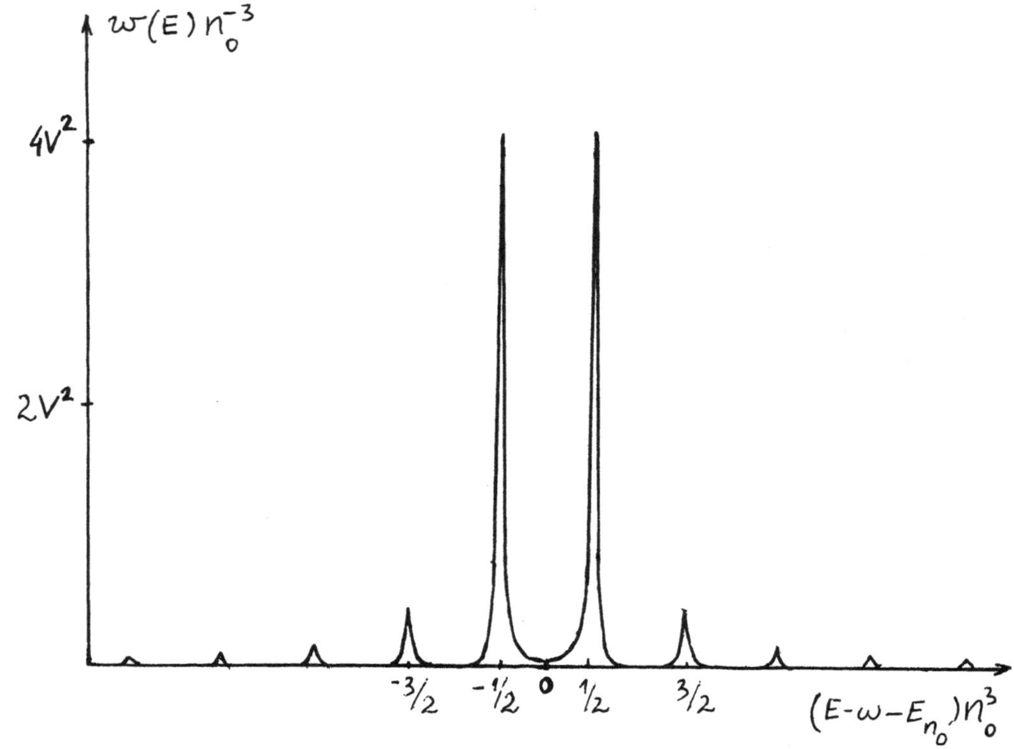

Figure 3. Strong field ($V \gtrsim I$) photoelectron energetic distribution.

The heights of the peaks and their widths are equal to

$$\left(w^{(n)}(E)\right)_{max} = \frac{V^2 n_0^3}{(n - n_0 + 1/2)^2} \quad , \quad \Gamma = \frac{2}{\pi^3 V^2 n_0^3} \tag{I2}$$

The stronger the field is the higher and narrower the peaks of $w(E)$ are. When $V > I$, $\Gamma \ll n_0^{-3}$, i.e. the widths of peaks at Fig.I are much less than the space between them, equal to n_0^{-3}.

2). Time-dependent photoionization probability in strong field has a form

$$w(t) = 1 - \exp(-\Gamma t) \tag{I3}$$

where Γ is given by equation (I2). Formally, equation (I3) coincides with the well-known exponential decay formula for ionization from an isolated level [I3] . But a deep difference is in a definition of the ionization rate constant Γ . In our case Γ (I2) is falling down with increasing V . When $V > I$ $\Gamma \ll n_0^{-3}$, i.e. the ionization rate constant is small in comparison with the classical Kepler frequency. The time of ionization $t_i = 1/\Gamma = \frac{1}{2}\pi^3 V^2 n_0^3$ is a growing function of V and when $V > I$ t_i is much longer than the Kepler period $t_i > 2\pi n_0^3$. These results show that equations (I2), (I3) predict field induced stabilization of the Rydberg atom in a strong field, or trapping of populations at Rydberg levels.

3). Time-dependent amplitudes $A_n(t)$ to find electron at discrete levels E_n in the case of strong field ($V > I$) have been found [7] to be

$$A_n(t) = \pi^{-2} \sum_{n'} \frac{\exp\{-itn_0^{-3}(n'+\frac{1}{2}) - \frac{1}{2}\Gamma t\}}{(n'+1/2)(n'+n_0-n+1/2)} \tag{I4}$$

When $\Gamma^{-1} = t_i \gg t \gg n_0^3$, $A_n(t)$ (I4) oscillate with a period $4\pi n_0^3$ and slowly decay with the rate constant $\frac{1}{2}\Gamma$ (I2).

3. INITIAL COHERENT POPULATION

Thus, there are two groups of interference effects described in the Introduction and in the previous section. The first of them occur in the weak field region ($V \ll I$) when there is an initial coherent population of Rydberg levels. The second kind of interference effects is connected with secondary coherent population of Rydberg levels in strong ionizing field ($V > I$) with the only Rydberg level E_{n_0} initially occupied. Comparison of these effects give rise at least to two questions: (i) what happens with effects connected with initial coherent population when the ionizing field becomes strong ($V > I$) and, (ii) how does the initial coherent population change the effects connected with secondary coherent population in strong field? In other words the task is to combine initial coherent population and strong field effects. This problem has been solved in ref. [8] . Initial coherent population has been taken in the form of equation (5). The resulting long-time strong field photoelectron spectrum has the same multipeak structure as described above. Again the peaks are localized at the points $E^{(n)} + \omega$, where $E^{(n)}$ is given by equation (I0). Each peak has the same Lorenz structure as described by equation (II)

$$w^{(n)}(E) = \frac{w_n}{\pi^4 V^2 n_0^3 \left[(E-E^{(n)}-\omega)^2 + (\pi^3 V^2 n_0^3)^{-2}\right]} \tag{I5}$$

Again the width of each peak Γ is given by equation (I2) and is a falling down function of V. But the envelope of these peaks given by the factors w_n in equation (I5) is different from that described in the previous section.

Now

$$w_n \equiv \int dE\, w^{(n)}(E) = \pi^{-2}\Big[\sum_{n'} \frac{C_{n'}}{(n-n'+1/2)^2}\Big]^2 \qquad (16)$$

As a whole the photoelectron spectrum in strong field in the case of initial coherent population (4) is described at Fig.4.

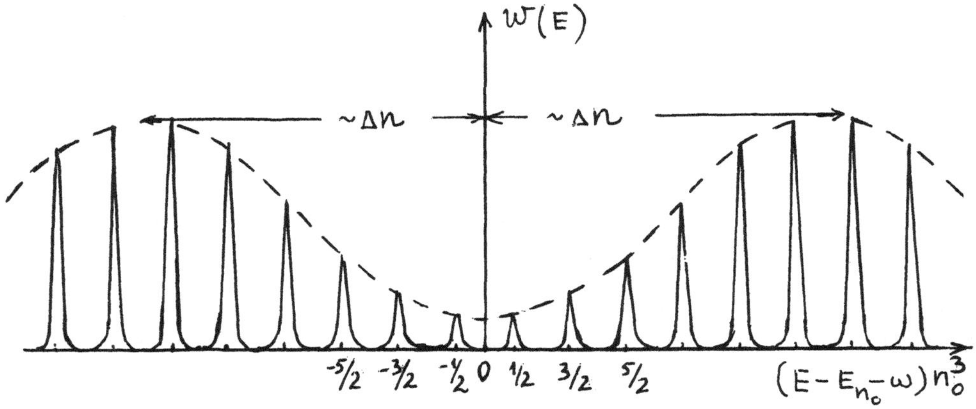

Figure 4. Photoelectron spectrum in strong field ($V > 1$) in the case of coherent initial population (4).

As for the time-dependent total photoionization probability $w(t)$ it has been reduced [8] to the form

$$w(t) = 1 - \frac{2i}{\pi^3 V^2} \sum_{n,k} \frac{C_n C_{n-k} \exp\{itkn_0^{-3} - \Gamma t\}}{k + \frac{2i}{\pi^3 V^2}} \qquad (17)$$

Analysis of this formula [8] shows that $w(t)$ is described by the same curve as at Fig.2, but with the parameters different from those which occur in the case of weak field [5] . Now in the time intervals $t < \Delta E^{-1} \equiv n_0^3 (\Delta n)^{-1}$ and $\Delta E^{-1} < t < n_0^3$ $w(t)$ has a form

$$w(t) = \frac{4t\Delta n}{\pi^{5/2} V^2 n_0^3} \qquad \text{and} \qquad w(t) = const. \sim V^{-2} \qquad (18)$$

respectively. The rate of initial linear growth and the level of plateau at Fig.2 are falling down functions of V when $V > 1$. In the region $t > n_0^3$ $w(t)$ is described by the previously derived formula (13) with Γ given by equation (12). Hence, in the region $t > n_0^3$ there is no influence of initial coherent population at all. In the region $t < n_0^3$ initial coherent population provides the same interference effect which exists in the case of weak field but with modified parameters (18) of the curve $w(t)$. As a whole the described analysis shows that if there is an initial coherent population of Rydberg levels and if the field is strong ($V > I$) two kinds of interference effects described above can coexist and each of them find its reflection both in the dynamics of ionization and in the photoelectron spectrum.

4. COMPLEX QUASIENERGIES

If unperturbed system under consideration has a continuous spectrum, there are two possibility to formulate the problem about quasienergies in an external periodically time-dependent field. In the first version continuum is considered exactly, the resulting quasienergies are real and quasienergetic spectrum is continuous [I3] . In the second version continuum is considered approximately as a reason of slow decay. The problem about quasienergies is formulated in the basis of discrete levels with ionization taken into account as the ionization width. In such a formulation guasienergetic spectrum becomes discrete but quasienergies γ become complex. Their real and imaginary parts determine a position and width of the quasienergetic levels. The wave functions corresponding to complex quasienergies are quasistationary quasienergetic wave functions. They are slowly decaying. In such a formulation the rate of decay is supposed to be much less than the space between any adjacent quasienergetic levels.

In the case of Rydberg atom in strong ionizing field ($V > I$) quasienergies γ can be shown [8] to obey the equation

$$1 + i\pi \sum_n \frac{|V_{n,\gamma+\omega}|^2}{\gamma - E_n} = 0 \tag{I9}$$

This equation has two solutions [8]

$$\gamma' = E_n , \quad \gamma'' = -\pi \left| V_{n, E_{n_0}+\omega} \right|^2 = -\frac{\pi V^2}{n_0^3} , \; V \ll 1 \tag{20}$$

$$\gamma' = E^{(n)} , \; \gamma'' = -(\pi^3 n_0^3 V^2)^{-1} , \; V > 1$$

Quasienergetic spectrum is shown as a whole at Fig.5.

In the region $V < I$ quasienergetic levels are localised near unperturbed Rydberg levels E_n, their width $|\gamma''|$ is a growing function of V (20), $|\gamma''|$ is much less than the space between adjacent levels n_0^{-3} so long as $V < I$. When V becomes as large as 1 the width $|\gamma''|$ becomes of the order of the space between levels $|\gamma''| \sim n_0^{-3}$ In this case rigorously a concept of quasistationary wave functions

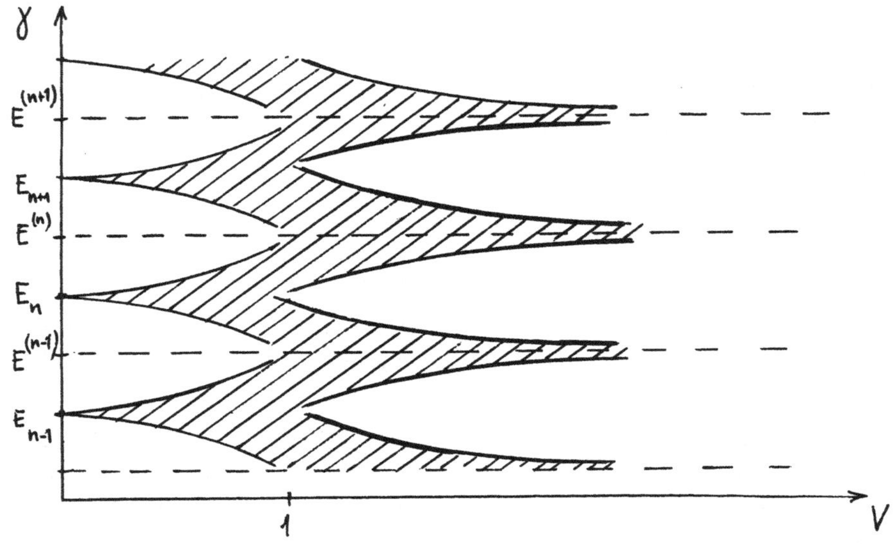

Figure 5. Quasienergetic zones γ versus field strength amplitude $\varepsilon_0 \propto V$

becomes inapplicable because the atom decays during one Kepler period. At the point $V \sim 1$ all the discrete spectrum at Fig.5 is smeared out and is replaced by quasicontinuum. At stronger field $V > 1$ that quasicontinuum disappears and instead of narrow levels localised near $E^{(n)}$ appear. In accordance with equation (20) their width is falling down function of V, and $|\gamma''|$ becomes again much less than n_0^{-3} as soon as V becomes larger than one.

5. A.C.STARK SPLITTING

Although the energy of Rydberg levels in the hydrogen atom does not depend on the angular momentum quantum number ℓ, a.c.Stark effect can split these levels removing their degeneration. Of course there is no splitting in the approximation used above when all the a.c.Stark effect is characterized by a general n- and l-independent shift $\varepsilon_0^2/4\omega^2$, common for all Rydberg levels. In accordance with estimates of ref. [9] this shift results from the zero-th order diagonal part of the a.c.Stark matrix $Q_{n\ell,n'\ell'}$. Corrections to this part have been estimated in ref.[9] to be of the order of ionization broadening considered above

$$\delta Q_{n\ell,n'\ell'} \sim V^2 n_0^{-3} \tag{21}$$

δQ matrix (21) is determined mainly by virtual transitions via levels separated from E_{n_0} by a distance $\sim\omega$. Matrix elements of such transitions can be described in a semiclassical approximation (2), (3) and they slowly depend on principal quantum numbers n , n', n'' of initial, final and intermediate states. For this reason the δQ -matrix (21) can be considered as slowly depending on n and n' and approximately it can be replaced by some n, n' independent (complex) matrix $\delta Q_{\ell\ell'}$ whose complex eigenvalues are q_α , $\alpha = 0, 1, 2, \dots \ell_{max}$. Such a generalization makes the theory much more complicated. But it does not change the main physical conclusions of section 2 [9]. The most interesting qualitative changes occur in quasienergetic spectrum. Strong field asymptotics of quasienergies is given by

$$\gamma_{n\alpha} = E^{(n)} - (\pi^2 n_0^6 q_\alpha)^{-1} \tag{22}$$

where $0 < |q_\alpha| \lesssim V^2 n_0^{-3}$ and the number of eigenvalues q_α is of the order of $\ell_{max} \sim n_0^{2/3}$. In the model without degeneration and a.c.Stark splitting quasicontinuum occurs in a single point $V \sim 1$ (Fig.5) Now quasicontinuum occurs for any $V > 1$. For any $V > 1$ there are some quasienergetic levels (with $|q_\alpha| \sim n_0^{-3}$) whose width is $\sim n_0^3$ and these levels form a quasicontinuum. Besides, there are two groups of narrow levels: those for which $|q_\alpha| \ll n_0^{-3}$ (they are weakly perturbed and broadened) and those for which the given V is large $(|q_\alpha| > n_0^{-3})$ and which became narrow because of interference effects in strong ionizing field. In this more general model photoelectron spectrum contains components similar to those described in section 2, with multipeak structure, with peaks localized near $E^{(n)} + \omega$, with narrowing and growing peaks. Again the time of total ionization t_i is determined by inverse width of these narrow peaks (12) though in details the dynamics of ionization deviates from that described by equation (13) [9]

6. ABOVE THRESHOLD IONIZATION

So a.c.Stark splitting does not change essentially the main results of section 2. This conclusion justifies the one-dimensional model formulated in the second section. In the frameworks of this model we will consider here another effect not included in the previous analysis. This effect is above threshold ionization or absorption of photons in the continuum. These processes have been considered in ref.[8] by a method close to that used by Deng and Eberly [10]. The main result of this generallization is a renormalization of the constant V^2 which, in general, is replaced by

$$\tilde{V}^2 = \frac{2V^2}{1+\sqrt{1+4\pi^2 V^2}} \tag{23}$$

In the limit of strong field $V > 1$ $\tilde{V}^2 = V/\pi$. This is the only change in equations (9)-(13) connected with above threshold ionization. All the qualitative conclusions of section 2 remain correct. Now the width (12) of narrow peaks in the photoelectron spectrum and the ionization time t_i take a form

$$\Gamma = \frac{2}{\pi^2 V n_0^3}, \quad t_i = \frac{1}{2}\pi^2 V n_0^3 \tag{24}$$

Photoelectron distribution (9), (11) described at Fig.3 repeats itself in higher above threshold zones, separated from the main one by $s\omega$ (with the indicated changes of V^2 and Γ). The number of effectively populated zones is $S_0 \sim V$

7. TWO-LEVEL MODEL

The main idea of two previous sections was to take into account some really existing factors and to make the model more realistic. The results show that though the model and equations become more complicated, the main qualitative results hold good. This conclusion is a stimulus to make a contrary attempt to find as simple model as possible in which, however, the described features would occur. Such a model does exist. This is a model of two close levels connected with the same continuum by interaction with almost coinciding matrix elements. Let Δ be the space between the two levels, Q_{ij} (i,j=1.2) be the elements of the complex a.c. Stark matrix, $Q_{11} \simeq Q_{12} \simeq Q_{21} \simeq Q_{22}$.

One of two quasienergies of such a system is given by

$$\gamma = \frac{1}{2}\left\{\Delta + Q_{11} + Q_{22} - \sqrt{(\Delta + Q_{11} - Q_{22})^2 + 4Q_{12}Q_{21}}\right\} \simeq \tag{25}$$

$$\simeq \frac{1}{2}\Delta + \frac{1}{2}\left(Q_{11} + Q_{22} - 2\sqrt{Q_{12}Q_{21}}\right) - \frac{1}{8}\frac{\Delta^2}{\sqrt{Q_{12}Q_{21}}}$$

where the last term on the right hand side of equation (25) is obtained in the strong field limit $|Q_{12} \times Q_{21}| \gg \frac{1}{4}\Delta^2$. If all the matrix elements Q_{ij} coincide exactly and if $\Delta = n_0^{-3}$, $Q_{ij} \sim V^2 n_0^{-3}$, $-\mathrm{Im}\,\gamma''$ from equation (25) coincides with Γ (12). It means that equation (25) describes the field-induced effect of narrowing of quasienergetic levels described above. In strong field $V > 1$ each narrow quasienergetic Rydberg level $E^{(n)}$ (10), (20) appears mainly due to interference interaction of two adjacent field-free Rydberg levels E_n and E_{n+I}.

Although very simple, equation (25) is rich enough not only to describe the effect of narrowing itself but to determine the limits of its applicability either. Let us put

$$Q_{11}+Q_{22}-2\sqrt{Q_{12}Q_{21}} \sim n_0^{-1} \times V^2 n_0^{-3} \tag{26}$$

where the additional factor n_0^{-I} indicates a degree of incomplete compensation of terms on the left hand side. Now equation (25) yields

$$\operatorname{Im}\gamma \sim -n_0^{-3}\left(\frac{V^2}{n_0}+\frac{1}{V^2}\right) \tag{27}$$

Hence $|\operatorname{Im}\gamma|$ has its minimum when $V \sim n_0^{1/4}$ and

$$|\operatorname{Im}\gamma|_{min} \sim n_0^{-1/2} \times n_0^{-3} \quad \left(t_{imax} \sim \sqrt{n_0}\, n_0^3\right) \tag{28}$$

The factor $\sqrt{n_0}$ determines the minimum achievable width of quasi-energetic levels (in comparison with the Kepler frequency n_0^{-3}) and maximum degree of stabilization or increase of the ionization time t_i (in comparison with the Kepler period n_0^3). When the renormalization (23) is taken into account the field V at which $|\operatorname{Im}\gamma|_{min}$ and t_{imax} (28) are achieved is estimated as

$$V_0 \sim \sqrt{n_0} >> 1 \tag{29}$$

8. SECONDARY POPULATED RYDBERG WAVE PACKET: STRUCTURE AND STABILITY

Secondary populated Rydberg states form the wave packet

$$\psi_R = \sum_n A_n(t)\,\varphi_n \tag{30}$$

where the amplitudes $A_n(t)$ are given by equation (I4). These amplitudes and probabilities $w_n = |A_n(t)|^2$ are oscillating and slowly damping functions of the time t. Averaged probabilities $\overline{w_n}$ are equal to [6]

$$\overline{w}_{n_0} = \frac{1}{3}e^{-\Gamma t}, \quad \overline{w}_n = \frac{2}{\pi^2(n-n_0)^2}e^{-\Gamma t}, \; n \neq n_0 \tag{3I}$$

Field induced decreasing of the ionization rate constant Γ (I2) means that the wave packet (30) is stable with respect to ionization by the same field which constructs this packet. This state (30) is stable with respect to photoionization by any other field either [6]. To confirm this statement we can consider the wave packet (30) as an unperturbed atomic wave function. Let this bound state be ionized by electromagnetic field with some frequency $\omega' \neq \omega$, $\omega' > 1/2\, n_0^2$ and let W be the corresponding interaction operator. Let C_E be the amplitude of probability for the atom to make transition from the state (30) to the state of continuum φ_E with energy E. Found in the first order perturbation theory in W C_E has a form [6]

$$C_E = \frac{2}{\pi} \sum_m \frac{\delta(E-E'-E^{(m)})}{m-n_0+1/2} \sum_n \frac{W_{En}}{m-n+1/2} \tag{32}$$

If in accordance with the approximation used and discussed above $W_{En}(n) \simeq$ const the last sum on the right hand side of equation (32) is zero [I4] and $C_E = 0$ confirming stability of the wave packet (30).

Additional and rather interesting information about the wave packet (30) can be deduced from analysis of its time-coordinate structure. This investigation has been done in ref.[6] with the aid of equation (I4) for $A_n(t)$ and semiclassical expressions for $\varphi_n(r)$ [I]. The resulting function $\psi_R(r,t)$ (30) has a form [6]

$$\psi_R = C(2ir)^{-1}(P_{n_0}(r))^{-1/2}\left\{e^{i\theta(r)} f_+(r,t) - e^{-i\theta(r)} f_-(r,t)\right\} \tag{33}$$

where $\theta(r)$ and $P_{n_0}(r)$ are given by equations (7), and $|f_\pm| = I$. If the functions $f_\pm$ would be equal to one identically there would be no difference between ψ_R (33) and $\varphi_{n_0}(r)$. But as a matter of fact f_+ and f_- change their signs and the points where $f_+(r,t)$ changes its sign do not coincide with the corresponding points for $f_-(r,t)$. These functions are described at Fig.6.

The described result shows that owing to interference semiclassical wave function (33) has sudden jumps of its phase at the points $\tau(r)$, $2\pi n_0^3 \pm \tau(r)$, $4\pi n_0^3 - \tau(r)$. As a whole, the functions $f_\pm(r,t)$ and $\psi_R(r,t)$ (33) are periodical functions of the time t with the period $4\pi n_0^3$ equal to the doubled Kepler period. The wave packet of secondary populated Rydberg states (30), (33) is not close to the earlier described Gaussian packet (4)-(6). The wave packet (30), (33) does not describe anything like classical localization.

Sudden jumps of the phase of the semiclassical wave function (33) are very important for explanation of stability of the wave packet (30), (33) with respect to its photoionization.

The amplitude of probability C_E for the atom to be ionized can be determined as previously. But using the explicit form of the unperturbed wave function (33) we can investigate details of the time

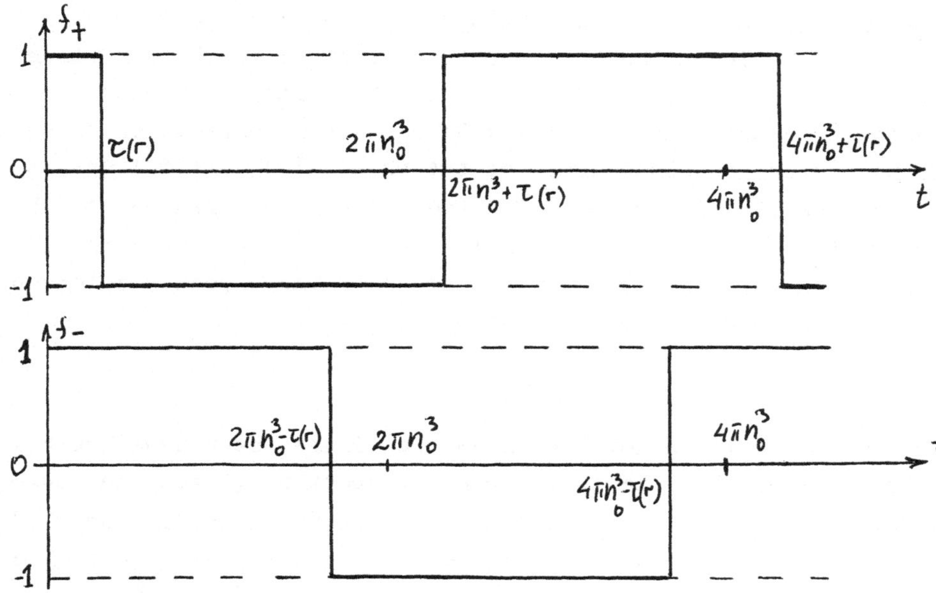

<u>Figure 6.</u> Functions $f_\pm(r,t)$ in equation (33), $\tau(r)$ is given by equation (7).

behaviour of $C_E(t)$. In the first order of the perturbation theory in W $C_E(t)$ is given by [6]

$$C_E = \int dr\, \varphi_E\, W\, C(2ir)^{-1}(P_{n_0}(r))^{-1/2}\, F(r,t) \tag{34}$$

where the real and imaginary parts of the function F are described at Fig.7.

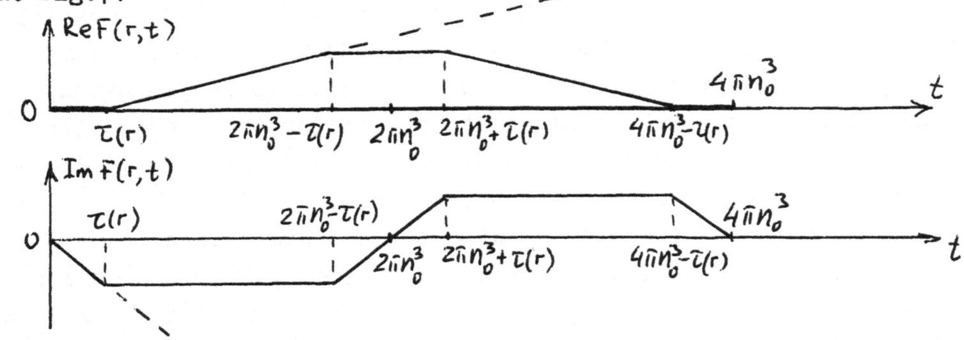

<u>Figure 7.</u> Real and imaginary parts of the function F(r,t) in equation (34). Dotted lines describe F(r,t) for the semiclassical wave function $\varphi_{n_0}(r)$ without jumps of the phase.

The curves at this figure show that transitions from the state (30) to the continuum (growing C_E) are replaced at $t = \pi n_0^3 \pm \tau$ by inverse transitions, from the continuum to discrete spectrum (falling down C_E). These two kinds of transitions cancell each other completely at $t = 4\pi n_0^3$ and then, all the pictures repeats itself with the doubled Kepler period $4\pi n_0^3$. These effects occur owing to the described above jumps of the phase. The dotted lines at Fig.7 show the linear growth of ReF and ImF which correspond to the linear growth of C_E or irreversable ionization. This process is interupted and replaced by inverse process at those points which correspond to jumps of the semiclassical phase.

9. EXPERIMENT

Two main results described in section 2 are the unusual multipeak structure of photoelectron energetic spectrum (Fig.3) and stabilization of the atom in the limit of strong field (equations (I2) (I3)). If $n \sim 10^2$ adjacent Rydberg levels and peaks of the photoelectron distribution function (9) are separated by $n_0^{-3} \sim 10^{-5}$ev. Probably this energy is too small for the peaks of photoelectron spectrum to be resolved experimentally. So the multipeak structure of $w(E)$ hardly is observable at present.

As for the effect of field-induced stabilization, its experimental observation seems to be more feasible though of course there are some problems either. The main assumption which has been done in the above consideration and which can seem to be nonrealistic is the assumption about suddenly switching on interaction. In a contrary case of very slow switching on interaction the parameter V becomes a slow function of t. If $V_{max} \gg 1$ at some time t_x in the process of switching on $V(t_x) \sim 1$. In this region according to the results described above the atom is totally ionized during a rather short time $\sim n_0^3$. If this time is much shorter then the pulse duration no atoms at Rydberg levels achieve the region (or time) where (or when) $V \sim V_{max}$, all these atoms are ionized and there is no object to observe the described interference effects.

So the assumption of suddenly switching on interaction is necessary for experimental realization of the described models and corresponding results. Usually in optics the assumption of suddenly switching on interaction is considered as nonrealistic. But in the range of microwave radiation this assumption can be realized with the aid of laser excitation [6]. The scheme of experiment is described at Fig.8.

A beam of protons (P) with the velocity $v \sim 3 \times 10^9 cms^{-1}$ in the process of charge exchange in a target (T) is transformed into a beam

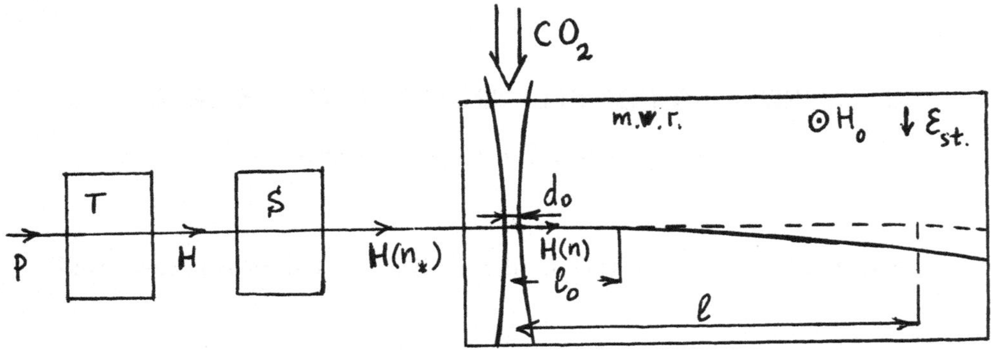

Figure 8. Scheme of experiment for observation field-induced stabilization of Rydberg atoms.

of hydrogen atoms (H). In a selector (S) only the atoms excited at the level $n_* = 5$ are selected to enter into the region occupied by microwave radiation (m.w.r.) In the same region a beam of radiation from CO_2 laser is focused, d_0 is the diameter of the focal region. Laser radiation excites the Hydrogen atoms at Rydberg levels with $n \sim 100$ during a time $\sim d_0/v$ which is much shorter than Kepler period $2\pi n_0^3$ if $d_0 < 2$ cm. In this case interaction of microwave field with Rydberg atoms is switched on almost suddenly. Then the length passed by Rydberg atoms before their total ionization depends on the field strength of the ionizing microwave radiation. If $V \sim 1$ this length is of the order of $\ell_0 \simeq 2\pi n_0^3 v \simeq 2$ cm. If $V \sim n_0^{1/2}$ the length of ionization is 10 times longer $\ell \sim 20$ cm. This difference in the length of ionization can be detected with the aid of transversal static magnetic (H_0) or electric ($\mathcal{E}_{st}$) field as it is shown in Fig. 8.

REFERENCES

1. L.D.Landau, E.M.Lifshitz. "Quantum Mechanics".FIZMATGIZ,Moscow, 1963
2. I.Ya.Bersons, Izv.Acad.Nauk, USSR, Ser.fiz., 1981,v.45, p.2289
3. S.P.Goreslavsky, N.B.Delone, V.P.Krainov. Zh.Eksp.Teor.Fiz.,1982, v.82, p.1789; J.Phys.B, 1983, v.16, p.2369.
4. G.Alber, P.Zoller. Phys.Rev.A, 1987, v.37, p.377.
5. M.V.Fedorov, A.M.Movsesian, JOSA B, 1988, v.5, p.850.
6. M.V.Fedorov, M.Yu.Ivanov, A.M.Movsesian, J.Phys.B, 1989
7. M.V.Fedorov, A.M.Movsesian. J.Phys.B, 1988, v.21, p.L153.
8. M.V.Fedorov, A.M.Movsesian, JOSA B, 1989, v.6
9. M.V.Fedorov, A.M.Movsesian, JOSA B, 1989, v.6
10. Z.Deng, J.Eberly, JOSA B, 1985, v.2, p.486.
11. S.I.Marmo, Voronez University, USSR, Theses, 1975.
12. K.Rzazewski, M.Trippenbach, M.V.Fedorov, A.E.Kazakov,J.Phys.B,1989
13. A.E.Kazakov, V.P.Makarov, M.V.Fedorov, Zh.Eksp.Teor.Fiz.,1976, v.70, p.38
14. I.S.Gradstein, I.M.Ryzik, Tables of Integrals, Series and Products, New York, Academic, 1966.

OBSERVATION OF ELECTRONIC WAVE PACKETS WITH SHORT LASER PULSES

A. ten Wolde, L.D. Noordam, H.G. Muller and H.B. van Linden van den Heuvell

FOM-Institute for Atomic and Molecular Physics,
Kruislaan 407, 1098 SJ Amsterdam, The Netherlands

1. INTRODUCTION

Under the influence of the development of lasers producing tunable ultra-short pulses, there has been an increasing interest in the behavior of atomic electron wave packets during the last few years. In general, deeper insight in quantum mechanics can be obtained by investigating the transition from the quantum mechanical description to the limit of classical mechanics. A striking difference between the two is, that the usual starting point in quantum mechanics is a stationary wave function extending over a relatively large region of space, while classical mechanics describes the behavior of moving point objects. The concept of a wave packet forms a bridge between these two extremes: localization in space, together with motion, can be described by quantum mechanics as a coherent superposition of stationary wave functions.[1] This means that the classical limit of a quantum mechanical system with a quantum number n is not obtained by simply letting $n \rightarrow \infty$, but by making a superposition of n-states. In the classical description of a one-electron atom, the electron moves around the atomic core in a Kepler orbit. In quantum mechanics, localization and motion on a Kepler ellipse can be achieved by a superposition of Rydberg states where the summation goes over all three quantum numbers n,l and m. In this way a wave packet localized in all space coordinates r,ϕ and θ can be created.[2] The motion of an atomic electron can be separated in a radial and an angular part. In the present contribution we discuss the behavior of radial electronic wave packets. Recently, we reported the experimental observation of a radial wave packet.[3,4] Theoretical attention to this type of wave packet has been paid by several authors.[5,6,7] Observation of angular wave packets has also been reported.[8] An unavoidable consequence of the Coulomb potential is dispersion of the radial wave packet. Such dispersion is absent in a potential which leads to energy levels with a constant spacing, such as a harmonic oscillator. Indeed more than six oscillations of a vibrational wave packet were observed in NaI.[9] Another dispersion-free system is that of an atom in a dc electrical field, where the linear Stark-shift leads to a constant energy spacing. Very recently we observed such a "parabolic" electron wave packet.[10]

2. CLASSICAL PICTURE

In order to estimate the spreading of a radial wave packet in time we can use simple classical arguments.[11] If we assume a classical ensemble of highly excited electrons with energies corresponding to Rydberg states around $\bar{n}$ (Bohr-quantization) they will describe Kepler orbits with different classical orbit times, given by $\tau_n = 2\pi n^3$ (atomic units are used throughout this text). The variation of the classical orbit time equals $\Delta\tau_n = 6\pi\bar{n}^2\Delta n$ ($\bar{n} >> \Delta n$). If we assume that the bandwidth of the exciting laser pulse is given by $1/\tau_p$, the number of involved Rydberg states Δn is approximately equal to τ_n / τ_p. This means that a very short exciting pulse leads to a narrow distribution at $t = 0$, but the spreading of this distribution is very rapid. On the other hand, a relatively long pulse leads to little dispersion, but the initial width is already large. If we minimize the total width of the electron distribution over the Kepler orbit at $t = 2\pi\bar{n}^3$, this leads to an optimal pulse duration $\tau_p = \sqrt{12\pi\bar{n}^5}$. In that case the expression for the total spreading becomes $\Delta t = \sqrt{12\pi\bar{n}^2 t}$. This means that the distribution is fully spread out over one Kepler orbit at $t = \frac{\bar{n}}{6}\tau_{\bar{n}}$, so the number of orbits during which the electron distribution remains localized on the Kepler orbit is of the order of $\bar{n}/6$. After this spreading, an approximate restoration of the localization will occur. This happens at the moment that one electron overtakes another one which has made one orbit less in the same time: $(N+1)\tau_n = N\tau_{n+1}$ gives $N \approx n/3$, which means that the distribution starts to reassemble itself as soon as it has spread. The further time evolution consists of periodic spreading and higher-order revivals. For further analysis of the system we will turn to the quantum mechanical description.

3. QUANTUM MECHANICAL DESCRIPTION

The energy levels in a Coulomb potential are given by the well-known relation $\omega_n = -1/2n^2$, so the spacing between the levels is given by $\Delta\omega_n = 1/n^3$. If two neighboring n-states are coherently excited, a quantum beat results with a period $\tau_n = 2\pi/\Delta\omega_n = 2\pi n^3$, which equals the classical orbiting time of an electron in a Kepler orbit. If more than two levels are coherently excited, different energy spacings will lead to different beating periods (see also Refs. 5,6). Therefore the resulting wave packet, which oscillates between the classical turning points ($r = 0$ and $r = 2n^2$ for $l = 0$), will start to spread. In order to calculate the evolution of a radial wave packet we expand the radial part of the wave function on the basis of the stationary Rydberg wave functions $R_{nl}(r)$:

$$\Psi(r,t) = \sum_n a_n(t)\ e^{i\omega_n t}\ R_{nl}(r) \quad , \tag{1}$$

where the expansion coefficients $a_n(t)$ are the population coefficients of the various n-states. We will now assume the following approximations in order to simplify the calculations: We adopt the rotating wave approximation. Spontaneous emission from the Rydberg states, which is small indeed on a picosecond time scale, is neglected. The intensity of the laser pulse is assumed to be low so that both ionization by absorption of an extra photon and depletion of the ground state can be neglected. The exciting pulse is taken to be a transform limited Gaussian pulse with a FWHM-duration τ_p and a central frequency ω_L : $E(t) = \cos \omega_L t \cdot \exp\{-2 \ln 2\, (t/\tau_p)^2\}$. Under these assumptions, the population coefficients after the exciting pulse will take the values

$$a_n = \frac{1}{n^{3/2}}\, e^{-\Delta_n^2 \tau_p^2 / 8 \ln 2} \quad , \qquad (2)$$

where $\Delta_n = \omega_L - \omega_n$ is the laser detuning with respect to level n. The pre-exponential factor comes from the dipole matrix element for the transition form the ground state to the Rydberg state, $\mu_{n1} = \langle n | r | 1 \rangle \propto 1/n^{3/2}$. For a given pulse duration τ_p and laser frequency ω_L, the

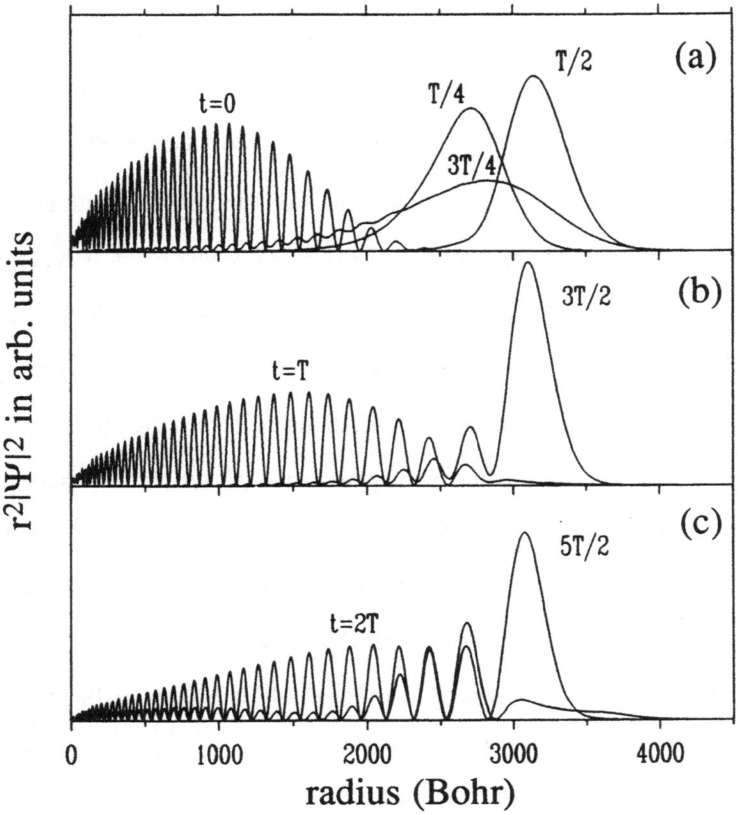

Fig. 1. The calculated time evolution of a radial wave packet excited with a pulse of 1.4 ps around $n = 40$. The probability density is plotted at $t = 0$, $T/4$, $T/2$ and $3T/4$ in (a), at $t = T$ and $t = 3T/2$ in (b), and at $t = 2T$ and $5T/2$ in (c) ($T = 2\pi n^3$). Note that the localization is almost lost after 5/2 orbits.

populations of the Rydberg state are fixed by Eqn. 2 and the time evolution of the wave packet is fully determined by Eqn. 1. Since the stationary Rydberg wave functions are known, we can now easily calculate the shape of the wave packet at time t (see also Refs. 5,6,7).

The probability to find the electron at a radius r at a time t is proportional to $r^2 \,|\Psi(r,t)|^2$. We calculated this probability as a function of r at various times for a coherent superposition of states around $n = 40$ ($t_n = 10$ ps) created with a 1-ps laser pulse. The results of the calculation are plotted in Fig. 1. In the upper figure we see that the initial distribution at $t = 0$ is localized around $r = 1000$. The oscillations indicate that the electron velocity in this region is high. It should be noted however, that the expression for the coefficients a_n in Eqn. (2) becomes valid only after the exciting pulse has vanished, so in this calculation $t = 0$ does not coincide with the maximum of the exciting pulse as one could intuitively think. The created wave packet starts to oscillate between the core and the outer classical turning point. The localization is the strongest at the outer classical turning point, because the electron velocity is low in this region. As time goes by, the wave packet starts to spread and after several oscillations the localization has almost disappeared. The effect of the spreading can be studied more quantitatively if we plot the expectation value of the distance to the core $\langle r \rangle$ (Fig. 2a) and the width of the packet Δr (Fig. 2b) as a function of time (in units of the classical orbit time, denoted by T). Here Δr is defined according to $\Delta r = 2\sqrt{\langle r^2 \rangle - \langle r \rangle^2}$. Initially, $\langle r \rangle$ varies from 1000 to more than 3000, while Δr is relatively small. The outer classical turning point for $n = 40$ is 3200 ($l = 0$). After several orbits, the wave packet has spread. One could expect that the mean radius of the dispersed wave packet equals that of the mean stationary wave function: $\langle r \rangle = \frac{3}{2}n^2 = 2400$ for $n = 40$. The width of this wave function is given by $\Delta r = n^2 = 1600$. In fact we see that $\langle r \rangle$ and Δr of the dispersed wave packet are oscillating around these values. Note that the width of the wave packet can be considerably smaller than the stationary value, especially if the wave packet is near the outer turning point: maxima in $\langle r \rangle$ go coincide with minima in Δr, because the low electron velocity near the outer classical turning point leads to a small width, and *vice versa*. The period of the oscillations around the stationary-state values corresponds to the sum-frequency of the energy spacings around $n = 40$, $\Delta\omega_+ \equiv \Delta\omega_n + \Delta\omega_{n-1} = 2n^{-3}$.

After this dispersion, we see that the wave packet shows an approximate revival after about 13 oscillations. The difference frequencies, $\Delta\omega_- \equiv \Delta\omega_n - \Delta\omega_{n-1} = n^{-3}, 3n^{-4}, 12n^{-5}, \ldots$ lead to beating periods of $2\pi n^3$ (the classical orbit time), $2\pi n^4/3$, $2\pi n^5/12$, etc. This is the same result as in the classical picture, and explains the first approximate revival at $t = 2\pi n^4/3 = \frac{1}{3} n \cdot \tau_n$, *i.e.* after 13.3 classical orbits.

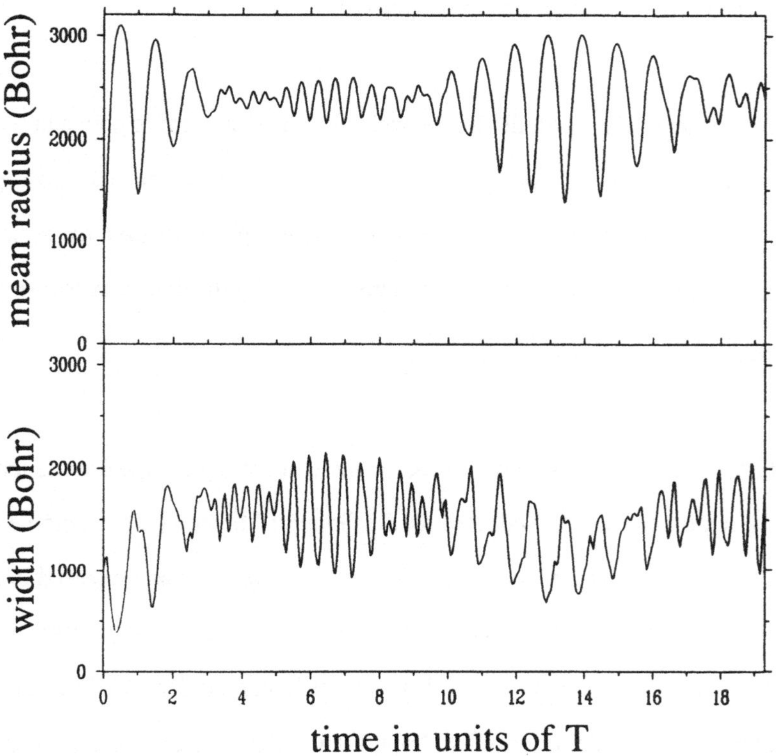

Fig. 2. The calculated time evolution of the mean radius $\langle r \rangle$ and the width Δr of the wave packet. The sum-frequency is visible in the oscillations after about 6.5 orbit times. The difference-frequency introduces an approximate revival of the wave packet after about 13 orbit times.

The dispersion and relocalization of the wave packet are good observables. For a time resolved wave packet experiment, apart from the creation of a radial wave packet we also need an experimental tool to detect it. This detection method should be r-sensitive in order to observe the oscillations. Photoionization is such a method, since it is only sensitive for the amount of wave function near the core. The photoionization signal of a wave packet by a delayed probe pulse has been calculated by the authors of Refs. 5 and 6. However, we want to investigate the r-dependence of the ionization process itself in somewhat more detail.

4. PHOTOIONIZATION OF THE WAVE PACKET

When a radial wave packet is oscillating between the classical turning points in the presence of an electromagnetic field, one can ask the question in what regions of space the absorption of photons leading to ionization actually takes place. In the dipole approximation, the ionization probability of a bound state $|i\rangle$ to a continuum state $|f\rangle$ is given by the dipole matrix element

$$\mu_{fi}(t) = \langle f|\hat{D}|i\rangle = \int d\vec{r}\ \Psi_f^*(\vec{r},t)\ \hat{D}\ \Psi_i(\vec{r},t) \tag{3}$$

where $\hat{D}$ is the dipole operator. As is well known, the form of the dipole operator depends on the gauge in which the matrix element is evaluated. In the Goppert-Mayer gauge the dipole operator is $\vec{E}\cdot\vec{r}$ (length form), where $\vec{E}$ is the electrical field. This form leads to an integrand in Eqn. (3) which grows with increasing r. In the radiation gauge we have $\vec{A}\cdot\vec{p}$ (velocity form), where $\vec{A}$ is the vector potential and $\vec{p}$ the momentum. In this form, the integrand depends less strongly on r, so the wave functions are probed more evenly. Applying the Kramers transformation (which is not a proper gauge transformation, but can be treated as such) to the system, the interaction is described by $\partial V/\partial r$ (acceleration form), where V is the atomic potential. For a typical atomic potential this form leads to an integrand which is heavily concentrated near the atomic core.

We see that although the value of the integral is invariant under these gauge transformations, the integrands are very different. This indicates that the integrand cannot simply be interpreted as "the contribution to the ionization signal at a distance *r* from the core". The subtlety lies in the fact that the influence of a perturbation implies the definition of an unperturbed system, and the zero-order Hamiltonian is chosen in different ways in the various forms of the theory. This also implies that the unperturbed final state $|f\rangle$ is written in a different form. The less the unperturbed final state resembles the perturbed one in the limit for $r \rightarrow \infty$, the more the integrand describes virtual photoabsorption and -emission processes which do not contribute to the ionization.

In this respect the length form is especially unfavorable. The integrand grows with increasing r. However, this does not imply at all the photoabsorption leading to ionization increases with *r* as well. In this form, the perturbed wave function contains a time-dependent phase-factor $\exp\{i\,\vec{A}(t)\cdot\vec{r}\}$, which in not incorporated in the unperturbed zero-order wave function. Therefore the main effect of the perturbation is the introduction of this phase-factor, which has no physical consequences whatsoever. In the calculation of the matrix element as an integral over all *r*, all contributions vanish except the one leading to photoionization. Therefore the dipole matrix element in the length form consists of an integral over a heavily oscillating function over a large region of *r*. Matrix-elements for a free-free transition can not be calculated in this form at all.

In the velocity form we are slightly luckier. The main inadequacy of this representation is that the zero-order wave function lacks the quiver motion of the electron in the electromagnetic field. The main effect of the perturbation at large *r* is therefore the introduction of this harmonic quiver motion through the unremitting absorption and emission of virtual photons from the field. The amplitude of this oscillation is given by E/ω^2. Note that no net energy is absorbed from the field in this way; again the dominant part of the integrand describes a process which has nothing to do with photoionization.

In the acceleration form things start to make sense. The Kramers transformation builds the quiver motion of a free electron into the zero-order description, so that the asymptotic form of an ionized electron approaches its true motion. In this case the perturbation vanishes as the electron approaches infinity, so that wave operators for the ionization process exist, and consistent scattering theory becomes possible. The perturbation in the case of a Coulombic potential now behaves asymptotically as r^{-2}. This supports the physical intuition that photoabsorption must take place close to the nucleus, because if the electron is far away from the core it is nearly free and can only scatter photons.

The acceleration form is not the best one can do, however. Although the zero-order is asymptotically correct, it is not optimal for finite r. If an electron quivering in the electromagnetic field feels the presence of a potential, in lowest order it experiences only a slight modification of the quiver amplitude. As long as higher derivatives of the potential can be neglected, the second derivative acts on the quivering electron as a harmonic potential. The response of such a driven harmonic oscillator differs from that of an unbound particle. Of course, the validity of this picture depends on the assumption that the amplitude of the oscillation remains small. If the frequency of the radiation field happens to be at resonance with the approximate harmonic oscillator, the amplitude grows without bound in response to the driving force, and the approximation breaks down. Note that the response of a harmonic oscillator to a harmonic driving force is still a periodic motion; the particle can not escape from the potential well and therefore ionization is impossible.

Once a wave packet is far away from the nucleus, its motion is very well described by

$$\Psi(r,t) = \chi(r - \alpha_0(r) \cos \omega t\, ,\, t) \cdot e^{i\, b(t)\, r} \tag{3}$$

where $b(t)$ is periodic in time, and $\chi(r,t)$ is the solution of the Schrödinger equation with a Hamiltonian which is now time-independent. The quiver amplitude α_0 is given by

$$\alpha_0(r) = \frac{E}{\omega^2 - \partial^2 V/\partial r^2} \quad . \tag{4}$$

The time-dependent wave function Ψ in the form of Eqn. (3) incorporates the quiver motion of the free electron, as well as the changing amplitude of this oscillating motion if the electron approaches the atomic core. The wave function does not incorporate the ionization itself, which can be deduced from the fact that apart from the phase factor, Ψ differs from χ only in a finite displacement $\alpha_0(r)\cos\omega t$. The ionization process is a higher-order effect, which is caused by the anharmonic terms in the expansion of the potential $V(r)$. As can easily be checked, substitution in the time-dependent Schrödinger-equation leads to terms of the order $\partial^3 V/\partial r^3 \propto 1/r^4$ for a Coulomb potential. This means that the terms leading to photoionization show a very rapid decrease with r indeed, in fact more than sufficient to prove our point that ionization takes place only when the wave packet passes close to the atomic core.

5. EXPERIMENT

For the measurement of radial wave packets in a photoionization experiment we used the following set-up (for more details, see Ref. 4). A schematic drawing of the experiment is given in Fig. 3. Rubidium atoms are contained in a vacuum vessel. With two laser pulses, atoms are excited to a coherent superposition of Rydberg states and subsequently ionized. The produced photoelectrons are detected by a set of channel plates. The advantage of measuring electrons instead of ions is that one can distinguish between photoionization, producing fast electrons, and ionization processes like atomic collisions, producing very slow electrons.

The laser system consists of a synchronously pumped linear dye laser (wavelength 594 nm). The pulses are amplified in a dye amplifier pumped by the second harmonic of a Q-switched Nd:YAG laser. After amplification the pulse duration is about 6 ps. The amplified pulses are split in a Michelson set-up. One of the mirrors of the Michelson is mounted one a translation stage, in order to vary the delay between the two pulses. With this experimental set-up the three-photon ionization signal of rubidium can be measured as a function of the time difference between the pump pulse (excitation of the wave packet) and the probe pulse (ionization of the wave packet near the core).

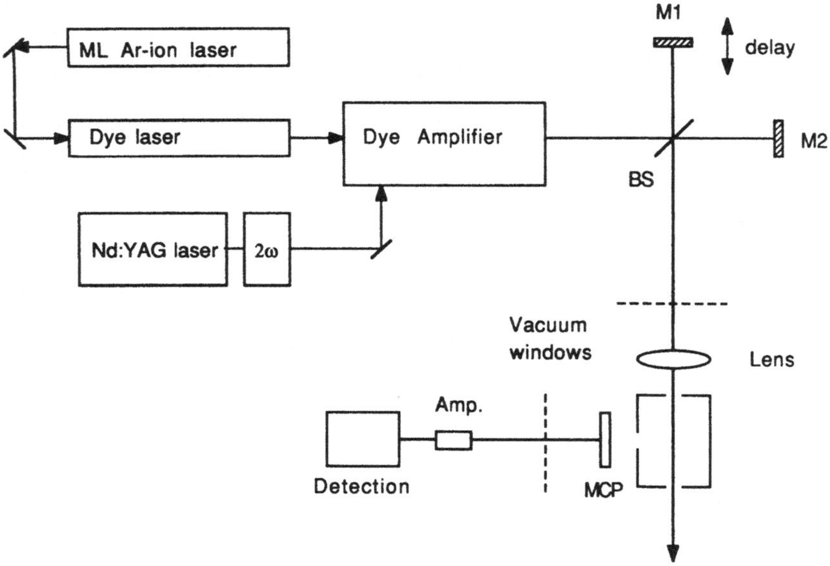

Fig. 3. A schematic drawing of the experimental set-up.

Instead of this single-laser set-up, in which the Michelson is used for the production of two identical pulses for the excitation and ionization, one could imagine a more sophisticated set-up in which two lasers are used. In that case the characteristics of the ionizing pulse could be chosen entirely different from the exciting pulse. We will shortly discuss the advantages of both schemes.

The used single-laser set-up is clearly the simplest solution, which is an advantage in itself. The generation of the pulses is relatively simple and there is no problem of synchronization of the pump and the probe pulse. A disadvantage is that photoelectrons created by both pulses have the same energy and can not be distinguished. Therefore the total photoionization signal will not only consist of the the ionization of the wave packet, but also of direct multiphoton-ionization (MPI) of the individual pulses. The energy of the probe pulse must be kept low, otherwise MPI of the ground state by the probe pulse will dominate the signal.

If the pump pulse and the probe pulse originate from different lasers, one can choose a probe pulse with optimal characteristics for the ionization of the wave packet. Firstly, the severe

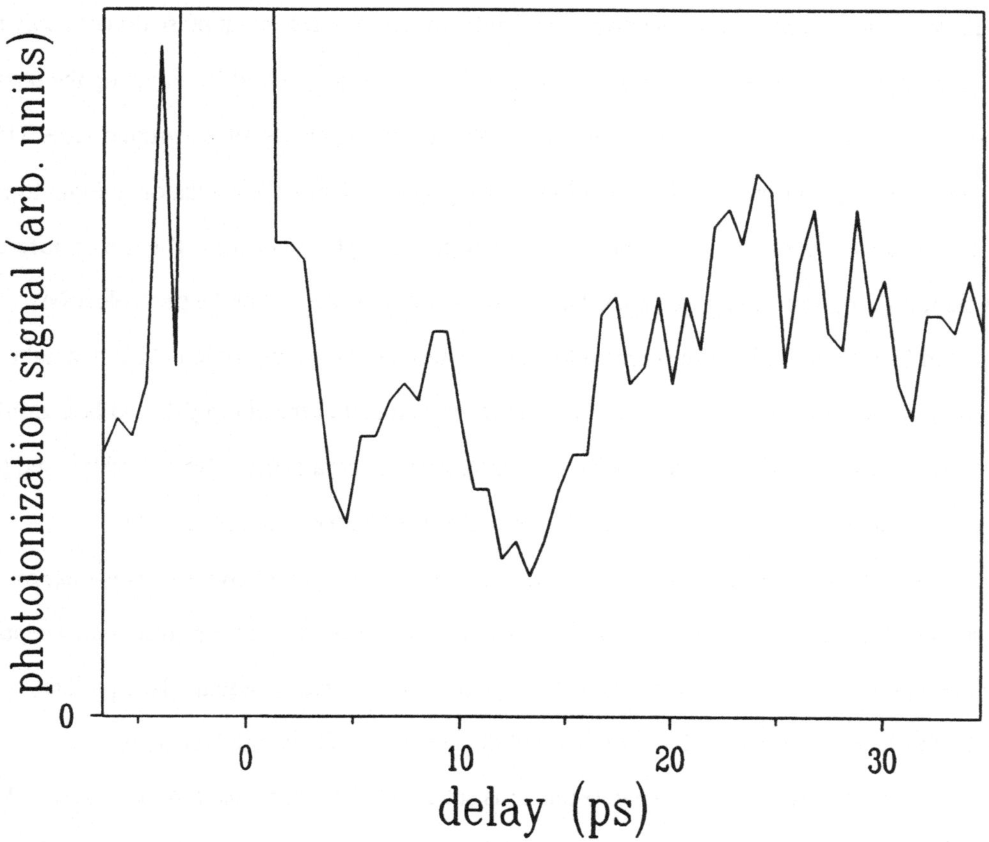

Fig. 4. The photoionization signal of the wave packet as a function of the delay between the pump- and probe pulse. The large peak in the ionization around delay zero is the result of the coherent spike in the light intensity. At 9 and 18 ps, the return of the wave packet to the core leads to an enhancement of the ionization signal. After two orbits, the dispersion of the wave packet has become so large that no further oscillations are observed.

requirements for the product of bandwidth and pulse duration of the pump pulse do not apply to the probe pulse. Furthermore, MPI by the probe pulse can be suppressed by choosing the photon energy of the probe pulse small in comparison with the binding energy of the ground state. This also enhances the ionization probability of the wave packet, because the oscillator strength from a Rydberg state to a continuum state increases with decreasing photon energy. The use of different wavelengths for pump- and probe pulse has the additional advantage that the photoelectrons produced by each pulse have different velocities, so they can be distinguished. Also, the light intensity of the probe pulse can be made much higher than that of the pump pulse without driving the MPI process, so the signal is actually increased. Finally, ultra-short pulses could be used for the ionization in order to measure the evolution of the wave packet more accurately.

In the actual experiment we used the Michelson set-up described above. A wave packet around the 42*d* state of Rb was created by two-photon excitation with a 6-ps pulse with a central wavelength of 594.8 nm. The classical orbit time of this wave packet equals 10.2 ps. The results for the ionization signal as a function of delay between the two pulses is given in Fig. 4. The ionization signal due to the separate pulses is subtracted from the signal. Around zero delay, the wave packet has not yet left the core and MPI takes place by absorption of photons of both pulses. The so-called coherent spike in the light intensity due to temporal and spatial overlap of the pulses gives rise to the extra large peak in the ionization yield. For increasing time delay we see two peaks in the ionization yield. The first one occurs at 9.4 ps: this corresponds to the first return of the wave packet to the core, predicted at 10.2 ps. A second return is observed around 18 ps. After two oscillations the width of the dispersing wave packet becomes too large to observe oscillations in the ionization with the relatively long probe pulse of 6 ps.

6. CONCLUSIONS

A coherent superposition of Rydberg states leads to a radially localized wave packet. This wave packet oscillates between the turning points of the classical Kepler orbit. The energy spacings between the Rydberg states determines the oscillation time. The differences in the spacings cause the effects of dispersion and revival of the wave packet. Photoionization is a sensitive probe to the amount of wave function near the core, because in a Coulomb potential ionization takes place only near the atomic core. In the experiment, a wave packet is created by coherent excitation of Rydberg states of Rb with a short laser pulse, and detected by photoionization with a delayed probe pulse. In this way two returns of the wave packet are observed.

REFERENCES

[1] E. Schrödinger, Die Naturwissenschaften **28**, 664 (1926).

[2] M. Nauenberg, to be published.

[3] L.D. Noordam, A. ten Wolde, H.G. Muller, A. Lagendijk and H.B. van Linden van den Heuvell, J. Phys. B **21**, L533 (1988).

[4] A. ten Wolde, L.D. Noordam, A. Lagendijk, and H.B. van Linden van den Heuvell, Phys. Rev. Lett. **61**, 2099 (1988).

[5] G. Alber, H. Ritsch, and P. Zoller, Phys. Rev. A **34**, 1058 (1986).

[6] J.A. Yeazell and C.R. Stroud, Jr., Phys. Rev. Lett. **56**, 716 (1986).

[7] J. Parker and C.R. Stroud, Jr., Physica Scripta **T12**, 70 (1986).

[8] J.A. Yeazell and C.R. Stroud, Jr., Phys. Rev. Lett. **60**, 1494 (1988).

[9] T.S. Rose, M.J. Rosker, and A.H. Zewail, J. Chem. Phys. **88**, 6672 (1988)

[10] A. ten Wolde, L.D. Noordam, A. Lagendijk, and H.B. van Linden van den Heuvell, to be published.

[11] H.B. van Linden van den Heuvell, H.G. Muller and A. ten Wolde, Electronic and Atomic Collisions, Eds. H.B. Gilbody, W.R Newell, F.H. Read, A.C.H. Smith (North-Holland, Amsterdam, 1988).

MANY PHOTON ABSORPTION IN A NEAR-CLASSICAL REGIME

James E. Bayfield
Department of Physics and Astronomy, University of Pittsburgh
100 Allen Hall, Pittsburgh, Pennsylvania 15260 U.S.A.

INTRODUCTION

A highly excited hydrogen atom placed in a partially ionizing microwave field exhibits a rich quantum dynamics not observed in few level quantum systems exposed to electromagnetic fields. If the number of absorbed microwave photons is not small, then the field can couple together many atomic energy eigenstates under conditions where elementary two state photon absorption and emission proceed at rates comparable to both the initial classical electron orbit frequency $\omega_n = n_0^{-3}$ and to the microwave frequency ω itself. Experimentally the system then is found to be near-classical, with quantum modifications that become large when ω is greater than twice ω_n. These results are particularly interesting because of theoretical predictions that microwave ionization of the atom is a result of chaotic time evolution of the driven atomic electron, in the classical limit. A high frequency quantum suppression of the ionization is then explained as a destructive wave interference effect that arises from the underlying classical chaos affecting the evolution of the semiclassical quantum system. The electron is quantum localized because of the disordered character of chaos, rather than being delocalized during an ionization process. The situation appears similar to Anderson localization of a quantum particle placed in a lattice with randomized lattice site potentials. In this paper the experimental support for these ideas is discussed. A parallel theoretical paper by G. Casati is elsewhere in this volume.

The microwave ionization of weakly bound electrons was first discovered experimentally (1,2) and then explained by completely classical numerical calculations (3). With the realization that the classical electron dynamics was chaotic (4), renewed experimental and theoretical study has focused on this quantum optical system as a prototype for addressing the issue of an integration of nonlinear dynamics with quantum mechanics. As the classical dynamics of this driven nonlinear oscillator is nonintegrable, numerical studies are necessary for the development of theory, both quantal and classical.

When the atom is taken to be one dimensional, time-extended classical phase space has the minimum dimension of three needed for chaotic dynamics. This particularly simple situation also makes detailed numerical studies practical. As we shall see, some experiments have been carried out under conditions that closely approximate a one dimensional situation. These work with atoms having all three parabolic quantum numbers n, n_1, m initially well defined, with $n_1 = m = 0$. Such hydrogen atom Stark states are highly elongated, are of mixed parity, and have associated large permanent electric dipole moments. The one dimensional experimental configuration involves defining the dipole moment direction with a static electric field, and aligning a linearly polarized microwave electric field classically parallel to the moment.

Considerable experimental and theoretical evidence indicates that a one dimensional picture is surprisingly useful in interpreting experimental data taken with ensembles of atoms having broad initial distributions in n_1 or in both n_1 and m, the so-called two and three dimensional (2-D and 3-D) situations (5,6). The electron classical action changes primarily for the motion in the dimension defined by the microwave field direction, the direction of the applied oscillating classical force. Quantum mechanically this is reflected in the size of electric dipole coupling matrix elements being largest for small changes in n accompanied by no change in n_1 or m. The decoupling of the motions has been studied, even quantum mechanically (7). Thus it is reasonable to make comparisons of one dimensional theoretical predictions with 2-D or 3-D experimental data.

In a quantum picture of the 1-D situation, one is working with a ladder of energy levels for the stationary states (n,0,0), coupled together by the classical microwave field $F \cos \omega t$. The anharmonicity of the levels is that of the field free atom, modified by the presence of any static field F_s. At low values of F, all that occurs with measurable probability is transitions at resonant frequencies between isolated pairs of levels. Depending on ω, these may or may not be multiphoton transitions; in either case, these low F two state transitions are understood. However, at high F the two state resonances are power broadened. sufficiently for a coupling of many states to occur (8). Then our system can become semiclassical whenever n_0 and the net number k of absorbed photons for ionization are both large. The quantity k probably should include a contribution from above threshold ionization, which might be large (9). In this regard, one should keep in mind that the classical system in some large field, large action limit is a free electron oscillating in response to the microwave field and only slightly perturbed by the nuclear Coulomb field.

The nature of the classical trajectories is expected to play an important role in semiclassical systems via the classical action function. For the operating conditions of the experiments to be discussed, both regular and chaotic trajectories lie in the initially populated region of classical phase space. In a purely classical picture, only the chaotic trajectories contribute to ionization.

EXPERIMENTS

Microwave ionization of Rydberg alkalai atoms has been studied using thermal atomic beams and pulsed microwave systems (9,10). However, in the case of highly excited hydrogen, experiments have all utilized the fast atomic beam optical double resonance (FAB ODR) technique (11,12). Here most of the needed excitation of the atom is acquired in an initial electron transfer collision between a proton traveling at about 0.003 c and an electron donor atom such as xenon (12). Those atoms with quantum numbers (7,0,0) are then selectively excited first to (10,0,0) and then to (n_0,0,0) with n_0 to be chosen between 35 and 100. Each ODR transition is done in a static field to split the Coulomb degeneracy, in the first transition to within the sign of m (12). The use of CW carbon dioxide lasers results in the 5 ppm spectroscopic resolution needed for single state atom excitation. The resolution is usually Doppler limited by the 0.1% atom beam velocity spread, although static field nonuniformities can also be troublesome. The FAB technique has one advantage in that large coils can surround the apparatus to produce a uniform magnetic field all along the atomic beam axis (8,13). This field generates a uniform electric field in the atom's rest frame, even within microwave waveguide structures, as long as the latter is made of nonmagnetic materials. Such a motionally produced static electric field is shown by the arrow marked F_s in the experimental schematic shown in Figure 1. Atoms with quantum numbers (n_0,0,0)

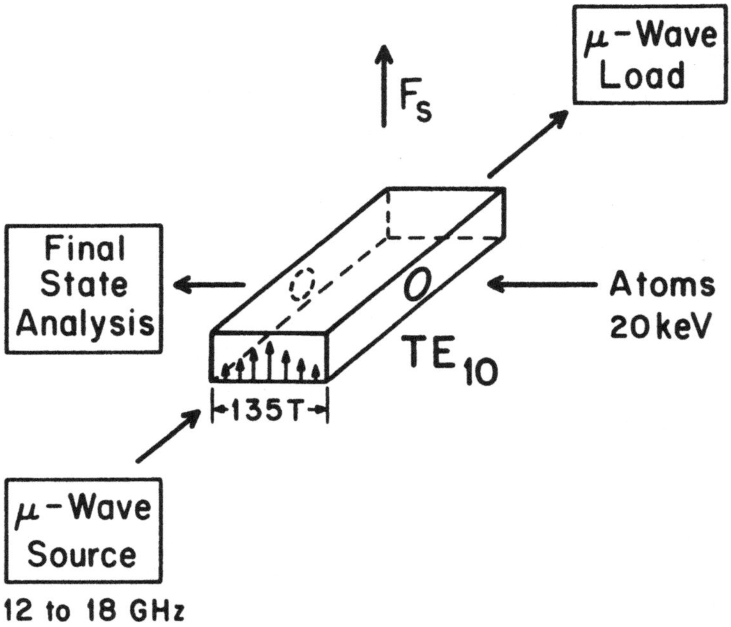

Figure 1. Schematic for recent one dimensional experiments exposing state selected highly excited hydrogen atoms to a short microwave pulse(5).

with reference to the direction of F_s that were produced by FAB ODR come from the right to pass through holes in the narrow sides of a rectangular TE_{10} waveguide. The microwave pulse seen by the atoms because of the passage has a pulse time determined by the atom velocity, with a pulse time spread of at least a few microwave cycles being set by transverse microwave field nonuniformities generated by the holes. Although this microwave interaction configuration has no uniform field region, it does produce the shortest practical pulse length. As quantum numerical calculations indicate that the central

1/3 of the pulse produces the bulk of the microwave photon absorption, the details of the early rise and late fall of the pulse, where the effects of the holes are greatest, are not very important for ionization measurements (14). The details do affect some early mixing of the initial state with nearby states. This can explain some deviations between experiment and theory for the bound state distribution in the beam after the pulse is over.

The final bound state distribution can be measured only in one dimensional experiments, using the state resolution for different (n,0,0) states provided by static electric field ionization (8,13). This has provided valuable details that can be stringent tests of theoretical predictions (15,16).

In actual experiments, static fields of one kind or another that are usually outside the microwave interaction region result in a definition of ionization that includes excitation to bound states with principle quantum number above a cutoff value n_c. The same definition must be used in calculations of quantities to be compared with the experimental data.

In one dimensional experiments the static field must be monodirectional and nonzero from the beginning of the ODR process through the microwave interaction and up to the point of final state analysis. Otherwise states with n_1 and/or m nonzero can be coupled in by nonadiabatic response of the atom dipole moment to the field changes. In some 2-D and 3-D experiments this state mixing mechanism is introduced deliberately to produce desired initial substate distributions (5,6).

In the 1-D experiments carried out to date, a sizable static field has been maintained in order to avoid field change state mixing. This produces several understood effects that can be handled in theoretical calculations, including shifts in classical and quantum resonance frequencies and a lowering of the threshold microwave field for classical chaos. However, a price to be paid for 1-D experiments has been the complication of quantum tunneling ionization of states with n above some number n_t, superimposed upon the effects of the microwaves in the interaction region. This effect has been difficult to include properly in quantum calculations, and has been modeled by the existence of a sink for state population flow past n_t. If n_t is large and well above n_c, the details of the sink do not appear very important as long as population is removed during the numerical time evolution on a time scale comparable to the classical orbit time for the state at n_t (5,14).

Very recently we have conducted 2-D ionization experiments for classically scaled microwave frequencies $\omega_0=n_0^3\omega$ between 0.5 and 2.7, using the short pulse configuration of Figure 1 (5). The static field was 0.87 V/cm. Figure 2a shows the classically scaled microwave field strength $F_0=n_0^4F$ for 10% ionization probability. The quantity ω_0 was varied both by varying ω between 12 and 18 GHz and by varying n_0 between 64 and 98. Values of n_c and n_t were measured for each data point, with $n_t = 165$ and n_c typically near 1.5 n_0. The dashed line is hand drawn through the data. As we shall see, this data contains a number of features that are explained by different aspects of the dynamics of the atom in microwaves system.

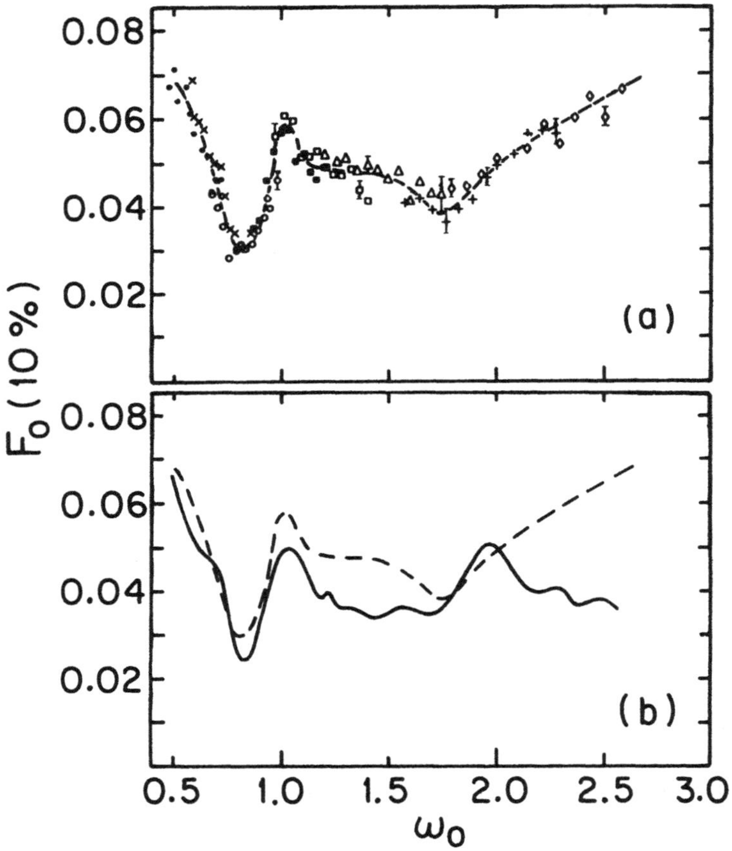

Figure 2. (a) Experimental data for the classically scaled microwave field strength for 10% ionization probability, as a function of the classically scaled microwave frequency (5). The different symbols for the data points correspond to different choices of initial atom principal quantum number. The scale for F_0 is experimentally uncertain by ± 10%. (b) A comparison of the experimental results of (a) with a smooth curve drawn through points obtained from classical one-dimensional numerical calculations (5).

EXPERIMENTAL TESTS OF CLASSICAL MODELS

In the classical limit, for fixed values of ω_0 and F_0 there is a threshold value of n_0 above which many electron trajectories are chaotic rather than all being regular (17,8). Alternatively, for fixed ω_0 and n_0, there is a threshold value of F_0 for chaos.

For an experimental initial condition of a quantum state of definite n_0, the corresponding ensemble of classical electron trajectories is definite in initial action $I_0=\hbar n_0$ and uniform in the conjugate angle variable. For F above the chaos threshold value, the time evolution of the chaotic portion of the ensemble involves a collection of random appearing walks within a portion of stroboscopically reduced phase space that adds up to a diffusion process in classical action space described by the Fokker-Planck equation (18). This diffusion must develop for some time before a given value for the ionization probability can be reached. In short pulse experiments, there may be insufficient time for all chaotic trajectories to contribute to ionization, that is, for all the available probability to reach the action corresponding to the cutoff quantum number n_c. In this case, experimentally defined ionization threshold fields will be larger than the threshold fields for classical chaos (19).

As long as theoretical definitions of quantities such as n_t, n_c, F_s, pulse shape and pulse time averaging are the same as experimental ones, classical numerical values for ionization threshold fields can be compared with experiment, as is done in Figure 2b for the data of Figure 2a.

At frequencies ω_0 below one, there is agreement between the 1-D classical and experimental curves. This reconfirms similar conclusions for earlier longer pulse data (3,6). At ω_0=1 the data displays a pronounced maximum that clearly is due to a classical resonance. Other such resonances have been seen at lower frequencies previously (2,6). However, the present data do not display the classical resonance at ω_0=2, where something else appears to be going on. In addition, away from the two classical resonances the classical prediction at ω_0>1 is a nearly constant value near 0.04, well below that observed. This confirms other recent experimental findings that the ionization is not classical at the higher frequencies, although large variations in the data precluded a conclusion on the reason for this (20).

Recently we have carried out 1-D experiments near ω_0=1 that determine final bound state distributions (16). On the basis of Figure 2b, one would expect that such distributions would be near-classical and would display effects arising from the ω_0=1 classical resonance. This is indeed the case, as can be seen in Figure 3. What is observed is a roughly exponentially decreasing distribution for n larger than n_0, with added maxima near n_0 and most prominently near n=66. As ω_0 is lowered to 0.74, the n=66 maximum disappears and another major maximum near n=80 appears instead. These maxima are believed associated with the ω_0=1 classical resonance, first because they are reproduced by classical numerical calculations, second because the ω_0=1 resonance island of regular trajectories in phase space is found numerically to extend from n=66 to n=80 (21) and third because the maxima become more pronounced as the field and hence the ionization probability is reduced to small values. Additionally, recent work with microwave noise instead of sinewave microwaves shows a final state distribution that uniformly fills the resonance island region of quantum numbers before dropping off for values above that region (22). The reasons for nonuniform filling of the island region in the sinewave case remain to be investigated.

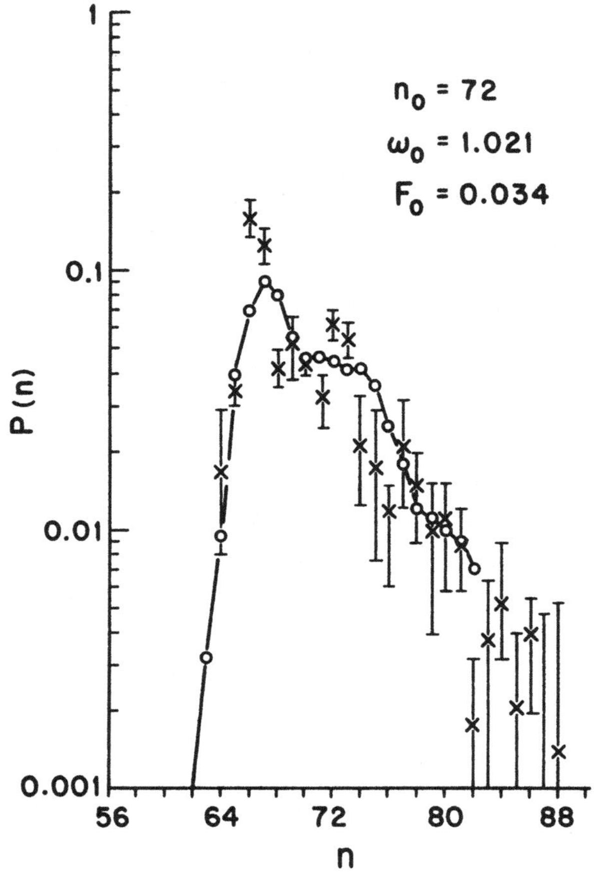

Figure 3. A comparison of experimental and classical numerical final bound state quantum number distributions for n_0=72 obtained using the apparatus of Figure 1. A static field strength of 6.25 V/cm was employed, producing a considerable reduction in the threshold microwave field for classical chaos of Figure 2a and a quantum tunneling cutoff value n_t=100.

QUANTUM MODIFICATIONS OF NEAR-CLASSICAL BEHAVIOR

The classical theory for the chaotic portion of the trajectory ensemble predicts diffusion in action space leading to ionization (18) and a roughly exponential bound state distribution for that part of the diffusing population density that doesn't reach the quantum number value defining ionization (18). Classically the characteristic length L of the distribution grows for a long time. However, quantum mechanically the growth in L is found to stop after a fairly short time, presumably because of destructive wave interference in the probability amplitudes for still larger values of n (24). In the language of quantum particle localization theory, the electron is said to be weakly localized in classical action or quantum number space. This is expected because of a mathematical analogy between the properties of the quantum kicked rotor and the Anderson problem of a quantum particle in a lattice with random site potentials (25). The kicked rotor and the sinusoidally driven bound electron are believed to be classically in the same class of driven nonlinear oscillator systems, and hence should exhibit the same qualitative behavior. There is no reason for their semiclassical quantum dynamics to not be similarly related. We take this line of reasoning seriously because of numerical verification (23,24), and now assume that the stopping time is reached. If the final bound state distribution is defined by the formula

$$P(n)=P(0)\exp(-2(n-n_0)/(L), \tag{1}$$

then L is approximately given by (23,24)

$$L=3.33\, n_0^2 F_0^2 \omega_0^{-7/3}, \tag{2}$$

which can be written as

$$L F_0^2 \omega_0^{-7/3} = 3.33 \left[F_0 n_0^{1/2} \omega_0^{-7/6}\right]^4. \tag{3}$$

This dependence is plotted as the solid line in Figure 4, along with numerical quantum values (23) and experimental values (16).

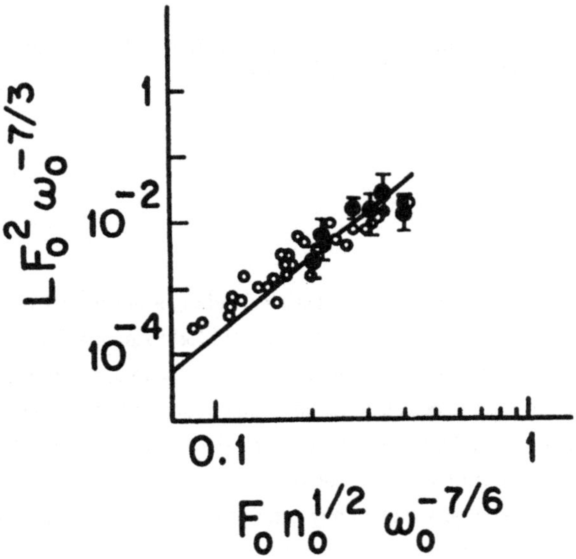

Figure 4. Experimental and numerical verification of the dependence of the characteristic exponent L on the parameters as given by equation (3). The experimental dots are from (16) and the quantum numerical circles from (23).

The quantum values are for quantum localization lengths, which are those values of L that are time-independent at long microwave pulse times (23). The agreement with experiment suggests that the experimental pulse time was long enough for the time-independence, while the agreement with classical results seen in Figure 3 suggest that the pulse time was just long enough. Numerical calculations verify this (14,19). Thus the effects of quantum localization just below and at ω_0 =l are small for the present experimental conditions, in accord with Figure 2b.

The consequences of quantum localization for the data of Figure 2a can be estimated by requiring the the amount of probability in the final state distribution above the cutoff value n_c be equal to the ionization probability (here 10%), and by assuming that the distribution is given by equations (1) and (2). The result is the solid curve of Figure 5, which is in agreement with the experimental results for ω_0 larger than two.

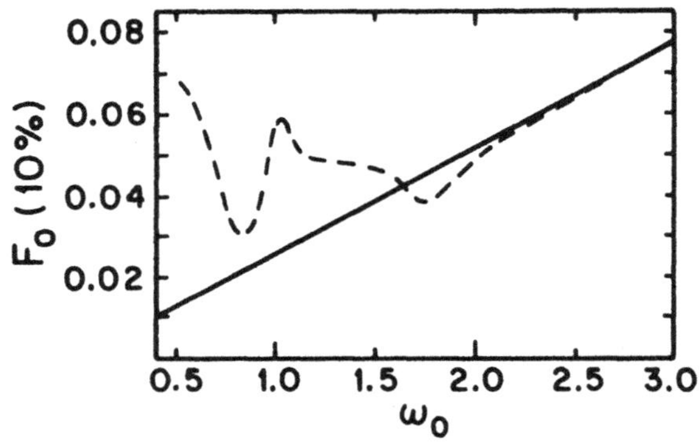

Figure 5. A comparison of the experimental microwave ionization curve of Figure 2a (dashed curve) with the predictions of a weak localization model (solid curve).

CONCLUSIONS

At ionizing values of microwave field strength, classical behavior plays a distinctive and major role in the excited hydrogen in microwaves system. The threshold field strength for chaos in the classical limit is closely related to threshold fields for ionization. The diffusion in action of the electron generated by the chaos in the classical limit produces characteristic final bound state distributions. One indirectly observed feature of the randomness in the chaos in the classical limit is quantum localization. However, the quantum system may only be mimicing the classical behavior in an approximate way. Thus the possibility of a direct experimental observation of any genuine randomness or deterministic chaos in any quantum system remains an open question.

The author thanks the U.S. National Science Foundation for continued support of the Pittsburgh laboratory.

REFERENCES

1. J. E. Bayfield and P. M. Koch, Phys. Rev. Lett. 33, 258 (1974).
2. J. E. Bayfield, L. D. Gardner and P. M. Koch, Phys. Rev. Lett. 39, 76 (1977).
3. J. G. Leopold and I. C. Percival, J. Phys. B 12, 709 (1979).
4. B. I. Meerson, E. A. Oaks and P. V. Sasorov, Sov. Phys.-JETP Lett. 51, 1259 (1979).
5. J. E. Bayfield, G. Casati, I. Guarneri and D. W. Sokol, submitted for publication, Nov. 1988.
6. K. A. H. van Leuwen, G. V. Oppen, S. S. Ewsnick, J. B. Bowlin, P. M. Koch, R. V. Jensen, O. Rath, D. Richards and J. G. Leopold, Phys. Rev. Lett. 55, 2231 (1985).
7. G. Casati, B. V. Chirikov, I. Guarneri and D. L. Shepelyansky, Phys. Rev. Lett. 59, 2927 (1987).
8. J. E. Bayfield, in Quantum Measurement and Chaos, edited by E. R. Pike and S. Sarkar (Plenum Press, New York, 1987), pp 1-33.
9. T. F. Gallagher, Phys. Rev. Lett. 61, 2304 (1988).
10. R. Blumel, R. Graham, L. Sirko, U. Smilansky, H. Walther and K. Yamada, Phys. Rev. Lett. 62, 341 (1989)
11. P. M. Koch, in Ryderg States of Atoms and Molecules, R. F. Stebbings and F. B. Dunning, editors (Cambridge University Press, New York, 1982).
12. J. E. Bayfield, Rev. Sci. Instrum. 47, 1450 (1976).
13. J. E. Bayfield and L. A. Pinnaduwage, Phys. Rev. Lett. 54, 313 (1985).
14. G. Casati, Private Communications.
15. J. N. Bardsley, B. Sundaram, L. A. Pinnaduwage and J. E. Bayfield, Phys. Rev. Lett. 56, 1007 (1986).
16. J. E. Bayfield and D. W. Sokol, Phys. Rev. Lett. 61, 2007 (1988).
17. R. V. Jensen, Phys. Rev. A 30, 386 (1984).
18. N. B. Delone, V. P. Krainov and D. L. Shepelyansky, Usp. Fiz. Nauk 140, 335 (1983) (Sov. Phys. Usp. 26, 551 (1983).
19. G. Brivio, G. Casati, I. Guarneri and L. Perotti, Physica D 33, 51 (1988).
20. E. J. Galvez, B. E. Sauer, L. Moorman, P. M. Koch and D. Richards, Phys. Rev. Lett. 61, 2011 (1988).
21. B. Sundaram, Private Communications.
22. J. E. Bayfield and D. W. Sokol, unpublished.
23. G. Casati, B. V. Chirikov, D. L. Shepelyansky and I. Guarneri, Phys. Reports 154, 77 (1987).
24. G. Casati, I. Guarneri and D. L. Shepelyansky, IEEE J. Quant. Electron. 24, 1420 (1988).
25. D. R. Grempel, R. E. Prange and S. Fishman, Phys. Rev. A 29, 1639 (1984).

LOCALIZATION PROPERTIES OF RYDBERG ATOMS IN THE PRESENCE OF NOISY MICROWAVE FIELDS

R. Blümel[a], R. Graham[b], L. Sirko[a,c],
U. Smilansky[d], H. Walther[a,e] and K. Yamada[a,f]

[a] Max-Planck-Institut für Quantenoptik, D-8046 Garching, FRG
[b] Fachbereich Physik, Universität Essen, D-4300 Essen, FRG
[c] Institute of Physics, Polish Academy of Sciences,
Al. Lotnikow 32/46, 02-668 Warszawa, Poland
[d] The Weizmann Institute of Science, 76100 Rehovot, Israel
[e] Sektion Physik, Universität München, D-8046 Garching, FRG
[f] Dept. of Physics, Keio Univ., 223 Yokohama, Japan

I. ABSTRACT

We present experimental and theoretical results on highly excited Rb-Rydberg atoms passing through a wave guide. The wave guide field consists of a coherent microwave field and a controlled component of technically generated colored noise. The presence of the noise field influences the localization properties of the Rydberg atoms and we show that the dynamics of Rydberg atoms subjected to a mixture of coherent and noisy fields can be classified into four dynamical regimes: (i) an initial classical diffusive regime, in which the initially prepared pure Rydberg state quickly broadens, (ii) a subsequent coherent localized regime, (iii) a transition, induced by the noise, in which coherence and localization are destroyed, and (iv), relaxation to equilibrium. The existence of the four dynamical regimes could be demonstrated in a clean atomic beam experiment in which excitation-, interaction- and analyzing regions are well separated from each other. The microwave interaction time is controlled by irradiating the Rydberg atoms with electronically shaped microwave pulses.

II. INTRODUCTION

Periodically driven non-linear dynamical systems may behave chaotically, in a classical description, giving rise to diffusion of an appropriately chosen action variable[1]. A simple model system is provided by the standard map[2]; a physical example is a Rydberg atom in a strong, coherent microwave field[3–7]. In a quantum description, these systems mimic the classical dynamics only over a short initial time-interval, after which quantum mechanical interference effects change the dynamics[8], leading in some cases to localization with respect to the action variable[7,9,10]. For the kicked rotor, this localization effect may occur for weak coupling as a perturbative effect[11], or, for strong coupling, it may be interpreted as a form of Anderson localization[9]. In both cases it crucially depends on the coherence of the wave function.

The presence of a small stochastic contribution to the kicking force of the rotor was shown to be sufficient to destroy localization and to lead back to diffusion on long time scales[12]. Recently the effects of dissipation and fluctuations on the localization of the angular momentum of the kicked rotor have been investigated[13,14]. It was shown that the coherence of the initial state and the accompanying localization are destroyed on a time scale inversely proportional to the dissipation rate or noise-intensity. For longer times it was found that the system is governed by noise driven diffusion.

Most of the material presented in this report has already been discussed in Ref. 15, where it is shown both theoretically and experimentally that the main features of the interplay between noise and coherence, derived from the study of the kicked rotor, also hold for the intensively studied case of Rydberg atoms in strong microwave fields. The present paper represents a more detailed description of our experimental setup with which we study the influence of noise on coherent quantum effects. So far, the influence of a stochastic component could only be demonstrated through the change of the ionization probability of Rydberg atoms[16]. The particular setup of the experiment in ref. 16 only allows to study a fixed exposure time of the atoms to the microwave field. Therefore, dynamical transitions between different localization regimes were not observed.

In the present paper we present a complete theory of the interaction of Rydberg atoms with a superposition of a strong coherent microwave field and a weak noise field. We also report on experiments which allow one to study the response of the Rydberg atoms to noisy microwave fields for variable exposure times. The experimental findings are in accord with the theoretical predictions.

III. THEORY

We consider Rydberg atoms traversing a wave guide which is excited by a superposition of a coherent signal in its TE_{01}-mode and a noisy signal. We assume that the Rydberg atoms, before they enter the wave guide in the x-direction (i.e. parallel to the electric microwave field), have been prepared in a parabolic substate in the x-direction such that a 1-dimensional approximation to the atomic dynamics is permissible[5,6]. The Hamiltonian then takes the form:

$$H = H_0 + H_{int} + H_R \tag{1}$$

with

$$\begin{aligned} H_0 &= \frac{p^2}{2m} - \frac{e^2}{4\pi\epsilon_0} \cdot \frac{1}{x} + e\,x\,F \sin(\omega t) \quad (x > 0) \\ H_{int} &= \hbar \sum_i g(\omega_i)\; x\; (b_i + b_i^+) \\ H_R &= \hbar \sum_i \omega_i\; (b_i^+ b_i + \frac{1}{2}) \end{aligned} \tag{2}$$

Here H_0 describes the 1-dimensional model of a hydrogen atom coupled to the electric microwave field with amplitude F and frequency ω. The coupling of the atom to the quantized modes of the wave guide is described by H_{int}. The coupling of the wave guide modes to the atom is given by the strength function $g(\omega_i)$. In the limit of infinite length of the wave guide, the sum over i is replaced by an integral over the density of modes of the wave guide. The field of the free wave guide is described by the Hamiltonian H_R. The operators b_i^+, b_i are the usual Boson creation and annihilation operators. The population of wave guide modes is determined by the mechanism that generates the noisy component of the field. In the present formalism we can treat any given noise distribution, such as *e.g.* electronic shot noise which is amplified and fed to the wave guide power supply[16], or thermally induced black body radiation in the wave guide. For the theoretical discussion we consider the latter source, and take the temperature T as a measure for the ratio between the coherent and the stochastic components of the driving field.

In order to proceed with the Hamiltonian (1), we introduce the interaction picture with respect to H_{int}, treating $H_0 + H_R$ as the unperturbed Hamiltonian. In the following we use the Floquet-basis $|\alpha>$ which exactly diagonalizes the atomic one-cycle propagator $U_0(\tau, 0)$:

$$U_0(\tau, 0)|\alpha> = e^{-i\mu_\alpha \tau}\, |\alpha> \tag{3}$$

$U_0(t',t)$ is the unitary time-evolution operator from t to t', generated by H_0, and $\tau = 2\pi/\omega$ is the microwave period. The μ_α are the quasi energies, which are chosen to lie in the interval $0 \leq \mu_\alpha < \omega$. H_{int} is now treated perturbatively applying the Born-Markoff approximation in a standard way[17)] to obtain a master equation for the reduced statistical operator of the hydrogen atom in the coherent microwave field. We emphasize that in the present approach the microwave field is treated exactly. Details of the derivation will be presented elsewhere.[18)]

In the interaction picture and basis, the master equation for the statistical operator $\tilde{\rho}$ reads:

$$< \alpha|\dot{\tilde{\rho}}|\alpha > \; = \; \sum_\beta \left(M_{\beta\alpha} < \beta|\tilde{\rho}|\beta > \; - \; M_{\alpha\beta} < \alpha|\tilde{\rho}|\alpha >\right) \tag{4a}$$

$$< \alpha|\dot{\tilde{\rho}}|\beta > \; = \; -\frac{1}{2}\sum_\nu \left(M_{\alpha\nu} + M_{\beta\nu}\right) < \alpha|\tilde{\rho}|\beta >, \quad \alpha \neq \beta \tag{4b}$$

The rate coefficients $M_{\alpha\beta}$ in (4) depend on the quasi-energy differences $\Omega_{\alpha\beta k} = |\Delta_{\alpha\beta k}|$, $\Delta_{\alpha\beta k} = \mu_\alpha - \mu_\beta + k\omega$, the mode-density $\bar{\rho}(\omega)$ and the coupling function $g(\omega)$ of the wave guide, and the Fourier components of the dipole matrix elements

$$x_{\alpha\beta k} = \frac{\omega}{2\pi}\int_0^{2\pi/\omega} e^{-i\,\Delta_{\alpha\beta k}\,t} < \alpha|U_0^+(t,0)\; x\; U_0(t,0)\;|\beta > dt \tag{5}$$

Explicitly:

$$M_{\alpha\beta} = \sum_{k=-\infty}^{\infty} \left[\gamma_{\alpha\beta k} + (\gamma_{\alpha\beta k} + \gamma_{\beta\alpha k})\cdot n_{\alpha\beta k}\right] \tag{6}$$

with

$$\begin{aligned} \gamma_{\alpha\beta k} &= 2\pi\,\bar{\rho}(\Omega_{\alpha\beta k})\,g^2(\Omega_{\alpha\beta k})\,|x_{\alpha\beta k}|^2\,\Theta(\Delta_{\alpha\beta k}) \\ n_{\alpha\beta k} &= \left[\exp(\frac{\hbar\Omega_{\alpha\beta k}}{k_B T}) - 1\right]^{-1} \end{aligned} \tag{7}$$

where Θ is the step function and k_B is the Boltzmann constant. The solution of (4) is given by single decaying exponentials for the coherences $< \alpha|\tilde{\rho}|\beta >$; the occupation probabilities $< \alpha|\tilde{\rho}|\alpha >$ of the quasi energy states, on the other hand, relax to a steady state as a superposition of exponentials. Returning to the Schrödinger representation, the statistical operator ρ at integral multiples $N\cdot\tau$ of the microwave period is given by:

$$< \alpha|\rho(N\tau)|\beta > \; = \; e^{-i(\mu_\alpha - \mu_\beta)N\tau} < \alpha|\tilde{\rho}(N\tau)|\beta > \tag{8}$$

In order to obtain some physical insight, we have solved (3)-(6) numerically for an ideal wave guide of rectangular cross section of side lengths $a = 1.9$ cm, $b = 0.95$ cm, and temperatures $T = 4$ K, 300 K, coherently excited in the TE_{01}-mode with amplitude $F = 4$ V/cm, 8 V/cm and frequency $\omega = 2\pi \cdot 10.6$ GHz. The initial state was chosen as the eigenstate of the 1-dimensional hydrogen model with principal quantum number $n_0 = 71$. Twelve atomic bound states, ranging from $n = 69$ to $n = 80$, were taken into account as a basis for the expansion of the Floquet states. This way, the final results

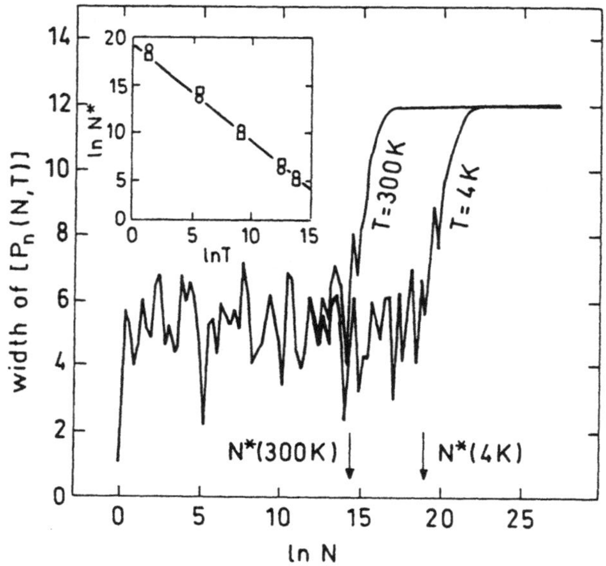

Fig. 1: *Width of the $\{P_n\}$-distribution as a function of time for two different temperatures and $F = 8$ V/cm. The inset shows the relation between temperature and break-time N^* for two different field strengths (squares: $F = 4$ V/cm; circles: $F = 8$ V/cm). The straight line interpolating the data was drawn to guide the eye, and reflects the relation $N^* \cdot T = const.$*

could be expressed in the basis of bound states of the hydrogen model. In Fig. 1 we plot the width function[10] $W(N,T) = \exp\{-\sum_n P_n(N,T)\ln P_n(N,T)\}$ where $P_n(N,T)$ is the probability to find the principal quantum number state $|n>$ in ρ at time $N\tau$ and temperature T. The width function W is a convenient measure for the number of hydrogen bound states contained in the ensemble ρ. Since initially only a single state is occupied, we have $W(0,T) = 1$. Fig. 1 actually consists of 100 snapshots of the width-function W at discrete times $N_j \cdot \tau$, $N_j = [10^{0.12 \cdot j}]$, $j = 1,...,100$, the square brackets denoting the integer part. Four regimes of the dynamics can be clearly distinguished. In the first regime, occurring for very short times, the width W rapidly increases from 1 to a value between 5 and 6 (not resolved in Fig. 1). During this initial stage, quantum effects due to the discreteness of the quasi energies are not yet relevant. In the second phase, W saturates and oscillates quasi-periodically around some finite avarage. The fact that W is bounded by a number which is significantly smaller than the total basis size, is the signature of localization. A transition which is sharp on the logarithmic scale of Fig. 1 follows the localization regime, in which nearly simultaneously the oscillations disappear and W increases abruptly, implying that coherence and localization are destroyed together at the transition time $N^*(T)$. A comparison of the two curves for $T = 4$ K and $T = 300$ K in Fig. 1 shows that the oscillations in the localized regime of the dynamics exactly reproduce each other i.e. they are independent of temperature (noise power), but that the critical time N^* increases for decreasing temperature. In fact, the simple relation $N^* \cdot T = const.$ is obeyed by our numerical results over a vast range of temperature, which is shown in the inset of Fig. 1. For $T \gg 300$ K, the "temperature" merely serves as a convenient parametrization of the noise power as was explained above. The relation $N^* \cdot T = const.$ indeed follows from (6) and (7) in the limit $n_{\alpha\beta k} \sim k_B T/\hbar\Omega_{\alpha\beta k} \gg 1$, where $M_{\alpha\beta k} \sim T$ holds. This limit is realized in our numerical examples. The fourth and final dynamical regime in Fig. 1 is the monotonic increase of W until equidistribution over the whole basis at $W = 12$ is attained. This increase is much slower than the initial classical spreading of W, and it is due to noise- induced population transfer between mutually incoherent quasi energy states. This process is described by the master equation (4a) for the diagonal elements of the density matrix. Accordingly, the time scale in this regime must again scale inversely proportional to T, i.e. $W(N,T) = \tilde{W}(N \cdot T)$. This scaling property is clearly seen in Fig 1, where the final rise of W follows nearly the same curve for different temperatures if the abscissa is shifted by the respective values of $\ln T$.

Summarizing our theoretical results, we have derived and solved a master equation for Rydberg atoms interacting with a strong microwave field in a noisy wave guide. Our predictions are that noise induced destruction of coherence and of localization onsets at a critical interaction time N^*. It is followed by the redistribution of populations of quasi energy states which approaches equipartition after very long time. This noise induced diffusion is characterized by a time scale, which, like N^*, is inversely proportional to temperature or noise intensity. In the following we shall show some experimental evidence which supports these predictions.

III. EXPERIMENT

The excitation and ionization behavior of a given Rydberg state strongly perturbed by a microwave field is essentially controlled by three parameters: the microwave field strength, the microwave frequency and the interaction time. While in existing

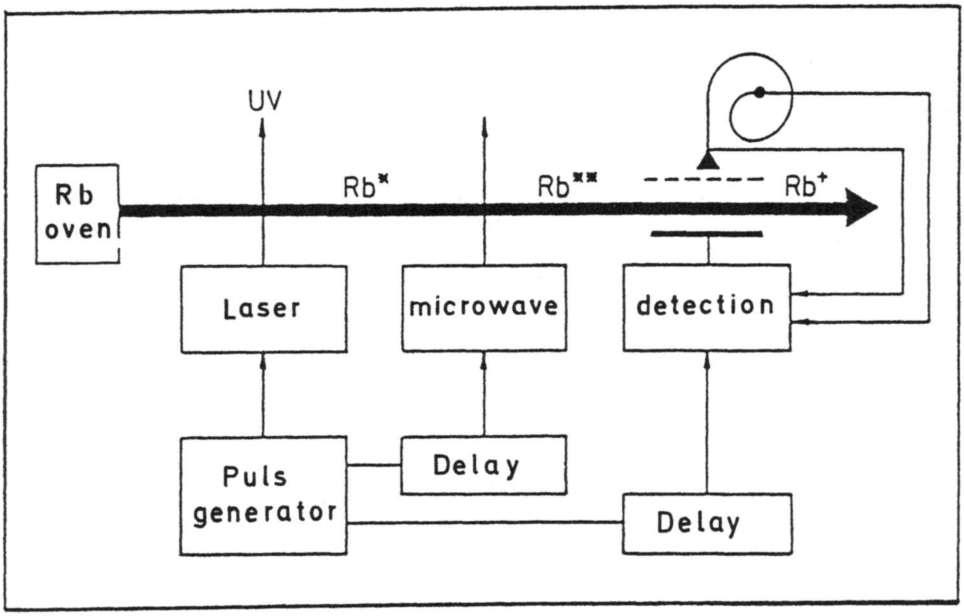

Fig. 2: *Sketch of the experimental setup*

experiments on quantum chaos in H-Rydberg atoms[16,19] at least one of the three parameters is hard to change (e.g. the interaction time), our experimental setup (see Fig.2) is "universal" in the sense that all three control parameters can be easily varied. Especially the variable interaction time in our experiments might help to clarify the notions of quantum localization and diffusion over Rydberg levels, since these phenomena can now be studied as a function of time. In contrast to the experiments of Bayfield[16] and Koch[19] who work with fast beams of highly excited H-atoms, our experiment uses a thermal beam of Rb-atoms, which can be laser excited to Rydberg states ranging from $n = 40$ to $n \approx 150$. Fig. 2 shows that our setup is electronically clean in the sense that laser-excitation, microwave interaction and analyzing regions are spatially well separated from each other. The Rb-atoms leave the oven as a thermal beam with a Maxwellian velocity distribution. A controlled microwave interaction time would, e.g., be possible by velocity selection and passing the atoms through a CW operated wave guide. A much cleaner experiment, however, which additionally avoids problems with stray fields, is possible in the following way: After leaving the oven, a group of atoms is laser excited by a short laser pulse (pulse duration $6\mu s$). The resulting spatially well defined group of atoms is now travelling into the microwave guide. Due to the velocity distribution, the packet is spreading, but the laser pulse is so short that during the travel time to the wave guide, the spread is smaller than the wave guide dimensions. As soon as the packet reaches the center of the wave guide, the atoms are irradiated with an electronically shaped microwave pulse whose duration, t, can be varied in steps of 10 microwave cycles from $t = 10$ ns to several tens of microseconds. The carrier frequency can be varied from 8 to 18 GHz. The microwave pulse induces n-changing transitions and ionization, depending on the duration, the carrier frequency and the field strength of the pulse. The packet of Rb-atoms, of which every single atom saw exactly the same pulse, is now travelling toward the field ionization region. Here, the atoms are ionized by an electric field ramp[20] ($1.8 \frac{\mathrm{V}}{\mathrm{cm}\cdot\mu s}$) and the electrons are detected with a channeltron multiplier. The exact timing of laser excitation, microwave interaction and field ionization is performed by a pulse generator and delay lines (see Fig. 2) such that independently of their particular velocity, the atoms all interact the same amount of time with the microwave field and the field ramp in the analyzing region.

In all experiments on microwave excitation the most difficult part is the gauge of the microwave field strength inside the interaction region. This might be done, e.g., by carefully monitoring the power flux through the wave guide and cavity structures assisted by a numerical calculation of the resonator modes. We choose a more direct

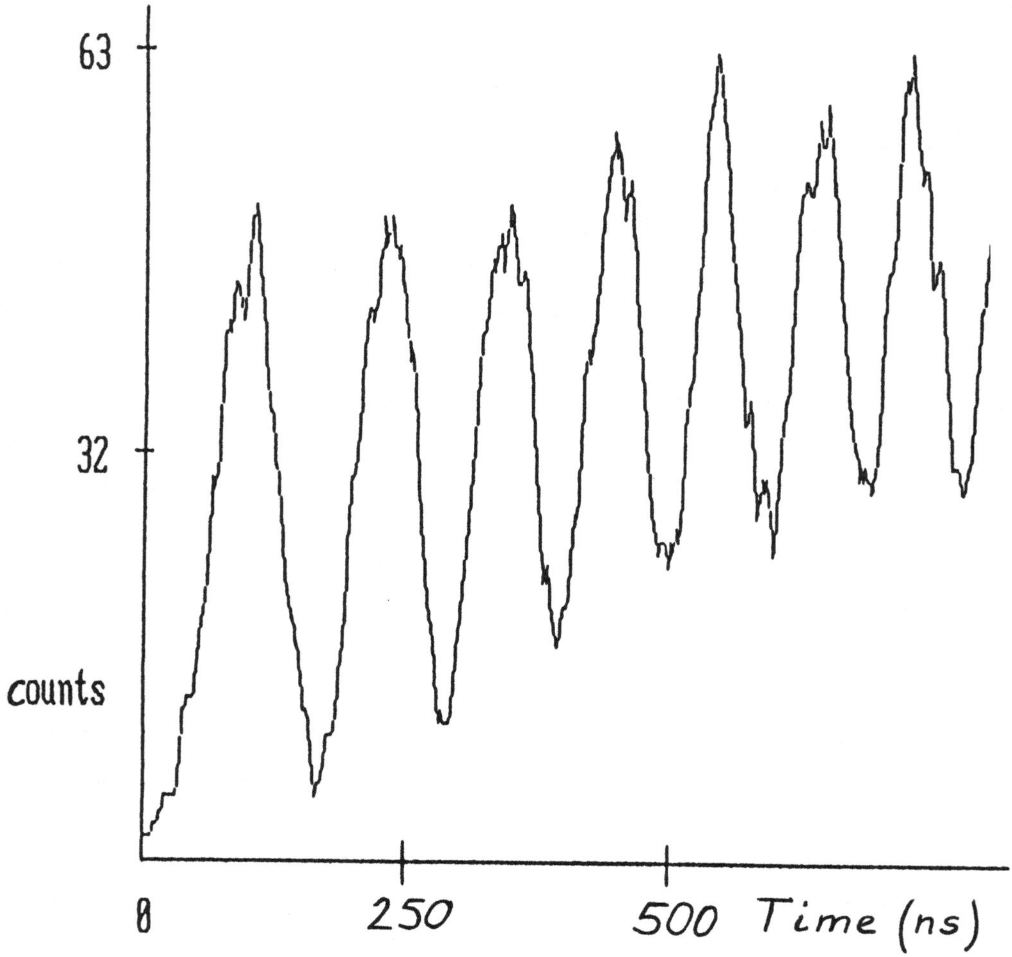

Fig. 3: *Rabi-flopping of the occupation probability between the states $^{67}P_{3/2}$ and $^{68}P_{3/2}$.*

and reliable procedure which determines the electric field strength inside our wave guide from direct observation of two-photon Rabi flopping between highly excited P-states, e.g., $67P_{3/2} \leftrightarrow 68P_{3/2}$ ($f_{1photon} = 12.05944$ GHz). Here we make use of the possibility of a variable interaction time and were able to measure the flopping pattern directly as a function of time (see Fig. 3). The field strength can now be evaluated from the known matrix elements[21)] and the Rabi frequency which is obtained simply by counting the number of Rabi cycles.

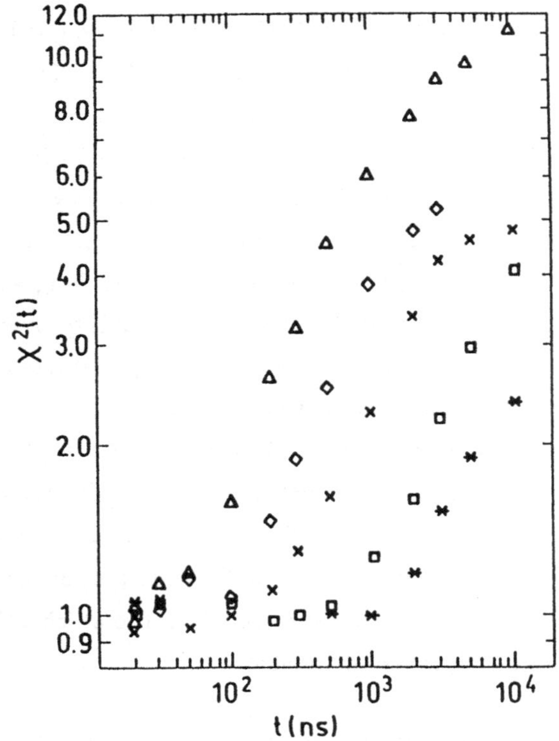

Fig. 4: *χ^2 - deviation (in arbitrary units) of the field ionization response functions $P^{(c)}(E;t)$ and $P^{(s)}(E,t)$ of Rb-atoms exposed to pulses of coherent and noisy microwaves as a function of the exposure time and several different noise powers. Triangles: 10 μW, diamonds: 4 μW, crosses: 1.6 μW, squares: .63 μW, stars: .25 μW. Signal power (= 3.2 mW) and noise-powers were measured before entering the wave guide.*

For the experimental investigation of the effect of noise on the localization properties of highly excited Rydberg atoms, we prepared the atoms in the the 84 $P_{3/2}$ state. Electronic shot noise (obtained from an idling travelling wave tube amplifier[16]) and ranging from 8 to 18 GHz) can be admixed to the coherent microwave pulses (variable time duration t, carrier frequency 9.654 GHz) in arbitrary ratios. In the field ionization region, the different Rydberg states are ionized at different times within the field ramp. Therefore $P(E;t)$ – the ionization probability at ramp field E of atoms having interacted a time t with the microwave field – will be a unique function of the final n-state distribution of the Rydberg atoms. If we denote by $P^{(c)}(E;t)$ the response function of atoms which have interacted with the coherent microwave alone, and by $P^{(s)}(E;t)$ the response to a superposition of coherent signal and shot noise, then

$\chi^2(t) \sim \int \left\{(P^{(s)}(E;t) - P^{(c)}(E;t))/\sqrt{P^{(s)}(E;t)}\right\}^2 dE$ is a good measure for the effect of the noise. Fig. 4 shows the $\chi^2(t)$ - function for five different noise powers as a function of the exposure time t. The fast (classical) diffusive broadening of the initially prepared pure Rydberg state (regime (i)) occurs on a scale of a few field cycles ($t \ll 10$ ns) and is not shown in Fig. 4. Classical diffusion stops abruptly and results in the localization plateau (region (ii)) which is followed by a rise of $\chi^2(t)$ (regime (iii)) where the noise takes over and a diffusive behavior prevails. For very long exposure time, $\chi^2(t)$ saturates which corresponds to equidistribution of probability over the atomic states (regime (iv)). The critical times t^* can be unambiguously extracted from Fig. 4. They are inversely proportional to the noise power (exponent: -1.07) which is in excellent agreement with the theoretical predictions. Moreover, for $t > t^*$, the slopes of the χ^2 - curves in regime (iii) are nearly parallel which is consistent with the claim that the time scale of redistribution of population in the diffusive regime is inversely proportional to the noise power. As a result, the existence as well as the proper scaling of a transition from the localization regime (short exposure time) to the stochastic diffusion regime (long exposure time) is quantitatively demonstrated.

One of us (R. G.) wishes to acknowledge support by the Deutsche Forschungsgemeinschaft through Sonderforschungsbereich 237. This work was supported in part by a grant from Stiftung Volkswagenwerk.

V. REFERENCES

1) A. J. Lichtenberg and M. A. Lieberman, *Regular and Stochastic Motion*, (Springer, Berlin, 1983).

2) B. V. Chirikov, Phys. Rep. **52**, 263 (1979).

3) J. G. Leopold and I. C. Percival J. Phys. **B12**, 709 (1979).

4) N. B. Delone, B. A. Zon and V. P. Krainov, Sov. Phys. JETP **48**, 223 (1978); B. I. Meerson, E. A. Oks and P. V. Sasarov, JETP Lett. **29**, 72 (1979).

5) R. V. Jensen, Phys. Rev. **A30**, 386 (1984).

6) D. L. Shepelyansky, in *Chaotic Behavior in Quantum Systems*, ed. by G. Casati, p. 187 (Plenum, New York, 1985).

7) G. Casati, B. V. Chirikov, D. L. Shepelyansky and I. Guarneri, Phys. Rep. **154, 77** (1987).

8) G. Casati, B. V. Chirikov, F. M. Izraelev and J. Ford, *Lecture Notes in Physics*, Vol. **93**, p. 334, (Springer, Berlin, 1979).

9) D. R. Grempel, R. E. Prange and S. Fishman, Phys. Rev. **A29**, 1639 (1984).

10) R. Blümel and U. Smilansky, Phys. Rev. Lett. **58**, 2531 (1987); R. Blümel, J. Goldberg and U. Smilansky, Z. Phys. **D9**, 95 (1988).

11) E. V. Shuryak, Zh. Eksp. Teor. Fiz. **71**, 2039 (1976).

12) E. Ott, T. M. Antonsen Jr. and J. D. Hanson, Phys. Rev. Lett. **53**, 2187 (1984).

13) S. Adachi, M. Toda and K. Ikeda, Phys. Rev. Lett. **61**, 655 (1988).

14) T. Dittrich and R. Graham, Europhys. Lett. **4**, 263 (1987), and Europhys. Lett. **7**, 287 (1988).

15) R. Blümel, R. Graham, L. Sirko, U. Smilansky, H. Walther and K. Yamada, Phys. Rev. Lett. 62, 341 (1989).

16) J. E. Bayfield and D. W. Sokol, in *Atomic Spectra and Collisions in External Fields II*, to be published in *Physics of Atoms and Molecules*, ed. by K. T. Taylor, M. H. Nayfeh and C. W. Clark, (Plenum, New York, 1988);

17) H. Haken, *Encyclopedia of Physics*, Vol. XXV/2c, ed. S. Flügge, (Springer, Berlin, 1970); W. H. Louisell, *Quantum Statistical Properties of Radiation*, (Wiley, London, 1973).

18) R. Blümel, A. Buchleitner, R. Graham, L. Sirko, U. Smilansky, H. Walther and K. Yamada, in preparation.

19) E. J. Galvez, B. E. Sauer, L. Mooreman, P. M. Koch, and D. Richards, Phys. Rev. Lett. **61**, 2011 (1988); L. Mooreman, E. J. Galvez, B. E. Sauer, A. Mortazawi-M., K. A. H. van Leeuwen, G. v. Oppen and P. M. Koch, Phys. Rev. Lett. **61**, 771 (1988).

20) J. A. C. Gallas, G. Leuchs, H. Walther and H. Figger, in *Advances in Atomic and Molecular Physics*, Vol. 20, p. 413, and references therein.

21) A. R. Edmonds, J. Picart, N. Tran Min and R. Pullen, J. Phys. **B12**, 2781 (1979).

Diamagnetism of the Hydrogen Atom in the Quasi-Landau-Regime

A. Holle, J. Main, G. Wiebusch, H. Rottke, and K.H. Welge
Fakultät für Physik, Universität Bielefeld
D-4800 Bielefeld 1
Federal Republic of Germany

INTRODUCTION

Since the discovery of quasi-Landau resonances by Garton and Tomkins [1] the physics of highly excited atoms in static homogenous magnetic fields has attracted much attention [2]. Substantial progress particularly with the hydrogen atom has been achieved in recent years by theoretical [3-11] and experimental [10-14] work. The interest in atomic diamagnetism arises from the fact that even for the hydrogen atom with its purely Coulombic field the Schrödinger equation is nonseparable in any coordinate system. In its simplest form, containing the Coulomb and diamagnetic term only, the Hamiltonian is given (in atomic units; magnetic field $\vec{B}$ in z-direction) by

$$H = \frac{1}{2} p^2 + \frac{1}{8} \gamma^2 \rho^2 - (\rho^2 + z^2)^{-1/2} \tag{1}$$

$$\gamma = B/(2.35 \cdot 10^5 \text{ Tesla})$$

This system has gained particular importance because of its classically chaotic dynamics that is where the Coulomb and diamagnetic interactions are of comparable strength [15]. Until recently it had been generally accepted that the dynamics and structure of highly excited magnetised atoms were determined solely by the Garton-Tomkins type quasi-Landau resonance. This resonance was explained first by Edmonds [16] as resulting from a two-dimensional bound motion of the electron on closed classical orbits in the (z = 0)-plane perpendicular to the magnetic field axis. However, the experiments with the hydrogen atom have revealed the existence of further, essentially new quasi-Landau resonance types correlated to three-dimensional closed orbits through the proton as origin [12, 13].

EXPERIMENTAL

Hydrogen atoms are excited at the center of magnetic field by tunable pulsed laser light (vuv, vacuum ultraviolet and uv, ultraviolet) in two steps,

$$H(1s) + h\nu(vuv) \rightarrow H(2p) + h\nu(uv) \rightarrow H^{*}$$

in a crossed atom-laser beam arrangement (Fig. 1). The beam intersection point is located between two flat, parallel, fine-mesh grid electrodes (7 mm apart) with their surfaces perpendicular to the magnetic field axis.

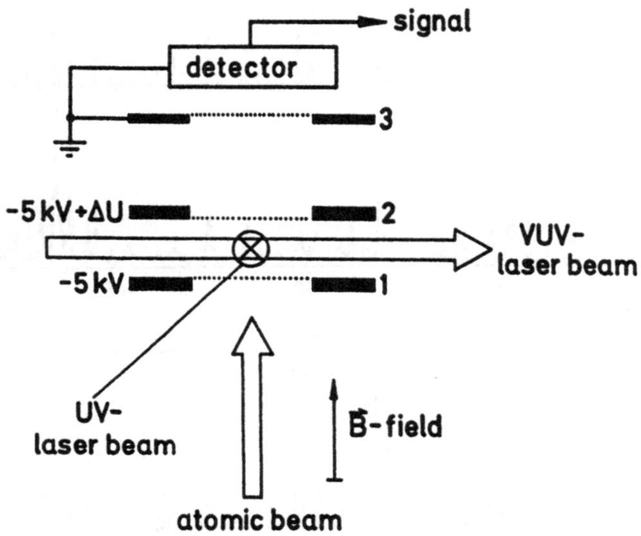

Fig. 1 Experimental crossed laser-atom beam set-up.

A surface-barrier diode located 30 mm behind the second electrode serves to monitor electrons from H^{*}-atoms ionized spontaneously or by the electric field between the second and third electrode. At the magnetic field strength employed ($2T \lesssim B \lesssim 6T$), the Lyman-$\alpha$ transition $(n = 1) \rightarrow (n = 2)$ is fully governed by the Paschen-Back effect. Three p-orbital type sublevels with magnetic quantum numbers $m = -1$, 0 or + 1 can be selectively excited by tunable laser light in the region of the Lyman-α wavelength (121.6 nm), linear polarized parallel or perpendicular to the field. From there final states with even parity and magnetic quantum numbers $|m| = 0$, 1, or 2 are

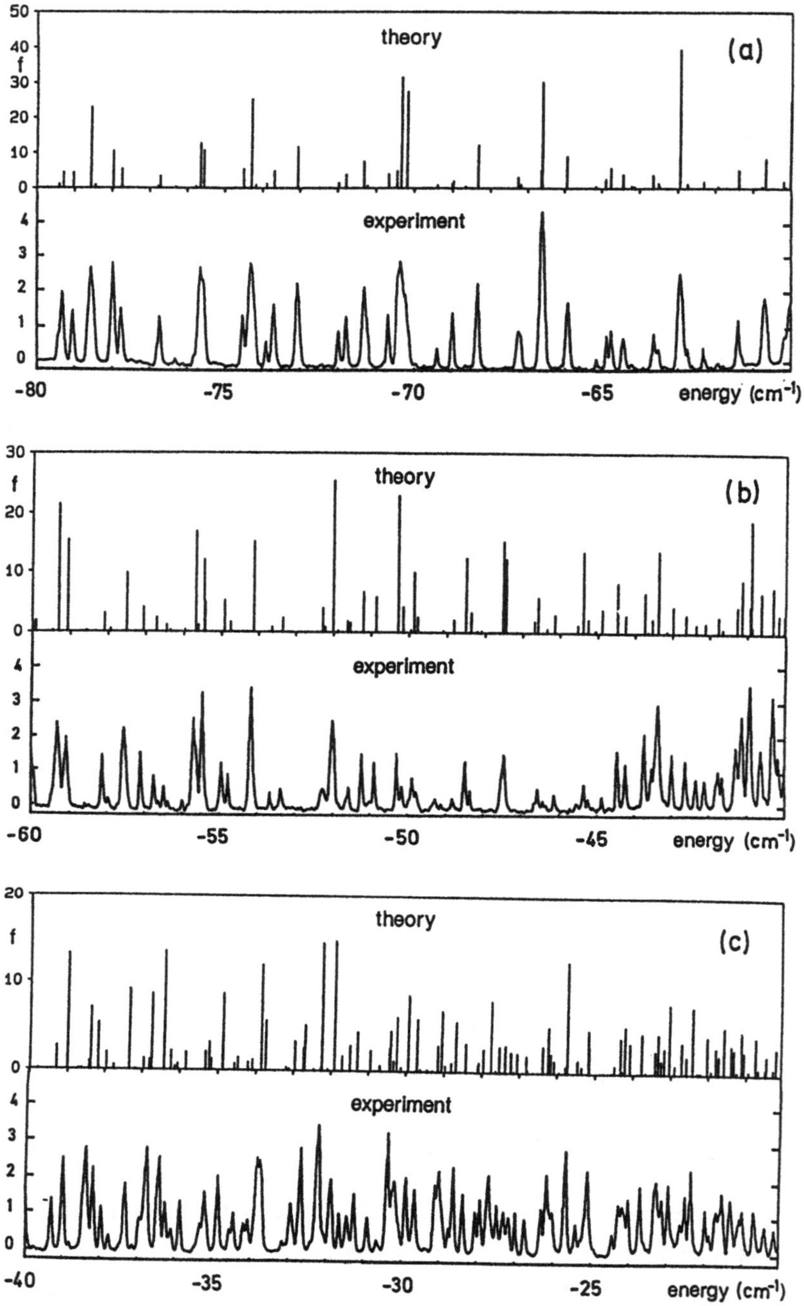

Fig. 2 Rydberg atoms in a magnetic field of 5.96 Tesla: comparison between the theoretical oscillator strength and the experimental photoabsorption spectrum for $\Delta m = 0$ Balmer transitions to $m = 0$, even parity final states. Oscillator strength in units of 10^{-6}; experimental signal in arbitrary units.

excited around the ionization limit by scanning the uv-laser around 365 nm.

RESULTS AT CONSTANT MAGNETIC FIELD STRENGTH

Fig. 2 shows a spectrum taken at a magnetic field strength B = 5.96 T and excited to final even parity states with magnetic quantum number m = 0. Also shown in Fig. 2 is a theoretical stick spectrum obtained by diagonalization of the Hamiltonian (eqn. 1) in a complete basis set [9, 11]. Within the precision and resolution limits of the experiment (1.5 GHz) the theoretical and experimental spectra are in excellent agreement for both line positions and intensities as may be verified by the reader by going through the figure line by line. We mention that the energy region from $- 80\ cm^{-1}$ to $- 20\ cm^{-1}$ shown here covers most of the transition region from complete regular to complete chaotic classical motion [9].

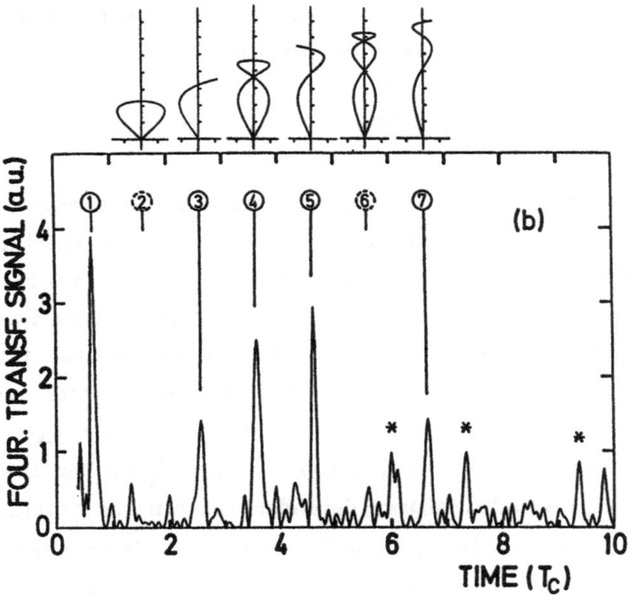

Fig. 3 Fourier-time spectrum at B = 5.96 T. Transformed is the experimental energy spectrum with m = 0, even parity final states from $-30\ cm^{-1}$ below to $+20\ cm^{-1}$ above the ionization threshold. Correlated to resonances two to seven calculated closed classical orbits of electron motion are shown in (ρ, z)-projection.

At the resolution achieved in this experiment the spectra have seemingly lost the oscillatory structures of quasi-Landau resonances discovered at lower resolution [12]. Periodic molulations in line density and/or oscillator strength can, however, be recovered by Fourier transformation in time-domain. Employing the semiclassical periodic orbit theory developed by Gutzwiller [17, 5] such energy spectrum oscillations can be rationalized by long-range periodic orbits of the electron motion, also in the classically chaotic regime. According to the general relation $\Delta E \cdot T = 2\pi\hbar$ periodic modulations (with spacing ΔE_i) in energy quantum spectra show up as resonances at time T_i in the Fourier-transform spectra, where i denotes the respective quasi-Landau resonance. T_i is envisaged as time period (or recurrence time) of the closed classical electron orbit [18]. On the basis of this theory a multitude of new quasi-Landau resonances correlated to closed classical orbits has been discovered [13]. Fig. 3 shows, as an example, a time spectrum obtained by Fourier-transformation of the energy spectrum (B = 5.96 T, magnetic quantum number m = 0) around the ionization limit. Also shown in Fig. 3 are calculated closed classical orbits correlated to resonances two to seven. The first resonance is the long known Garton-Tomkins resonance [1] with correlated periodic electron motion perpendicular to the magnetic field axis.

CONSTANT-SCALED-ENERGY SPECTROSCOPY

The discovery of a multitude of new quasi-Landau resonances with correlated closed classical orbits raises the question as to the "entire" manifold of quasi-Landau resonances resulting from final states with given parity and magnetic quantum number m as the system evolves from regularity in the ℓ/n - mixing regime to the chaotic regime around the ionization limit. Time spectra like in Fig. 3 resulting from Fourier-transformation of energy spectra taken at constant magnetic field strength can deliver resonances T_i only when the oscillation spacing ΔE_i in the energy spectrum is not, or at most weakly, energy dependent within the integration intervall. We have solved this problem by employing "constant-scaled-energy spectroscopy", which allows to observe also strongly energy dependent quasi-Landau resonances. The technique is based on the scaling property of the classical Hamiltonian with respect to the simultaneous dependence on the excitation energy (E) and the magnetic field strength (B). Following theoretical work [19] the classical Hamiltonian (eqn. 1) can be transformed to scaled variables. Using the scaling relations

$$\tilde{\vec{r}} = \gamma^{2/3}\,\vec{r}; \quad \tilde{\vec{p}} = \gamma^{-1/3}\,\vec{p} \tag{2}$$

the scaled Hamiltonian $\tilde{H} = \gamma^{-2/3}\,H$ is given by

$$\tilde{H} = \frac{1}{2}\tilde{\vec{p}}^2 + \frac{1}{8}\tilde{\rho}^2 - (\tilde{\rho}^2 + \tilde{z}^2)^{-1/2}. \tag{3}$$

In scaled form, the Hamiltonian does no more depend on E and B independently but on the scaled energy

$$\tilde{E} = E\gamma^{-2/3} \tag{4}$$

only. According to the scaling relations (eqn. 2) the semiclassical Bohr-Sommerfeld quantization condition for the two nonseparable coordinates ρ and z is transformed to

$$\frac{1}{2\pi}\oint_i (\tilde{p}_\rho d\tilde{\rho} + \tilde{p}_z d\tilde{z}) = n\gamma^{1/3} \equiv C_i \tag{5}$$

where i denotes a closed classical orbit and n is a positive integer quantum number.

For given orbit type i the scaled action integral (eqn. 5) depends on the scaled energy only. If $E\gamma^{-2/3}$ is kept constant $C_i = n\gamma^{1/3}$ describes a spectrum of *equidistant* lines on a scale $\gamma^{-1/3}$. The Fourier transformation of such a spectrum shows in the conjugate coordinate a resonance at the value C_i, which is correlated to the respective orbit type i. According to the scaling conditions spectra have been taken at constant scaled energy, that is the magnetic field strength was varied linearly on the scale $\gamma^{-1/3}$ while the excitation energy E was adjusted (via scanning uv laser wavelength) such that $\tilde{E} = E\gamma^{-2/3}$ = const was obeyed. Apart from this novel spectroscopic technique the experiments have been carried out as previously.

RESULTS AND DISCUSSION

Fig. 4a shows three examples of constant-scaled-energy spectra taken on a scale linear in $\gamma^{-1/3}$ and at scaled energies $\tilde{E} = -0.45$, -0.30, and -0.11. The ranges of E and B are indicated at the top of the figures. Also shown (Fig. 4b) are the respective Fourier-transform action spectra on $n\gamma^{1/3}$-scale. Classically, a resonance C_i in the action spectrum corresponds to one (or more) closed orbits. Some resonances are marked by V_μ^ν, R_μ^ν or X_μ with μ and ν being small integer values. This notation corresponds to Fig. 6 where individual closed classical orbits are shown in (ρ, z)-projection. The relation of this notation with the systematics of classical orbits is discussed later. Constant-scaled-energy spectra like in Fig. 4a have been measured over a range of $\tilde{E}$ from $\tilde{E} = -0.5$ to $\tilde{E} = +0.2$.

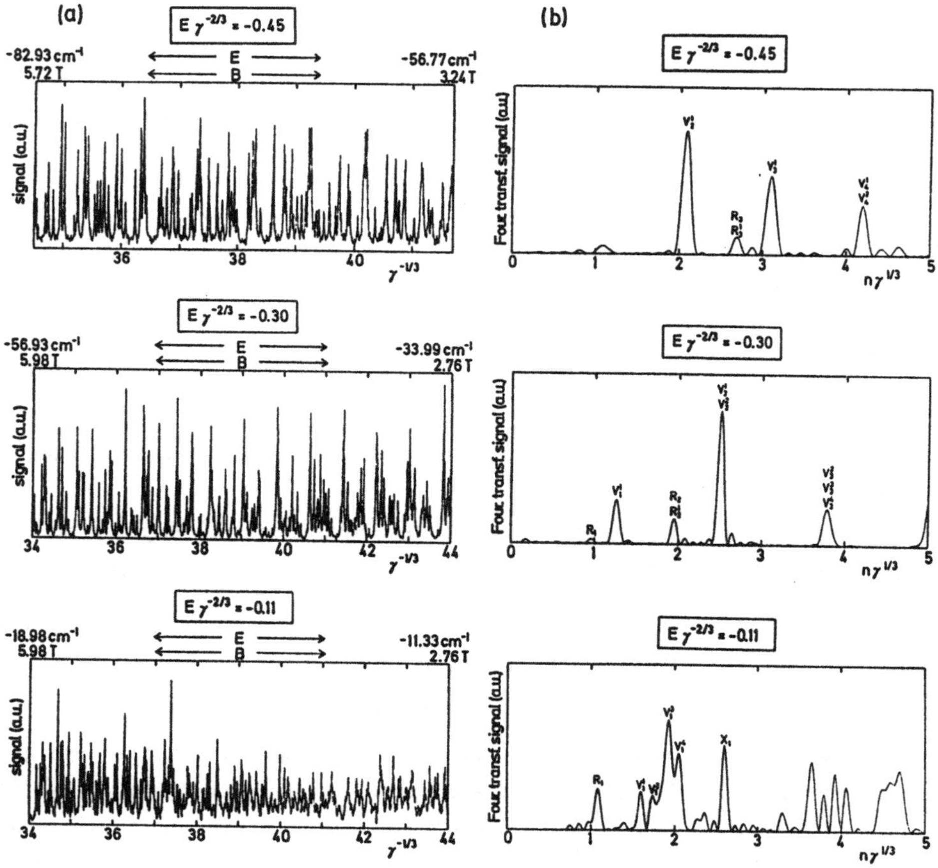

Fig. 4: (a) Scaled-energy spectra at $\tilde{E} = -0.45$, -0.30 and -0.11. (b) Fourier-transformed action spectra of (a). Some resonances are classified according to correlated closed classical orbits (see Fig. 6).

Fig. 5a shows in concise overlay fashion a set of action spectra taken at different scaled energies in steps $\Delta\tilde{E} = 0.01$, representing the evolution of the entire quasi-Landau spectrum as function of $\tilde{E}$ and C from the regular into the chaotic regime. Seen here for the first time it evolves as a remarkably well-structured, ordered system of clustered branches of resonances, not previously predicted or anticipated.

To rationalize the experimental results, we have calculated the complete semiclassical action spectrum, that is the positions of all resonances in the ($\tilde{E}$, C)-plane correlated to closed orbits with the proton as origin by numerical integration of eqn. 5. The result is shown in Fig. 5b in the range $-0.5 \lesssim \tilde{E} \lesssim +0.04$ with $\tilde{E}$ in steps of 0.01. Viewing the figure at glancing angle, one recognizes (in regions of sufficiently low resonance density) strings of dots belonging to given resonances i and showing their ($\tilde{E}$, C)-dependence. Some examples are indicated by solid or broken lines. Evidently there exists a striking resemblance of Figs. 5a and 5b in the overall distribution of resonances in the ($\tilde{E}$, C)-plane. Both spectra evolve fanlike in an obviously similar systematic fashion of well-distinguished clusters and branches.

As the calculations show, the entire manifold of resonances can be traced to roots of three basic resonance types with correlated closed orbits: (a) "vibrators", one-dimensional orbits parallel to the magnetic field axis; (b) "rotators", two-dimensional orbits perpendicular to the field; and (c) "exotics" with genuine three-dimensional orbits. From them evolves the whole quasi-Landau spectrum by sequential bifurcation into a system of complex branches of higher generation resonances with respective three-dimensional orbits.

<u>Vibrators:</u> The basic vibrators, V_μ ($\mu = 1, 2, \ldots$) in Fig. 5b form a harmonic series with V_1 the fundamental. Correlated to the V_μ's are the one-dimensional orbits along the z-axis. From them bifurcate directly first-generation vibrators, indicated by V_μ^ν ($\mu, \nu = 1, 2, \ldots$) in Fig. 5b. The first series, V_1^ν, is just the one previously discovered in experiments at constant magnetic field strength (Fig. 3). The systematics of the whole first-generation vibrators, V_μ^ν, is strikingly mirrored by the geometrical nature of the correlated closed classical orbits, which form a topologically homologous matrix set (Fig. 6a).
As to the further evolution of vibrators, the calculations show that sequential bifurcation of resonances into higher generations increases rapidly in the higher clusters. However, any topological systematics of correlated orbits gets quickly lost with increasing resonance density.

<u>Rotators:</u> The basic rotators, R_μ ($\mu = 1, 2, \ldots$) in Fig. 5b also form a harmonic series with the fundamental, R_1, being the original Garton-Tomkins resonance with correlated two-dimensional closed classical orbit in the ($z = 0$)-plane. From the harmonics bifurcate directly first-generation rotators, R_μ^ν ($\mu > \nu$) correlated to a topologically homologous set of

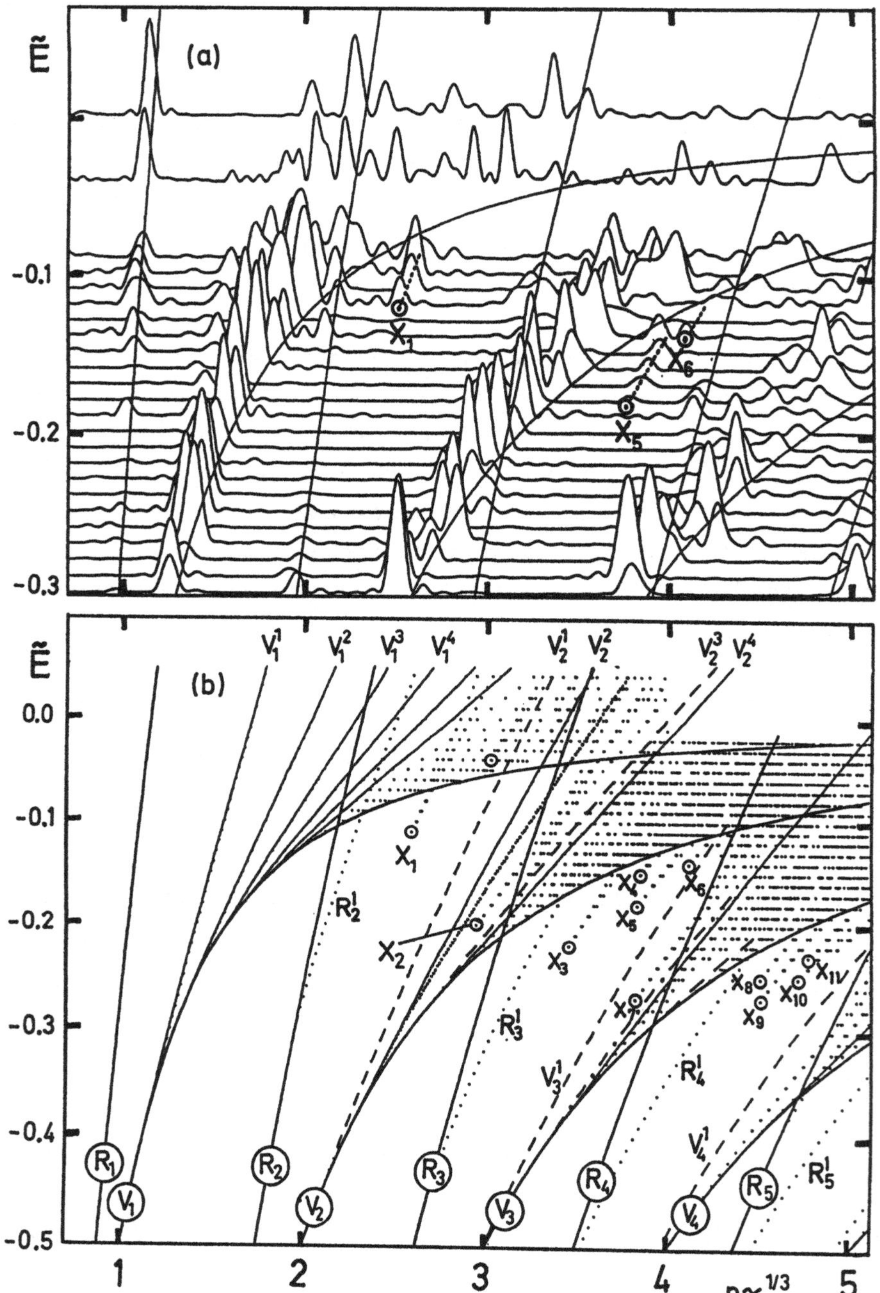

Fig. 5: (a) Experimental quasi-Landau resonance action spectrum as a function of scaled energy $\tilde{E}$ in overlay form. Even parity, magnetic m = 0 final state. (b) Semiclassically calculated ($\tilde{E}$, C)-spectrum of quasi-Landau resonances correlated to closed classical orbits through origin.

closed classical orbits (Fig. 6b). Only a few rotators are observed in this experiment with m = 0 final states, the strongest being R_1.

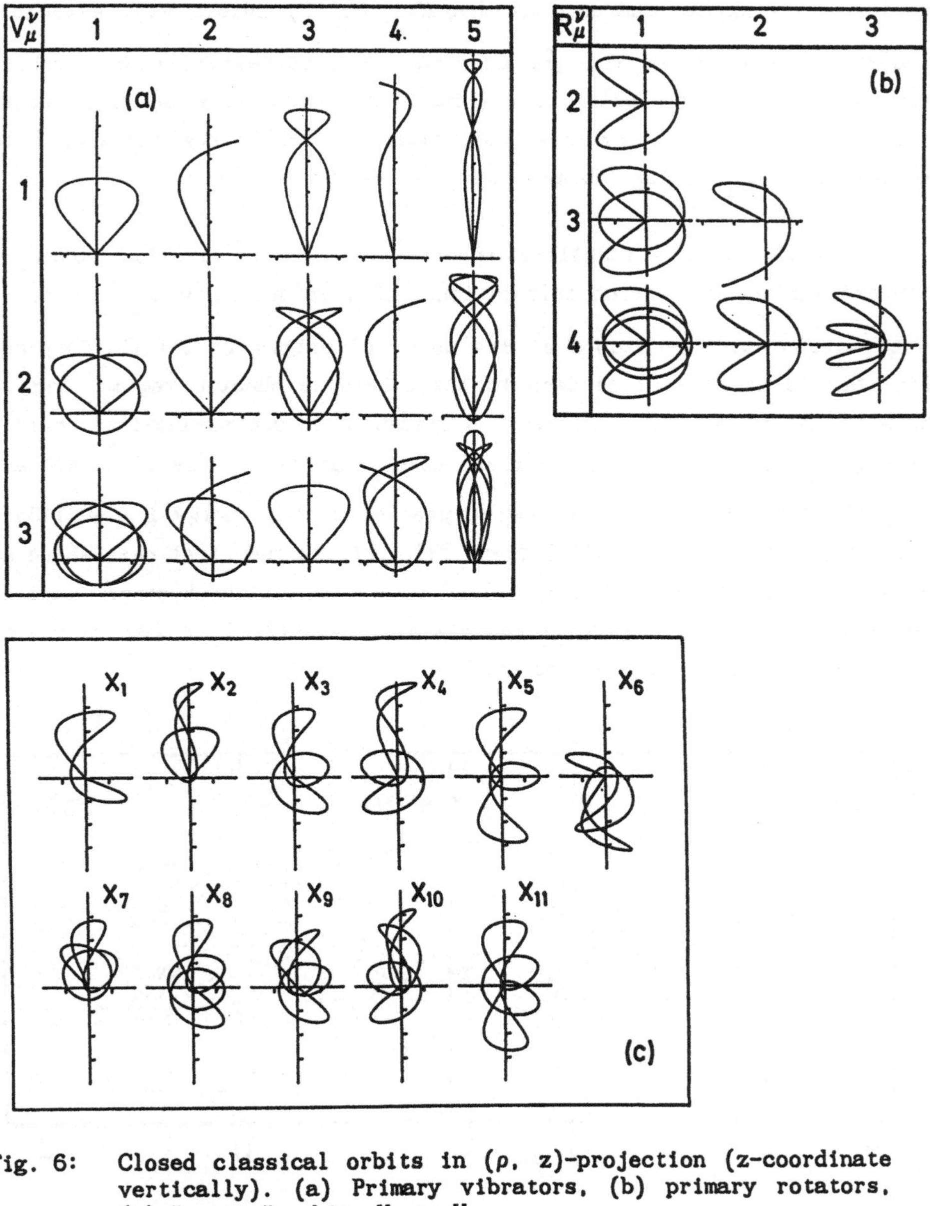

Fig. 6: Closed classical orbits in (ρ, z)-projection (z-coordinate vertically). (a) Primary vibrators, (b) primary rotators, (c) "exotic" orbits X_1 to X_{11}.

<u>Exotics:</u> Surprisingly, some resonances are observed in regions where neither vibrators nor rotators exist that originate from the basic ones by

direct or sequential bifurcation. Examples of such "exotic" resonances are indicated in Fig. 5b (circles and dotted lines). Consistent with the experimental observation, the calculations also reveal the existence of exotics. They appear suddenly at singular $(\tilde{E}, C)$ points with evidently random distribution in the $(\tilde{E}, C)$-plane. The correlated closed orbits bifurcate right at the point of origin. They do not show any topological systematics like the vibrators and rotators. The orbits correlated to exotics X_1 to X_{11} in Fig. 5b are shown in Fig. 6c.

We have analysed the stability specifically of the exotic orbits by Poincaré surface of section calculations [20]. To our surprise we find X_1 to be microscopically stable at its point of origin in the $(\tilde{E}, C)$-plane (Fig. 5b) although it lies deep in the otherwise chaotic regime. Fig. 7 shows the Poincaré surface of section presented in semiparabolical coordinates [21]. However, the stable phase space region is extremely small and observable only with a corresponding precise scaled energy $\tilde{E} = -0.11544$. Predicted by theoretical calculations [3], this is the first case where an isolated microscopical stability island of the magnetized hydrogen atom in the otherwise chaotic regime is experimentally verified by observation of the correlated quasi-Landau resonance.

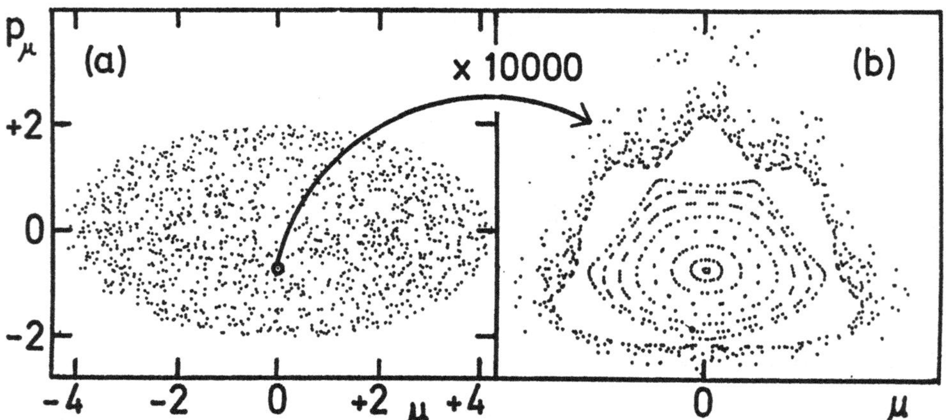

Fig. 7: Poincaré surface of section at scaled energy $\tilde{E} = -0.11544$. Representation in semiparabolical coordinates $(p_\mu, p_\nu, \mu, \nu = 0)$. (a) complete phase space; (b) magnification of microscopic stability island around $\mu = 0$, $p_\mu = -1.1522$.

Concerning the intensity of quasi-Landau resonances, this is generally determined by the stability of the correlated classical orbits [8] and by the angular dependence of the excited final wave function [12]. In the present case, where the final state orbital $|d; m = 0\rangle$ is mostly oriented in the z-direction, one expects preferential excitation of resonances correlated to orbits generally centered along the z-direction, that is, orbits originating from basic vibrators, V_μ. Consistent with this model, in the observed spectrum (Fig. 5a) vibrators are, on a whole, more strongly excited than rotators.

In summary, studies during the last few years have revealed a much advanced picture of the physics of highly excited, magnetized atoms. Essentially new insight has been gained in the structure and dynamics of these systems under classically chaotic conditions by the discovery of a multitude of new quasi-Landau resonances with correlated closed classical orbits and their evolution from the regular into the chaotic regime.

REFERENCES

[1] Garton, W.R.S. and Tomkins, F.S., Astrophys. J. 158, 839 (1969)

[2] Kleppner, D., Littmann, M.G., and Zimmermann, M.L., in *Rydberg States of Atoms and Molecules*, edited by Stebbings, R.F. and Dunning, F.B. (Cambridge Univ. Press, Cambridge, England, 1983), pp. 73-116

[3] Delande, D. and Gay, J.C., Phys. Rev. Lett. 57, 2006 (1986), and 59, 1809 (1987)

[4] Wintgen, D. and Friedrich, H., Phys. Rev. Lett. 57, 571 (1986) and Phys. Rev. A 36, 131 (1987)

[5] Wintgen, D., Phys. Rev. Lett. 58, 1589 (1987), and 61, 1803 (1988)

[6] Du, M.L. and Delos, J.B., Phys. Rev. Lett. 58, 1731 (1987)

[7] Al-Laithy, M.A., O'Mahony, P.F. and Taylor, K.T., J. Phys. B 19, L773 (1986)

[8] Wintgen, D., J. Phys. B 20, L511 (1987);
Schweizer, W., Niemeier, R., Friedrich, H., Wunner, G., and Ruder, H., Phys. Rev. A 38, 1724 (1988)

[9] Wunner, G., Woelk, U., Zech, I., Zeller, G., Ertl, T., Geyer, F., Schweizer, W., and Ruder, H., Phys. Rev. Lett. 57, 3261 (1986)

[10] Wintgen, D., Holle, A., Wiebusch, G., Main, J., Friedrich, H., and Welge, K.H., J. Phys. B 19, L557 (1986)

[11] Holle, A., Wiebusch, G., Main, J., Welge, K.H., Zeller, G., Wunner, G., Ertl, T., and Ruder, H., Z. Phys. D 5, 279 (1987)

[12] Holle, A., Wiebusch, G., Main, J., Hager, B., Rottke, H., and Welge, K.H., Phys. Rev. Lett. 56, 2594 (1986)
[13] Main, J., Wiebusch, G., Holle, A., and Welge, K.H., Phys. Rev. Lett. 57, 2789 (1987) and Z. Phys. D 6, 295 (1987)
[14] Holle, A., Main, J., Wiebusch, G., Rottke, H., and Welge, K.H., Phys. Rev. Lett. 61, 161 (1988)
[15] Rau, A.R.P., Phys. Rev. A 16, 613 (1977)
[16] Edmonds, A.R., J. Phys. Paris 31 Colloque C4, 71 (1970)
[17] Gutzwiller, M.C., J. Math. Phys. 8, 1979 (1967), and 10, 1004 (1969), and 11, 1791 (1970), and 12, 343 (1971)
[18] Reinhardt, W.P., J. Phys. B 16, 635 (1983)
[19] Hasegawa, H., Adachi, S., and Harada, H., J. Phys. A 16, L503 (1983)
[20] Robnik, M., J. Phys. A 14, 3195 (1981);
Harada, A. and Hasegawa, H., J. Phys. A 16, L259 (1983)
[21] Moon, P. and Spencer, J.J. in *Field Theory Handbook*, Springer-Verlag, New York, 1961

PART V: Other Laser Interactions

ATOMIC INTERACTIONS IN SUPERINTENSE, HIGH-FREQUENCY LASER FIELDS

M. Gavrila*, M. Pont*, J. van de Ree**

* FOM-Institute for Atomic and Molecular Physics, Kruislaan 407, 1098 SJ Amsterdam, The Netherlands

** Physics Départment, Eindhoven University of Technology, Eindhoven, The Netherlands

I. Introduction

The generation of superintense fields, in excess of one atomic unit of intensity ($3.51\ 10^{16}$ W/cm^2), has been a continuous goal of laser technology over the last years. It is now possible to generate such fields, in the form of laser pulses of 1 ps or less, at frequencies ranging from the infrared to the ultraviolet (for recent advances see Ref. 1). Intensities several orders of magnitude higher than the atomic unit are underway.

Atomic phenomena acquire new features under these conditions (e.g. see Ref. 2). It now becomes insufficient to describe them by means of standard perturbation theory to lowest order in the electron-field interaction. Indeed, perturbation theory yields the transition amplitudes as expansions in powers of the intensity I, and at these intensities the higher order (untractable) terms of the expansion become comparable to the lower order ones and should be included, let aside the uncertainties concerning the convergence of the summation. The alternative is to devise direct, nonperturbative methods of solution of the Schrödinger equation describing the interaction.

One such theory, valid at sufficiently high frequency ω, was presented a few years ago by Gavrila and Kaminski for the case of scattering[3], and then extended to atomic structure and decay[4]. The theory is nonperturbative in the intensity I, but proceeds by successive approximations which are of increasing order in (essentially) $1/\omega$. It has been developed so far for the case of one electron systems. Firstly, exploratory calculations were carried out for laser modified scattering from potentials. The approximation made consisted in a simplification of the laser induced effective ("dressed") potential acting on the electron, which reduced substantially the computations. Calculations were done for the Coulomb[5] and Yukawa[6] potentials. The same type of approximation was then applied to calculate the structure of atomic hydrogen[7,8].

Recently, however, some very accurate computations have been performed within the high-frequency limit of the theory for the structure of atomic hydrogen[9,10], and for Coulomb scattering[11]. The former[9,10] have revealed an unprecedented radiative stretching of the atom placed in superintense fields, culminating in its splitting into two parts ("dichotomy"). The latter[11] have shown the emergence of peculiar interference patterns in the scattering and of an unexpected symmetry.

In the following we shall briefly review the present status of the high-frequency theory. In Sec. II we recall the basic formalism. In Sec. III we consider the radiative distortion of the hydrogen atom, and in Sec. IV the laser modified Coulomb scattering. Finally, in Sec. V an outlook is given on the further development of the theory, and its possible experimental verification.

II. Basic formalism

The starting point of the high-frequency theory[3,4] is the "space-translated" version of the semiclassical Schrödinger equation for a particle moving under the combined forces of a potential $V(\mathbf{r})$ and a radiation field, which was derived by Kramers[12], and independently by Henneberger[13]. It assumes that the radiation field can be described by its dipole approximation in the physically important region of space; we denote the corresponding vector potential by $\mathbf{A}(t)$. By applying the time dependent translation $\mathbf{r} \rightarrow \mathbf{r} + \vec{\alpha}(t)$, with

$$\vec{\alpha}(t) = -\frac{e}{mc} \int_0^t \mathbf{A}(t')\, dt' , \qquad (1)$$

to the one-electron Schrödinger equation in the momentum gauge (for the laboratory frame of reference), the space-translated version emerges:

$$\left[\frac{1}{2m}\mathbf{P}^2 + V(\mathbf{r} + \vec{\alpha}(t))\right] \psi = i\hbar \frac{\partial \psi}{\partial t} . \qquad (2)$$

Note that in passing from the momentum gauge form of the Schrödinger equation to Eq.(2), we have removed the term $e^2\mathbf{A}^2(t)/2mc^2$ by a phase transformation. Thereby, the connection between the wave function ψ entering Eq.(2) and the solution ψ_L of the Schrödinger equation in the momentum gauge for the laboratory reference frame is:

$$\psi_L(\mathbf{r},t) = \psi(\mathbf{r} - \vec{\alpha}(t)) \exp\left[-(ie^2/2\hbar mc^2) \int_0^t \mathbf{A}^2(t')\, dt'\right] . \qquad (3)$$

We are using in this section the Gauss system of units.

As easily seen, Eq.(1) represents the field driven part of the motion of a classical free electron. Thus, Eq.(2) characterizes the dynamics of our problem in a moving frame of reference which follows the field driven motion of the classical electron (i.e. in which it moves with constant velocity). This we shall call the "Kramers reference frame". In it the center of force (the proton) has a motion represented by $-\vec{\alpha}(t)$, as apparent also in Eq.(2).

From now on we shall assume that the radiation field can be represented by a *monochromatic plane wave* of frequency ω (i.e. with constant amplitude in space and time). For *linear polarization*, the vector potential of the field in the laboratory frame can be written as $\mathbf{A}(t) = a\mathbf{e}\cos\omega t$, where $\mathbf{e}$ is the real unit vector of the polarization. Eq.(1) then gives

$$\begin{aligned} \vec{\alpha}(t) &= \alpha_0 \mathbf{e} \sin \omega t \ , \\ \alpha_0 &= -(ea/mc\omega) \ , \end{aligned} \tag{4}$$

which describes the quiver motion of a classical electron. In atomic units (Bohr radii) α_0 becomes

$$\alpha_0 = I^{1/2}\, \omega^2 \ , \tag{5}$$

where I is the time averaged beam intensity in atomic units $I_0 = 3.51 \cdot 10^{16}$ W/cm^2, and the photon energy is given in Hartrees (2 Ry).

By inserting Eq.(4) into Eq.(2) one gets a differential equation with periodic coefficients, which can be handled according to the Floquet method[3,4]. This leads to a system of coupled differential equations in coordinate space for the Floquet components of the wave function ψ, containing a quasi-energy parameter E. The system has to be supplemented by appropriate boundary conditions to describe the physical situation considered (e.g. steady decay of the atom by multiphoton ionization; scattering from a center of force). An iterative method of solution was devised, proceeding essentially in inverse powers of ω. To lowest order in the iteration, i.e. the *high-frequency limit*, the set of coupled equations reduces to a single one

$$\left[\frac{1}{2m}\mathbf{P}^2 + V_0(\alpha_0,\mathbf{r})\right]\psi_0 = E\psi_0 \ . \tag{6}$$

This has the form of a time independent Schrödinger equation for the zeroth Floquet component ψ_0, the only one in the expansion of ψ to survive in this limit. Thus:

$$\psi(\mathbf{r},t) \cong \psi_0(\mathbf{r}) \exp(-iEt/\hbar) \ . \tag{7}$$

Eq.(6) contains the "dressed potential" $V_0(\alpha_0,\mathbf{r})$ which is the time average over one period of the oscillating $V(\mathbf{r}+\vec{\alpha}(t))$:

$$V_0(\alpha_0,\mathbf{r}) = \frac{1}{2\pi}\int_0^{2\pi} V(\mathbf{r}+\alpha_0\mathbf{e}\sin\zeta)\, d\zeta \ . \tag{8}$$

The result has a simple intuitive meaning: in the high-frequency limit the electron is not capable of feeling the rapidly oscillating potential $V(\mathbf{r}+\vec{\alpha}(t))$, but only its time average V_0. In the Coulomb case, $V(r) = -e^2/r$, Eq.(8) can also be looked upon as the electrostatic potential created by a linear (inhomogeneous) distribution of charges extending from $-\alpha_0$ to $+\alpha_0$ along the line of oscillation of direction $\mathbf{e}$ passing through the center of force (see Eqs.(16) and (17) of Ref. 4). Note that, in the high-frequency limit, ω and I enter the theory only through the parameter α_0 contained in V_0.

The dressed Coulomb potential can be expressed as:

$$V_0(\alpha_0,\mathbf{r}) = -(2e^2/\pi)(r_+r_-)^{-1/2}\, K\left[2^{-1/2}\,(1-\hat{\mathbf{r}}_+\hat{\mathbf{r}}_-)^{-1/2}\right] , \tag{9}$$

where $\mathbf{r}_\pm = \mathbf{r} \pm \alpha_0\mathbf{e}$ (the origin of the coordinates is kept at the center of V(r)) and K is the complete elliptic integral of the first kind. V_0 has $r^{-1/2}$ singularities at the endpoints of the charge distribution $\pm\alpha_0\mathbf{e}$, and a logarithmic singularity along the segment inbetween (due to the behavior of $K(x)$ for $x \to 1$). A graphical representation of V_0 was given in Fig. 1 of Ref. 3 and in Fig. 1 of Ref. 4.

The reduction of the exact Schrödinger equation, Eq.(2), in the high-frequency limit to Eq.(6) has important consequences. The eigenvalues $E(\alpha_0)$ of Eq.(6) are obviouly real, which means that an atom placed in the field has truly stationary states in this limit, i.e. does not decay by multiphoton ionization. When referring to collisions, Eq.(6) can, obviously, describe only elastic scattering, showing that free-free transitions are suppressed in this limit. Thus, in the high-frequency regime the absorption (emission) of photons is inhibited.

Radiative transitions become possible only in the *second order* of the iteration. In the case of atomic structure, complex values are now found for the quasi-energies E, and, at the same time, expressions for the absorption amplitudes. Similarly, in the collisions case, expressions are obtained for the free-free transition amplitudes. In both cases, the n-photon transition amplitudes can be expressed in terms of the solutions of Eq.(5) as[3,4]

$$f_n = -\frac{m}{2\pi\hbar^2} <\psi^{(-)}_{\mathbf{p}_n}|\, V_n |\, \psi_0> \tag{10}$$

Here $\psi^{(-)}_{\mathbf{p}_n}$ is the scattering solution for asymptotic momentum $\mathbf{p}_n$ and having spherically converging waves, which describes the final state; V_n is the n-th component of the Fourier expansion of the oscillating potential $V(\mathbf{r}+\vec{\alpha}(t))$ appearing in Eq.(2); ψ_i is the solution representing the initial state. For ionization ψ_i is a bound state solution of Eq.(6), whereas for scattering, a solution corresponding to initial momentum **p**, and having spherically diverging waves. The continuum waves are normalized to unity asymptotic amplitude, and the bound state to unity normalization integral.

It is argued elsewhere[3] that when dealing with atomic structure the *frequency condition* under which the iteration procedure should converge (in some pragmatic sense) is

$$\hbar\omega >> |E_0^m(\alpha_0)| \ , \tag{11a}$$

where $E_0^m(\alpha_0)$ represents the lowest eigenvalue of the manifold belonging to the same magnetic quantum number m as the initial state. Note that this condition refers to the atom in the field, characterized by a given value of α_0. In the collision problem, the frequency condition is somewhat more restrictive, requiring that

$$\hbar\omega >> |E_0^0(\alpha_0)| \ ; \qquad \hbar\omega >> E \ , \tag{11b}$$

where $E_0^0(\alpha_0)$ is the energy of the ground state (m=0) for the dressed potential V_0, and E is the incoming electron energy.

There are no restrictions on α_0 (except for being kept finite), i.e. the theory applies as well for $\alpha_0 << 1$, which is the realm of perturbation theory[14], as for $\alpha_0 \geq 1$, where the theory is nonperturbative. It is in this latter case that we shall be primarily interested in.

The solution ψ of Eq.(2), and its approximate form for the high-frequency limit Eq.(6), pertain to the Kramers reference frame. From Eq.(3), in the laboratory frame the (momentum gauge) solution of the Schrödinger equation corresponding to Eq.(7) is

$$\psi_L(\mathbf{r},t) \cong \psi_0(\mathbf{r}-\vec{\alpha}(t)) \exp\left\{(-it/\hbar)\,[E+e^2a^2/4mc^2]\right\} \ . \tag{12}$$

(As customary, we have eliminated from the right-hand side of Eq.(12) the rapidly oscillating phase factor $\exp[-(ie^2a^2/8\hbar\omega mc^2)\sin 2\omega t]$.) ψ_L and the electronic charge density $|\psi_L|^2$ oscillate without loss of normalization, like a free classical electron. From Eq.(12) it is apparent that the energy spectrum (including the continuum threshold) for the Schrödinger equation in the laboratory frame is shifted with respect to that given by Eq.(6) by the constant $e^2a^2/4mc^2$.

Eq.(6) was obtained earlier for the bound state problem via other methods by Henneberger[13], Gersten and Mittleman[15], see also Mittleman[16]. Gersten and Mittleman were the first to realize its high-frequency character.

After having recalled the main points of the high-frequency theory, we shall consider in the following its application to the structure of the hydrogen atom and to laser modified Coulomb scattering. From now on we shall be using *atomic units* ($|e| = \hbar = m = 1$).

III. The structure of atomic hydrogen[9,10]

For any spherically symmetric potential V(r), the dressed potential Eq.(8) has obviously axial symmetry around the axis of **e**, and is an even function with respect to its center. This reduction of the spherical symmetry implies that now only the projection of the angular momentum on the **e** axis (associated quantum number m) and the parity (quantum number P) are conserved quantities. In fact, as can be easily seen, the eigenvalues depend on |m|, rather than on m, which implies a twofold degeneracy for $m \neq 0$. The symmetry of the problem suggests therefore a *level classification scheme* for the states similar to that for diatomic molecules (although the form of the potential is quite different in the two cases). Thus, we will designate the states with $|m| = 0,1,2,...$ by $\sigma,\pi,\delta,...$ respectively, and even or odd (gerade or ungerade) parity by P = 0 or 1. In order to distinguish a particular state within a symmetry manifold with given |m| and P, we will label it by the quantum numbers (n,ℓ) of the unperturbed state it is connected to continuously in the field free limit ($\alpha_0 = 0$). The eigenfunctions will therefore be denoted by $\phi_{(n\ell)m}(\mathbf{r})$; their parity is necessarily that of ℓ.

It has been shown that at small and large α_0 the eigenvalue problem can be solved independently of the main computation, in part analytically. Thus, at *small* α_0 the value obtained for the displacement of the levels from their Bohr values ($\alpha_0 = 0$) is

$$(\Delta E)_{n\ell|m|} = \alpha_0^2 \left[\frac{1}{3n^3}\delta_{\ell 0} - \frac{\ell(\ell+1)-3m^2}{2n^3\ell(\ell+1)(\ell+1/2)(2\ell-1)(2\ell+3)}(1-\delta_{\ell 0})\right]. \quad (13)$$

At sufficiently *large* α_0, it was shown analytically that each eigenfunction splits into two practically nonoverlapping parts, an occurrence we designate by *"atomic dichotomy"*. When this takes place, the α_0 dependence of the eigenvalues is given by $\alpha_0^{-2/3}$. This happens for any bound state of V_0, at sufficiently large α_0. On the other hand, at fixed α_0, the high-lying states of each manifold $\{|m|,P\}$ are of the Rydberg type. This is due to the fact that for $r \geq \alpha_0$ the dressed potential Eqs.(8),(9) reduces to a Coulomb potential corrected by terms in $r^{-(2s+1)}$, with s = 1,2,... The Rydberg character will be manifest only for those states whose wave function lies to a good extent in the Coulombic region of the potential, i.e. at radial distances well beyond α_0.

The fact that we have knowledge of the small- and large-α_0 behavior of the eigenvalues $E(\alpha_0)$ enables the drawing of *correlation diagrams* for the eigenvalue curves

$E(\alpha_0)$, which connect the small- and large-α_0 eigenvalues lying within the same symmetry manifold. If all symmetries of the problem are fully accounted for, the von Neumann-Wigner noncrossing rule should be satisfied. This leads in general to avoided crossings of the eigenvalue curves at places where in an approximate calculation with a Hamiltonian of higher symmetry the curves would have crossed.

The *numerical solution* of the eigenvalue problem Eq.(6) was carried out by diagonalization of the Hamiltonian matrix in an orthonormalized Gaussian basis, analogous to the procedure used in structure calculations for diatomic molecules. Each symmetry manifold $\{|m|,P\}$ was considered separately. Since the potential V_0 is singular along the line of charges (see Sec. II), we have distributed the centers of the Gaussian basis functions $\chi_i(\mathbf{r})$ along this line, chosen as the z-axis. Powers of x and y were included in the $\chi_i(\mathbf{r})$ so as to account for the correct azimuthal angular dependence imposed by m. Finally the correct parity P was incorporated from the start in $\chi_i(\mathbf{r})$.

We have investigated the four symmetry manifolds σ_g, σ_u, π_g, π_u for values of α_0 ranging from 0 to 200. (Note that when the ArF* excimer laser, ω = 6.4 eV, is operated at intensity I = 1 a.u. we have α_0 = 9 a.u., and at I = 100 a.u., α_0 = 90 a.u.) Due to the fundamental importance of the hydrogen atom problem we have deemed it worthwhile to carry out a very accurate computation. Up to 10 different centers and a maximal number of 100 basis functions were used to obtain the lowest lying level in each symmetry class to 5 significant figures for the σ states and to 4 figures for the π states. The accuracy for the excited states is somewhat lower.

Our numerical *results for the eigenvalues* are displayed graphically in Fig. 1, where we present the correlation diagrams for the σ_g, σ_u, π_g, π_u states. A general feature of the correlation curves $E(\alpha_0)$ is their global tendency to increase to zero at large α_0 values. In some cases this increase is monotonical (e.g. $(1s)\sigma_g$, $(2p)\pi_u$), in others it may be preceded by a minimum (e.g. $(2s)\sigma_g$, $(3d)\sigma_g$, $(2p)\sigma_u$). In the range $\alpha_0 < 40$ the increase is rather sharp for the lower states. It is quite dramatic for the ground state $(1s)\sigma_g$ (not fully represented in Fig. 1), whose binding energy drops at $\alpha_0 = 40$ to about 1/10 its value at $\alpha_0 = 0$ ($|E(0)|$ = 0.5 a.u.). However, for the higher excited states, which start at $\alpha_0 = 0$ with an already small value of the binding energy, the increase is much slower and takes place over large α_0 intervals.

The small-α_0 behavior of the correlation curves can be compared with the analytic formula Eq.(13). The agreement is excellent, well within the α_0^3 corrective terms to Eq.(13). At large α_0 on the other hand, one can check the numerical results against the predicted $\alpha_0^{-2/3}$ behavior of the eigenvalues. The agreement is quite good for the ground state, and sets in at unexpectedly low values of α_0 (for example, at $\alpha_0 = 30$ the variance is of only 2%). For higher lying states the agreement sets in at larger values of α_0.

We shall now consider the *energy eigenfunctions*. In view of the φ dependence of the eigenfunctions $\phi_{(n\ell)m}(\mathbf{r})$, we may write

$$\phi_{(n\ell)m}(x,y,z) = e^{im\varphi}\,\zeta_{(n\ell)m}(r,\theta)\,. \tag{14a}$$

By setting here $\varphi = 0$, we find:

$$\phi_{(n\ell)m}(x,0,z) = \zeta_{(n\ell)m}(r,\theta)\,, \qquad \text{for } x > 0\,. \tag{14b}$$

Thus, knowledge of $\phi_{(n\ell)m}(x,0,z)$ is sufficient to determine $\phi_{(n\ell)m}(\mathbf{r})$ throughout space. Two of the $\phi_{(n\ell)m}(x,0,z)$ functions are represented in Fig. 2. The x,z axes define the horizontal plane in Fig. 2, and the vertical axis gives the values of $\phi_{(n\ell)m}(x,0,z)$.

Let us consider the *distortion of the eigenfunctions* with increasing α_0. In Fig. 2 this is shown for the states $(1s)\sigma_g$ and $(2p)\sigma_u$.

The case of the ground state wave function $(1s)\sigma_g$ was presented in more detail in Fig. 1 of Ref. 9. Note that the wave function does not change sign and maintains its value at reflections in the xz plane with respect to both the x or z axes. From both Fig. 2 and Fig. 1 of Ref. 9 it is apparent that, with increasing α_0, the wave function undergoes *radiative stretching* in the z direction, following the elongation of the line of charges (length $2\alpha_0$) generating the dressed potential V_0 (for the shape of the latter see Fig. 1 of Ref. 3). However, as α_0 approaches 20 a saddle appears, and by $\alpha_0 = 30$ two pronounced maxima are formed around the endpoints $\pm\alpha_0\mathbf{e}$ of the line of charges. Beginning with $\alpha_0 = 50$, *dichotomy* sets in, and is complete by $\alpha_0 = 70$. While splitting, the wave function also becomes more diffuse around the endpoints (note the change of scale in the pictures of Fig. 2 for growing α_0). At large α_0 (e.g. $\alpha_0 = 100$), the ground state wave function attains a truly Rydberg state size (some $2\alpha_0 = 200$ a.u. linear extension).

In Fig. 2 we also show the parallel evolution of the $(2p)\sigma_u$ eigenfunction. This can take both positive and negative values, in such a way that it remains unchanged at reflections in the xz plane with respect to the z axis, and changes sign at reflections with respect to the x axis.

The $(2p)\sigma_u$ and the $(1s)\sigma_g$ wave functions are totally dissimilar at lower α_0 values (e.g. $\alpha_0 = 10$). However, after dichotomy has fully set in for both of them (at about $\alpha_0 = 100$), it is apparent that the wave functions coincide for $z > 0$, whereas they differ in sign for $z < 0$. This is an intrinsic consequence of the large-α_0 behavior of the eigenfunctions of Eq.(6).[10]

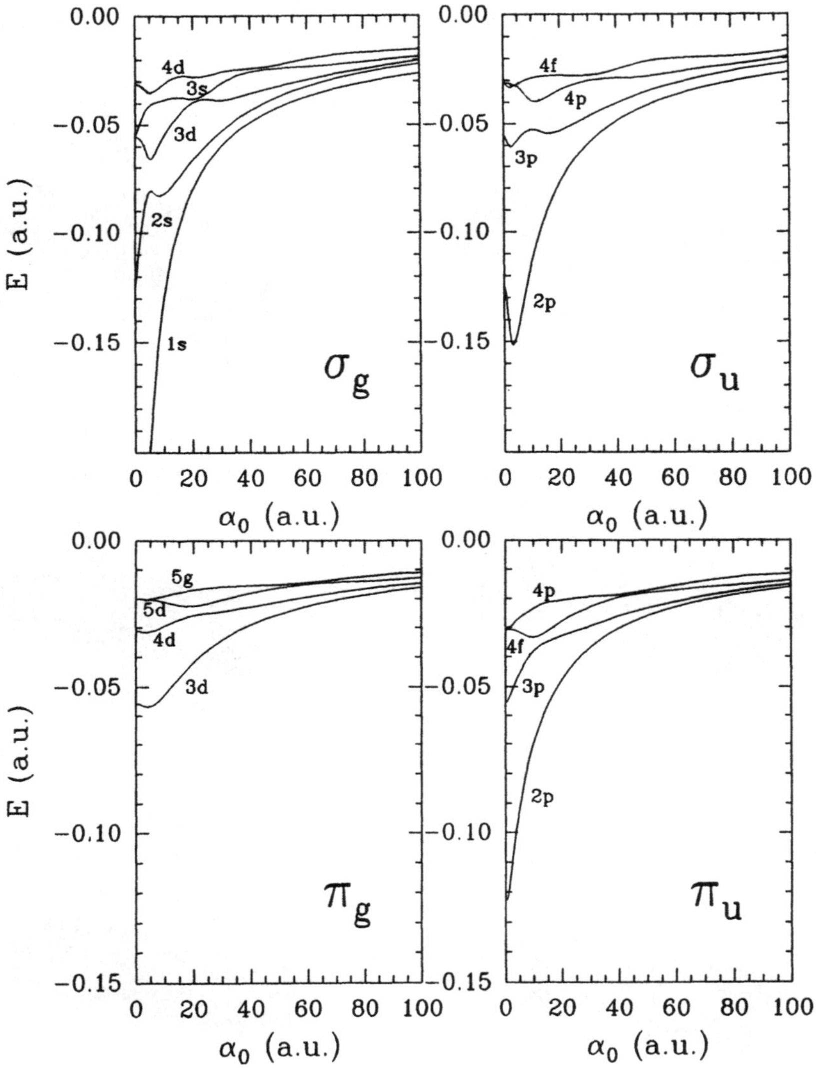

Fig. 1. The energies (in a.u.) of the first eigenvalues of the σ_g, σ_u, π_g and π_u manifolds as functions of α_0 (in a.u.).

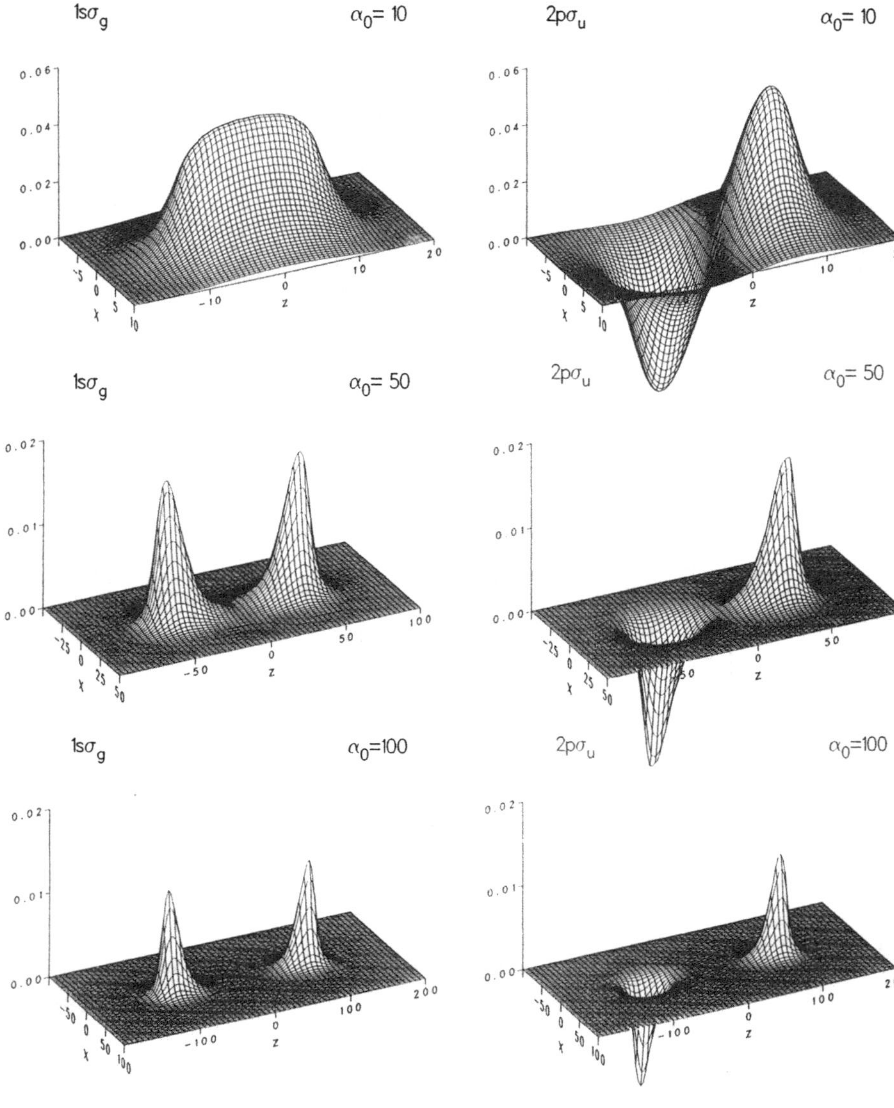

Fig. 2. The normalized eigenfunctions $\phi(x,0,z)$ of the $(1s)\sigma_g$ and $(2p)\sigma_u$ states in the Kramers frame of reference, for increasing α_0. The choice of the coordinate axes is defined in Sec. III. The length unit is the Bohr radius a_0, and the unit for $\phi(x,0,z)$ is $a_0^{-3/2}$. There is a change of scale in the figures for the various α_0. Note how dichotomy begins to set in at about $\alpha_0 = 50\, a_0$.

The stable, high-frequency limit eigenfunctions we have calculated and discussed pertain to the Kramers frame of reference. Their form in the laboratory frame can be obtained from Eq.(12).

We shall now discuss some of the *experimental implications* of our results.The atomic structure we have been discussing so far pertains to the high-frequency limit of our theory. In fact, ω will be finite no matter how large. As mentioned in Sec. II, this can be taken into account by passing to the *second approximation* of the theory, which allows for atomic ionization. The latter manifests itself in the fact that the eigenvalues become complex, i.e. there is a shift and a broadening of the physical levels with respect to the values obtained here, and, at the same time, the wave functions lose their square integrability while acquiring diverging spherical waves. The latter yield the decay amplitudes for the various channels of multiphoton absorption. Nevertheless, the better the condition Eq.(11a) is satisfied, the closer the exact results will be to the ones calculated here.

The energy spectrum of the photoelectrons depends on the binding energy $E_0(\alpha_0)$ of the initial (e.g. the ground) state of the atom at the time of emission, according to

$$E_\nu = -|E(\alpha_0)| + \nu\omega, \qquad (\nu \geq 1)\,. \tag{15}$$

(Recall that in our case one photon ionization, $\nu = 1$, is already possible, see Eq.(11a)). The decrease of the ground state binding energy $|E(\alpha_0)|$ with respect to its unperturbed value implies that the whole energy spectrum of the ionized electrons at the time they are *leaving the atom* is shifted towards *higher* energies than in the low intensity case. This is a prediction which can be compared to experiment, if the empirical data are appropriately interpreted. It contrasts with the findings from multiphoton experiments performed in the infrared or the visible at intensities up to 10^{-1} a.u. on noble gas atoms, e.g. see Refs. 17,18. (The fact that the effective potential for the noble gas atoms is not Coulomb, as in our case, should not play a role in the following.) Under those circumstances an *increase* of the ground state ionization potential occurs and the lowest energy ionization peak may be suppressed (see Ref. 19). There is no contradiction with our results, however, since the increase or decrease of $|E_0|$ will depend on ω and I, and our theory does not apply in their case (the frequency condition Eq.(11a) is not satisfied). Obviously, there can be no suppression of ionization peaks in our regime.

The previous remarks referred to predictions of our high-frequency theory. We now comment on one of its *underlying assumptions* and the extent to which it is fulfilled in experiment.

As all (semiclassical) theories of stationary decay based on the Floquet method[20,21], ours too uses a *plane wave description* for the radiation field. Obviously, this is an idealized situation, only approximately realized in experiment. What happens in practice is that high intensities can be achieved only in the form of very short laser pulses. Thus the amplitude of the laser pulse acting on the atoms has a space and time variation, and, besides, may vary from shot to shot (in an unknown way). Since the linear extension of the laser pulse, as determined by the focusing lenses, is larger than the wavelength of the light, the variation of the amplitude over the size of each particular atom can be disregarded and, therefore, in this respect plane wave results are satisfactory. (Nevertheless, when irradiating an assembly of atoms with the laser pulse, the different atoms, located at various points in it, will feel different intensities.) On the other hand, if the time dependence induced by turning on and off the pulse is too rapid, it may well happen that the plane wave results are not applicable. This difficulty can be avoided if one assumes that the amplitude varies sufficiently slowly ("adiabatically") during: (1) a light period, so that the notion of frequency is still meaningful; (2) a typical period of the (classical) electronic motion, so that the atom has time to adjust continuously to this variation, without making quantum transitions. Under these circumstances our plane wave results for the eigenvalues and eigenfunctions can still be expected to describe instantaneously the situation, if we replace the constant field amplitude by the slowly varying time function (on the atomic scale) it actually is. To accomodate for the existence of statistics in the photon beam would require a reformulation of the present theory.

IV. Modified Coulomb scattering[11]

We now consider the scattering of an electron from the dressed Coulomb potential V_0, Eq.(9). Because of the axial symmetry of V_0, the problem now bears similarity to that for radiationless electron scattering from a homonuclear diatomic molecule. Of the many approaches developed in that context we have adapted for our needs the "one-center expansion", which analyzes the scattering equation in a spherical harmonics basis. Since the azimuthal quantum number ℓ is no longer conserved, the scattering can no longer be characterized by phase shifts, but rather by a (nondiagonal) S matrix. However, because of conservation of the angular momentum along **e** (associated magnetic quantum number m), the S matrix is diagonal with respect to m, i.e. is of the form $S^m_{\ell\ell'}$.

We are seeking the scattering solution $u^{(+)}_{\mathbf{k}_i}(\mathbf{r})$ as

$$u^{(+)}_{\mathbf{k}_i}(\mathbf{r}) = \frac{2\pi}{kr} \sum_{\ell,\ell',m} i^{\ell'+1}\, e^{i\sigma_{\ell'}}\, g^{\ell'm}_{\ell m}(\alpha_0,k;r)\, Y_{\ell m}(\hat{\mathbf{r}})\, Y^*_{\ell' m}(\hat{\mathbf{k}}_i)\,. \tag{16}$$

We are denoting here by $\mathbf{k}_\alpha = k\hat{\mathbf{k}}_\alpha$ (α = i,f) the initial and final momenta of the electron ($k^2 = 2E$, where E is the scattering energy), and σ_ℓ are the Coulomb phases. The polar axis is taken along **e** and the summation indices satisfy: $\ell, \ell' \geq |m|$. $u^{(+)}_{\mathbf{k}_i}$ should satisfy the same asymptotic condition as for a usual Coulomb potential. This imposes the following asymptotic behavior on the functions $g^{\ell' m}_{\ell m}(r)$:

$$g^{\ell' m}_{\ell m}(r) \underset{r\to\infty}{\longrightarrow} \delta_{\ell\ell'} \exp\left\{-i\left(kr - \gamma \ln 2kr - \frac{\ell\pi}{2} + \sigma_\ell\right)\right\}$$

$$-S^{m}_{\ell\ell'}(\alpha_0,k) \exp\left\{+i\left(kr - \gamma \ln 2kr - \frac{\ell\pi}{2} + \sigma_\ell\right)\right\}. \qquad (17)$$

Eq.(17) contains the S-matrix elements $S^{m}_{\ell\ell'}(\alpha_0,k)$ in the spherical harmonics basis, and $\gamma = -i/k$.

To determine the S matrix we have to solve the equations for the functions $g^{\ell' m}_{\ell m}(r)$. These can be obtained by inserting Eq.(16) into the integral form of the Schrödinger equation (incorporating the boundary conditions), which results in a system of coupled integral equations. For the solutions of the latter we have followed a method used in Ref. 22. The S matrix elements were calculated with an absolute accuracy of 10^{-5} for any given number of coupled ℓ-channels. Their values for small α_0 were checked against an analytic formula obtained from perturbation theory.

As noted in Sec. III, the dressed potential V_0 is a Coulomb potential modified at finite distances. It is then useful to apply the "two-potential formalism"[23] and write the scattering amplitude in the form of the sum:

$$f_{\mathbf{k}_i}(\alpha_0;\hat{\mathbf{k}}_f) = f_c(\theta) + f'_{\mathbf{k}_i}(\alpha_0;\hat{\mathbf{k}}_f), \qquad (18a)$$

$$f_c(\theta) = \frac{(-\gamma)}{2k \sin^2(\theta/2)} \exp[-i\gamma \ln \sin^2(\theta/2) + 2i\sigma_0], \qquad (18b)$$

$$f'_{\mathbf{k}_i}(\alpha_0;\hat{\mathbf{k}}_f) = \frac{2\pi}{ik} \sum_{\ell,\ell',m} i^{\ell'-\ell} e^{i\sigma_\ell} [S^{m}_{\ell\ell'}(\alpha_0,k) - \delta_{\ell\ell'}] e^{i\sigma_{\ell'}} Y_{\ell m}(\hat{\mathbf{k}}_f) Y^{*}_{\ell' m}(\hat{\mathbf{k}}_i). \qquad (18c)$$

Here $f_c(\theta)$ is the Coulomb scattering amplitude, and f' is the extra contribution due to the departure of V_0 from the Coulomb form. The scattering amplitude $f_{\mathbf{k}_i}(\hat{\mathbf{k}}_f)$ has a number of symmetry properties, deriving from those of the dressed potential V_0.

The modified elastic scattering cross section is

$$(d\sigma/d\Omega) = |f|^2 = |f_c|^2 + 2\mathrm{Re}(f_c^* f') + |f'|^2. \qquad (19)$$

The first term on the right-hand side is the original Rutherford cross section, $d\sigma^c/d\Omega = |f_c|^2$, and the second represents the interference between the Coulomb and the extra scattering amplitude.

In order to calculate the amplitude Eq.(18) we need choose a set of independent angles to characterize the configuration of the momenta, in which we want in include the scattering angle θ. To this end we shall denote the reference system used above in Eqs.(16)-(18) by T, specifying that it has the x axis in the $(\mathbf{e}, \mathbf{k}_i)$ plane. We then introduce another reference system T', obtained by a rotation of T by an angle θ_i around the y axis in the direct sense, so that the z axis coincides with $\mathbf{k}_i$. The polar angles of $\mathbf{k}_i$ in T are $\theta_i, \varphi_i = 0$, and the polar angles of $\mathbf{k}_f$ in T' are θ and φ, where θ is the scattering angle; $\theta_i, \varphi, \theta$ are the angles we shall be using. By definition their range of variation is $0 \le \theta_i \le \pi$, $0 \le \theta \le \pi$ and $0 \le \varphi \le 2\pi$. In terms of these angles the symmetry properties of the amplitude $f(\theta_i; \theta, \varphi)$ may be expressed as:

$$f(\theta_i; \theta, \varphi) = f(\pi-\theta_i; \theta, \varphi) , \qquad f(\theta_i, \theta, \varphi) = f(\theta_i; \theta, 2\pi-\varphi) , \tag{20}$$

These allow the reduction of the ranges of the angular variables to $0 \le \theta_i \le \pi/2$, $0 \le \theta \le \pi$ and $0 \le \varphi \le \pi$. (Note that for symmetry reasons, at $\theta_i = 0$, f is independent of φ for all θ, whereas for geometrical reasons at $\theta = 180°$, f is independent of φ for any θ_i.) The expression of f', Eq.(18), can be obtained as function of $\theta_i, \theta, \varphi$ by using the transformation properties under rotations of the spherical harmonics.

The number of coupled ℓ-channels included in the calculation of the S matrix was increased until the result for the cross section Eq.(19) has converged to better than 1%. The number of channels required increases rapidly with growing α_0 or energy; for example, when $\alpha_0 = 2$ and $E = 0.01$, some 15 channels are needed.

We present our results for the laser modified scattering by giving $R = (d\sigma/d\Omega)/(d\sigma^c/d\Omega)$, the ratio of the cross section Eq.(19) to the original Rutherford one. Thus, any deviation of R from 1 indicates laser effects. A typical example is shown in Fig. 3 where we represent $R(\theta)$, the ratio R as function of θ, at given θ_i and φ. We have chosen the low energy of $E = 0.01$ Ry, and have taken for α_0 the nonperturbative value $\alpha_0 = 2$ a.u. The results present some remarkable features:

(1) *Forward scattering*. For $\theta \to 0$, $R(\theta) \to 1$ in all cases. This is due to the fact that $|f_c| \to \infty$, whereas f' stays finite, so that the relative contribution to $R(\theta)$ of the last two terms in Eq.(19) tends to zero. This can be explained in classical terms by the fact that small angle scattering occurs at large impact parameters, where the deviations of V_0 from the Coulomb form vanish.

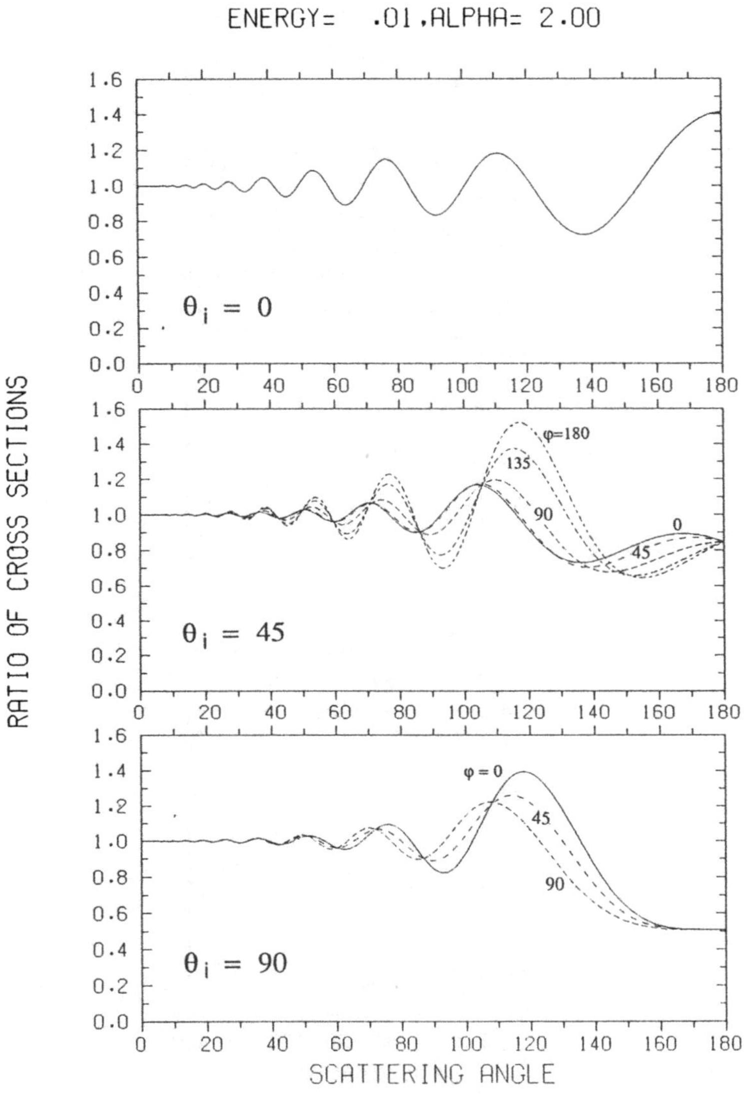

Fig. 3. Ratio R(θ) of the elastic scattering cross section for the dressed Coulomb potential, Eq.(19), to the Rutherford cross section, as function of the scattering angle θ. Values of θ_i and φ as indicated. For θ_i = 90°, the R curves for φ = 135° and 180° coincide with those for φ = 45° and 0°, respectively (see Ref. 11). The electron energy is E = 0.01 Ry and $\alpha_0 = 2\,a_0$.

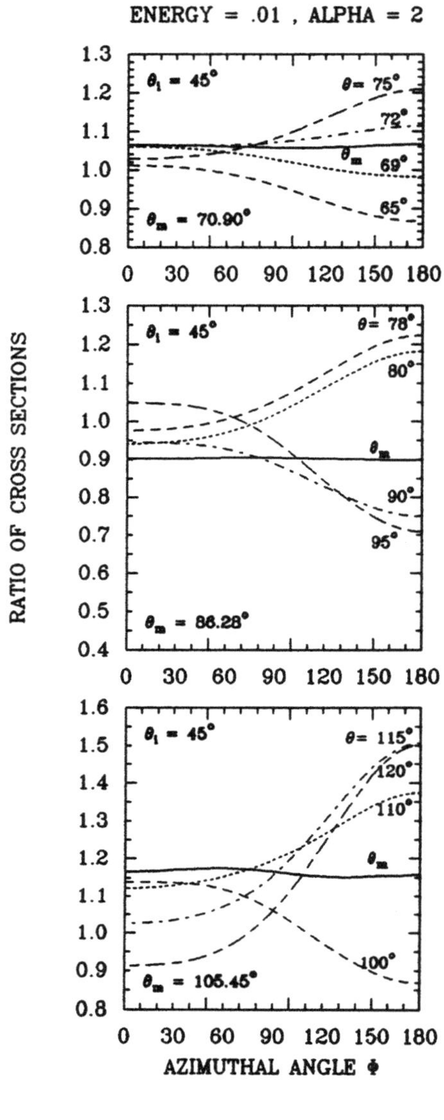

Fig. 4. Ratio $R(\varphi)$ as function of φ, at $\theta_i = 45°$, and fixed θ, as marked, in the vicinity of magic angles θ_m. The full curves represent $R(\varphi)$ for θ_m. The electron energy is $E = 0.01$ Ry and $\alpha_0 = 2\, a_0$.

(2) *Coulomb interference oscillations.* At small and for vanishing θ, $R(\theta)$ has an infinite number of oscillations (which cannot be distinguished in Fig. 3). These are due to the interference term in Eq.(19) and to the fact that f_c contains the oscillating θ-dependent phase factor, Eq.(18b). Their amplitude diminishes for $\theta \to 0$, when the relative importance of the interference term diminishes with respect to $|f_c|^2$, which is more singular. They are enhanced at small energies, by the increase of $\gamma = -1/k$ contained in the Coulomb phase factor, Eq.(18b). Moreoveor, they are strongly dependent on the initial configuration (i.e. on θ_i, and especially on φ). Physically, these oscillations originate in the fact that V_0 is a Coulomb potential with a short range distortion, which yields the split form of the amplitude, Eq.(18a). Whereas Coulomb interference oscillations may occur in other physical situations as well (e.g. high-energy scattering of ions from nuclei[24]) they have, to our knowledge, not yet been detected in atomic physics experiments. Note that, in our case, they appear within a range of parameters well accessible to experiment.

(3) *Backward scattering.* At large scattering angles, $\theta \geq 90°$, the Coulomb interference pattern is distorted and R may deviate substantially from 1. This is markedly dependent on the configuration of the momenta, in particular at $\theta = 180°$, when, at larger α_0 (e.g. Fig. 3), R may vary with θ_i by one order of magnitude.

(4) *Magic angles.* A striking feature in Fig. 3 is that the $R(\theta)$ curves for *all* φ (at given $\theta_i \neq 0$) tend to cross at certain scattering angles θ_m, which we shall call *"magic angles"*. E.g. for $\theta_i = 45°$, $\theta_m = 104.45°$; $86.28°$; $70.90°$, etc.). These curve crossings are not exactly pointwise; we have found them in all cases with $\alpha_0 k < 1$ we have considered. To better illustrate these statements, we consider in Fig. 4 again the case of $\alpha_0 = 2$ a.u., $E = 0.01$ Ry. We now represent the ratio $R(\varphi)$ at $\theta_i = 45°$ as function of φ, for fixed values of θ covering the angular spread between the first magic angles θ_m. Fig. 4 shows how φ-independent R can be at the magic angles, in contrast to its large variations at θ angles in between. Physically speaking, the occurrence of these angles means that, for any $\hat{k}_i$ (or θ_i), if the scattering angle θ is equal to θ_m the scattering is (approximately) axially invariant around $\hat{k}_i$. This represents a particle diffraction effect peculiar to the dressed Coulomb potential, which we were not able to trace back to the symmetry properties of the problem. On the other hand, it can be shown that when $\alpha_0 k >> 1$ (in first Born approximation) there are no magic angles left.

Thus, laser modified elastic scattering from a Coulomb potential at high frequencies (when multiphoton free-free transitions are quenched) may deviate considerably from the Rutherford formula and presents remarkable interference effects. Generalization of our results to a coherent pulse of radiation, with adiabatically varying field amplitude, can be made by adopting the procedure described in Ref. 25. However, a separate calculation would be required to take into account photon statistics.

V. Outlook

We have shown above that the high-frequency theory of laser-atom interactions leads to definite and rather unusual predictions. The results obtained so far refer, however, to the high-frequency limit, when multiphoton transitions are inhibited. As in reality ω will be finite, no matter how high, in order to improve on our results it is essential to consider the next order approximation which yields transition amplitudes and rates (see Eq.(10)). These formulas involve the wave functions of the high-frequency limit. Once the latter were obtained, we could pass to the calculation of transition amplitudes, now in progress.

On the other hand one may inquire how the predictions of the theory could be checked by experiment, in view of the limitations on both of them. For example, to date there are no intense lasers to satisfy the high-frequency condition Eq.(11a) for the ground state of atomic hydrogen (at all α_0 encountered during the rise of the laser pulse, starting from 0). In this respect one way out is to study ionization (e.g. the photoelectron energy spectrum) from excited atomic states for which the condition can be satisfied even by the infrared photons of the powerful existing lasers (e.g. Nd:YAG). An analysis shows that an experiment along these lines appears conceivable at the present time.

References

1. *"Ultrafast Phenomena VI"*, Editors: T. Yajima, K. Yoshihara, C.B. Harris and S. Shionoya (Springer Verlag, 1988), Parts I and II.
2. C.K. Rhodes, Phys. Scripta **T17**, 193 (1987).
3. M. Gavrila and J.Z. Kaminski, Phys. Rev. Lett. **52**, 614 (1984); and in preparation.
4. M. Gavrila, in *Fundamentals of Laser Interactions*, Ed. F. Ehlotzky (Lecture Notes in Physics, vol. 229; Springer, 1985), p. 3.
5. M.J. Offerhaus, J.Z. Kaminski and M. Gavrila, Phys. Lett. **112A**, 151 (1985).
6. M. Gavrila, M.J. Offerhaus and J.Z. Kaminski, Phys. Lett. A **118**, 331 (1986).
7. M. Pont and M. Gavrila, Phys. Lett. A **123**, 469 (1987).
8. M. Pont, M.J. Offerhaus and M. Gavrila, Zeits. f. Phys. **9**, 297 (1988).
9. M. Pont, N.R. Walet, M. Gavrila and C.W. McCurdy, Phys. Rev. Lett. **61**, 939 (1988).
10. M. Pont, N. Walet and M. Gavrila, submitted for publication.
11. J. van de Ree, J.Z. Kaminski and M. Gavrila, Phys. Rev. A **37**, 4536 (1988), and in preparation.
12. H.A. Kramers, *Collected Scientific Papers* (North-Holland, Amsterdam, 1956), p. 866.

13. W.C. Henneberger, Phys. Rev. Lett. **21**, 838 (1968).
14. M. Gavrila and J.Z. Kaminski, in preparation.
15. J.I. Gersten and M.H. Mittleman, J. Phys. B **9**, 2561 (1976).
16. M.H. Mittleman, *Theory of laser-atom interactions* (Plenum, New York, 1982), Chap. 3.
17. P. Kruit, J. Kimman, H. Muller and M. van der Wiel, Phys. Rev. A **28**, 248 (1983).
18. P. Agostini, F. Fabre and G. Petite, in: *Multiphoton Ionization of Atoms*, Eds. S.L. Chin and P. Lambropoulos (Academic Press, New York, 1984) p.133.
19. H. Muller and A. Tip, Phys. Rev. A **30**, 3039 (1984).
20. N.L. Manakov, V.D. Ovsiannikov and L.P. Rapoport, Phys.Reports **141**, 319 (1986).
21. Shih-I-Chu, Adv. Atom. Molec. Phys. **21**, 197 (1985).
22. J. van de Ree, J. Phys. B **15**, 2245 (1982).
23. A. Messiah, *Quantum Mechanics* (North Holland Publ. Co., Amsterdam, 1975), Secs. XI-11 and XIX-11.
24. D.M. Brink, *Semi-classical Methods for Nucleus-nucleus Scattering* (Cambridge University Press, Cambridge, 1986), Chap. 2.
25. H.A. Krüger and Chr. Jung, Phys. Rev. A **17**, 1706 (1978).

PHOTOIONIZATION AND DISSOCIATION OF THE TRIATOMIC HYDROGEN MOLECULE

H. Helm, L. J. Lembo, P. C. Cosby and D. L. Huestis
Molecular Physics Laboratory
SRI International, Menlo Park, CA 94025

ABSTRACT

Rotationless triatomic hydrogen molecules are synthesized in a fast molecular beam. Laser excitation of the molecules allows study of electric field induced ionization as well as autoionization of the ns, np, nd, and nf Rydberg series of H_3. Two photon ionization and double resonance techniques are applied to determine vibrational frequencies of trihydrogen.

Experimental analysis of the electron-core interaction shows that selective optical excitation of core rotation and core vibration can be achieved concurrent with excitation of the Rydberg electron.

Predissociation of low principal quantum number Rydberg states is observed to lead to the formation of both H_2 + H as well as H + H + H fragments.

INTRODUCTION

Triatomic hydrogen presents topics for experimental and theoretical investigation that are interesting from a variety of points of view. As the simplest polyatomic molecule it should have received extensive experimental attention, yet its electronic emission spectrum was first identified only ten years ago [He79]. From a chemist's perspective H_3 consists of H_2 with an extra hydrogen atom, or the unstable transition state in the most fundamental of chemical reactions,

$$H + H_2 \rightarrow H_2 + H \ .$$

The unstable nature of the ground state of H_3 easily explains the difficulties in its quantitative experimental investigation.

From a physicist's perspective we might think of H_3 as the combination of the exceptionally stable molecular ion H_3^+ with an extra electron. In our investigations we take this perspective literally. We begin with a beam of H_3^+ molecular ions, to which we attach a single electron by charge transfer in a Cs vapor cell. Most of the resulting H_3 neutrals are predissociated by the vibrational continuum of the ground electronic state. The few that remain are in a single rotational level of a single electronic state and survive only because they have a symmetry that is not present in the ground-state continuum [GP83,He86]

We detect the existence of these levels by laser photoabsorption, leading either to photodissociation, in which case the H_2 + H or H + H + H fragments are detected in coincidence by a position-sensitive detector, or by reionization, in which case reformed H_3^+ ions are detected. The resulting spectra are remarkably simple, appearing more like that of the hydrogen atom, than that of the hydrogen molecule. The objective of our studies is to quantitatively investigate the interactions among the relatively uncoupled electronic, vibrational, and rotational motions of H_3.

In this lecture we first develop the symmetry classification of molecular wavefunctions, and the selection rules that govern radiative transitions and configuration interaction in this molecule. We then discuss experimental examples of different types of interactions between the Rydberg electron and the H_3^+ core, and our observations of predissociation, autoionization, and field ionization.

SYMMETRY CLASSIFICATION

The "overall symmetry" classification of rovibronic wavefunctions for symmetric-top molecules is used to label the rovibrational levels of H_3^+ and D_3^+ and the rotational levels of the vibrationless excited states of H_3 and D_3 [CP77,Ho62,Bu79,HHW84]. These overall symmetry labels are the same as for the representations of the normal symmetry point group of the nonrotating molecule, but the rotational wavefunction also makes a contribution, since changing the numbering of the nuclei requires a reanalysis of the Euler angles. To develop selection rules for radiative transitions and for perturbations there are two things to accomplish (1) rovibronic wavefunctions must be combined with appropriate nuclear spin wavefunctions to achieve the correct Fermi-Dirac or Bose-Einstein permutation symmetry and (2) the parities of the levels must be established.

As might be expected, some levels have rovibronic wavefunctions that are antisymmetric with respect to nuclear exchange, which for H_3 must be combined with symmetric nuclear spin wavefunctions. These are called ortho levels, having a total nuclear spin of I = 3/2, and spatial symmetry labels A_2' and A_2'', with and even odd inversion parity, respectively. Since we cannot construct an antisymmetric nuclear spin wavefunction out of three spin 1/2 nuclei, exchange-symmetric rovibronic wavefunctions are forbidden (these would be labeled A_1' and A_1'', and are allowed for D_3). A third kind of rovibronic wavefunction is neither exchange-symmetric nor antisymmetric. These are to be combined in a direct-product sense with the similarly asymmetric I = 1/2 nuclear spin wavefunctions, resulting in so called para levels, with labels E' and E".

Symmetry Classification of Rovibronic Levels

To describe the symmetry properties of the rovibronic levels of H_3 we construct a basis set by multiplying together wavefunctions for the angular parts of degenerate vibration $\left(\chi_{v\ell}\right)$ and motion of a single electron $\left(Y_{L\lambda}\right)$ in the body-fixed frame and rotation of the entire molecule in the laboratory frame $\left(D^{(N)}_{KM}\right)$. Thus we write

$$\Psi^{vLN}_{\ell\lambda K} = \chi_{v\ell}\cdot Y_{L\lambda}\cdot D^{(N)}_{KM}$$

The vibrational-projection quantum number ℓ indicates the vibrational angular momentum, and ranges over v,v-2,...-v, where v is the total number of vibrational quanta in the ν_2 (bending or asymmetric stretch) mode (see [Bu79] pp196-200). The electronic orbital angular momentum quantum number L is expected to be rather well defined for the outer electron in Rydberg states of H_3^*. For the discussion below we only need to know if it is even or odd. For H_3^+ we can take $L = \lambda = 0$.

Classification of the states is accomplished using Hougen's three-part symmetry operations [Ho62],

$$Op = {}^1Op \cdot {}^2Op \cdot {}^3Op ,$$

where we each of the three operators, ${}^1Op$, ${}^2Op$, and ${}^3Op$, is chosen from the from the same symmetry group, D_{3h} in the case of H_3. The first type of operator, ${}^1Op$, affects only body fixed coordinates, i.e. $\chi_{v\ell}$ and $Y_{L\lambda}$. The second, ${}^2Op$, affects only the angles of the $D^{(N)}_{KM}$. The third, ${}^3Op$, affects only the center-of-mass coordinates in the laboratory frame. Hougen describes how to apply the operators C_n (n = 3 for H_3), σ_h, and σ_v on the rotation matrices $D^{(N)}_{KM}$. He also says that ^1C_n rotates by $-2\pi/n$ while ^2C_n rotates by $+2\pi/n$. We can easily work out the transformation properties of $\chi_{v\ell}$ and $Y_{L\lambda}$ in the body-fixed frame (supposing that the phases of $\chi_{v\ell}$ can be chosen to be the same as for $Y_{v\ell}$). Thus we can write the results of multiplying the group operators as

$$C_n\,\Psi^{vLN}_{\ell\lambda K} = e^{-2\pi i(\ell+\lambda-K)/n}\,\Psi^{vLN}_{\ell\lambda K} = e^{-2\pi iG/n}\,\Psi^{vLN}_{\ell\lambda K}$$

$$\sigma_h\,\Psi^{vLN}_{\ell\lambda K} = (-1)^{v+L-\ell-\lambda+K}\,\Psi^{vLN}_{\ell\lambda K} = (-1)^{v+L-G}\,\Psi^{vLN}_{\ell\lambda K}$$

$$\sigma_v\,\Psi^{vLN}_{\ell\lambda K} = (-1)^{N+\ell+\lambda-K}\,\Psi^{vLN}_{-\ell-\lambda-K} = (-1)^{N+G}\,\Psi^{vLN}_{-\ell-\lambda-K}\,,$$

where we have introduced the quantity $G = \ell+\lambda-K$, which Hougen calls the "convenient" quantum number. G corresponds to the projection onto the principal axis of the angular momentum associated with the mechanical rotation of the H_3 frame. It is therefore related to the nuclear-exchange symmetry of the level. We are now ready to begin classifying the total symmetry of our rovibronic wavefunctions.

C_3: If G is 0 or a multiple of 3, the wavefunction has a character of +1 under rotation. This means that it belongs to one of the one-dimensional representations: A_1', A_1'', A_2', or A_2''. If G is not a multiple of 3, then the representation is either E' or E".

σ_h: The operation of σ_h allows us to decide between symmetric reflection in the plane (A_1', A_2', and E' for v+L-G even) and antisymmetric reflection (A_1'', A_2'', and E" for v+L-G odd).

σ_v: Finally, we use σ_v to decide on the subscript for the one-dimensional representations (i.e. for G = 3t, t=0,1,2,...). If ℓ, λ, and K are all simultaneously zero then $\Psi^{vLN}_{\ell\lambda K}$ and $\Psi^{vLN}_{-\ell-\lambda-K}$ are identical, and the character for σ_v is $(-1)^N$ (+1 means A_1' or A_2'', -1 means A_1'' or A_2') otherwise we form linear combinations.

$$\left(\Psi^{vLN}_{\ell\lambda K} + \eta\,(-1)^{(N+G)}\,\Psi^{vLN}_{-\ell-\lambda-K}\right)/\sqrt{2}$$

and get both +1 and -1 (i.e. the value of η chosen) as characters. Thus two representations are obtained, either both A_1' and A_2', or both A_1'' and A_2''.

Nuclear-spin Labeling

Following equations (2-9), (7-204), (7-206), and (7-251) of Bunker [Bu79] we identify Hougen's σ_v operator as the overall-inversion or parity operator E^*. Similarly, Hougen's product $\sigma_h \sigma_v$ is identified as exchanging two protons. We follow Townes and Schawlow [TS55 p69-71] and construct a nuclear spin basis set consisting of three $m_I = 1/2$ states (G = 0, 1, 2) of the form

$$\Phi^G_{nspin} = \left[\beta\alpha\alpha + e^{-2\pi iG/3}\,\alpha\beta\alpha + e^{-4\pi iG/3}\,\alpha\alpha\beta \right]/\sqrt{3} \; .$$

A similar set of nuclear-spin basis functions can be constructed for $m_I = -1/2$ by replacing all the α's by β's and vice versa. By applying the nuclear raising and lowering operators (I_+ and I_-) we can show that these wavefunctions give $I_{total} = 3/2$ for G = 3t and $I_{total} = 1/2$ for G = 3t+1 and 3t+2, these two spins corresponding to ortho and para respectively.

We then construct normalized, combined rovibronic/nuclear-spin wavefunctions for both the ortho and para states of H_3 of the form

$$\Xi^{vLN}_{\ell\lambda K} = \left[\Psi^{vLN}_{\ell\lambda K}\cdot\Phi^{\ell+\lambda-K}_{nspin} - (-1)^{v+L+N}\,\Psi^{vLN}_{-\ell-\lambda-K}\cdot\Phi^{-\ell-\lambda+K}_{nspin} \right]$$

$$\times \left[(2N+1)/8\pi\left(1+\delta_{\ell 0}\delta_{\lambda 0}\delta_{K0}\right) \right]^{1/2} \; .$$

Such a wavefunction will transform into itself when we apply the C_3 operator (C_3 corresponds to cyclic permutation of all three nuclei, with respect to which the total wavefunction must be symmetric). Similarly, applying the operator $\sigma_h\sigma_v$ results in interchange of the $\ell+\lambda-K$ term with the $-\ell-\lambda+K$ term with the appropriate sign that it always is antisymmetric, thus satisfying Fermi-Dirac nuclear permutation rules. For $G = \ell - \lambda - K = 3t$ it has spatial symmetry classification A_2' if v + L - G is even and A_2'' if v + L - G is odd. Similarly, for G = 3t+1 and 3t+2, the spatial symmetries are E' and E" for v + L - G even and odd, respectively. The wavefunction vanishes identically for quantum numbers that prevent construction of an antisymmetric combination, i.e. v + L + N even and $\ell = \lambda = K = 0$.

Transformation of Ordinary Objects from the Laboratory Frame

The angular part of the wavefunction for a single particle that is uncorrelated with the molecular frame, say a photon or free electron, can be expressed in the laboratory frame as a spherical harmonic, $Y_{JM}(\text{lab})$. Following Chang and Fano [CF72] we can rewrite it, using rotation matrices, as a combination of spherical harmonics in the body-fixed frame,

$$Y_{JM}(\text{lab}) = \sum_{\lambda} Y_{J\lambda}(\text{body}) \cdot D^{(J)}_{\lambda M}$$

Such an object has $v = \ell = 0$, $L = N = J$, $G = 0$. We can immediately apply Hougen's symmetry operators, establishing that the symmetry classification is A_1' or A_1'' if J is even or odd, respectively.

Optical Selection Rules

A particularly interesting case is the dipole transition operator, which has J=1, and thus is of symmetry A_1'', and can induce the transitions $A_1' \leftrightarrow A_1''$, $A_2' \leftrightarrow A_2''$, and $E' \leftrightarrow E''$. Thus without even knowing about the existence of nuclear spin labeling we have divided the energy levels into three distinct classes, within which optical transitions are possible, but between which optical transitions are strictly forbidden.

The states of H_3^+

We construct normalized rovibrational wavefunctions (including nuclear spin) for the ground electronic state of H_3^+, with total angular momentum N^+ and figure-axis projection K^+, of the form

$$\theta^{vN^+}_{\ell K^+ M} = \left(\chi_{v\ell} \cdot D^{(N^+)}_{K^+M} \cdot \Phi^{\ell - K^+}_{\text{nspin}} - (-1)^{v+N^+} \chi_{v-\ell} \cdot D^{(N^+)}_{-K^+M} \cdot \Phi^{-\ell+K^+}_{\text{nspin}} \right) \times \left[(2N^+ +1)/8\pi \left(1+\delta_{\ell 0}\delta_{K^+0}\right) \right]^{1/2} .$$

Wavefunctions for H_3 in Hund's Case (d)

We follow Chang and Fano [CF72] in combining a core state of H_3^+ (with quantum numbers v, ℓ, N^+, K^+, and M) with a Rydberg electron (with laboratory angular quantum numbers L and m) using angular momentum addition to obtain states of well defined total rovibronic angular momentum N:

$$\gamma^{v\ell L N^+ K^+}_{NM} = \sum_{m} Y_{Lm}(\text{lab}) \cdot \theta^{vN^+}_{\ell K^+ M-m} \; (Lm, N^+ M-m \,|\, LN^+ NM) \; .$$

We use the frame transformation introduced above to write this Hund's case (d) wavefunction in terms of Hund's case (a/b) wavefunctions,

$$\gamma^{v\ell L N^+ K^+}_{NM} = \sum_{\lambda} \Xi^{vLN}_{\ell\lambda\lambda+K^+} \left[(2N+1)/8\pi \left(1+\delta_{\ell 0}\delta_{\lambda 0}\delta_{K^+0}\right) \right]^{-1/2} \times \left[(2N^+ +1)/8\pi \left(1+\delta_{\ell 0}\delta_{K^+0}\right) \right]^{1/2} (L\lambda, N^+K^+ \,|\, LN^+ N\lambda+K^+) \; ,$$

and to write Hund's case (a/b) wavefunctions in terms of Hund's case (d) wavefunctions,

$$\Xi^{vLN}_{\ell\lambda\lambda+K^+} = \sum_{N^+} \gamma^{v\ell LN^+K^+}_{NM} (-1)^{(L-\lambda)} (L\lambda,N-\lambda-K^+|LNN^+-K^+)$$
$$\times \left[\left(1+\delta_{\ell 0}\delta_{K^+0}\right)/\left(1+\delta_{\ell 0}\delta_{\lambda 0}\delta_{K^+0}\right)\right]^{1/2} .$$

Metastability

All of the excited states of H_3 lie in the vibrational continuum of the repulsive ground state, which correlates asymptotically to $H_2(X^1\Sigma_g^+)$ + H(1s). At the equilibrium geometry of H_3^+ this state is called $\tilde{X}$ 2p $^2E'$, which we identify as L = 1, λ = 1. The symmetries of the levels of the $\tilde{X}$ are classified, as functions of angular momentum, N, and bending vibration, v, as follows:

N = 0 We must have K=0. Thus v + L - G is always even, and the possible states have symmetry A_2' and E' (plus A_1' for D_3).

N > 0 K can be either even or odd, thus v + L - G is correspondingly either even or odd, giving states of symmetry A_2' and E' or A_2'' and E" (plus A_1' or A_1'' for D_3), respectively.

Thus we conclude that all of the excited states of H_3 and D_3 are predissociated for N > 0, since all possible symmetries are present in the ground state. For an excited state with L odd, predissociation is mediated by rotation, while for L even it is mediated by bending vibration.

Any states that are metastable with respect to predissociation are limited to N=0, and must have one of the symmetries A_2'' or E" (plus A_1'' for D_3). Since K = 0, v + L - G = v + L - ℓ - λ must be odd, which allows us to establish that L - λ must be odd, since v - ℓ is always even. The low-lying choices are only L = 1, λ = 0, and L = 2, λ = 1, which reduces to the single electronic state, $\tilde{B}$ 2p $^2A_2''$, if we insist that the state also be radiatively metastable. This state is predicted to be immune to predissociation for all rotationless vibrational levels.

EXPERIMENTAL RESULTS

Formation of H_3^*

We form the metastable species of H_3 by electron transfer from a stationary alkali target (Cs or Rb) to a mass selected H_3^+ ions in a fast beam. As mentioned above only a small fraction of charge transfer products are metastable [JB84]. Typically 10^7 to 10^9 metastable molecules are produced per second in a 1 mm diameter beam traveling at energies of several keV. Experimental schematics for the laser photoionization and photodissociation spectrometers have been presented in recent publications [LHH89a,HC87].

Photodissociation in the Visible

The lowest photoexcitation transitions from the $\tilde{B}$ 2p $^2A_2''$ are to the 3s $^2A_1'$ and 3d$^2E''$ states at 16694 and 17297 cm^{-1} respectively, as reported by Herzberg et al. [DH80,HHW84]. In our experiments, the initial state is unique, having N"=0, thus the one-photon absorption spectrum can have only a single rotational branch with N'=1. These same transitions are easily observed by collecting H_2 + H photodissociation products, as shown as peaks labeled 3sν_0 and 3dν_0 in Figure 1 [CH88]. The energy analysis capability of the position-

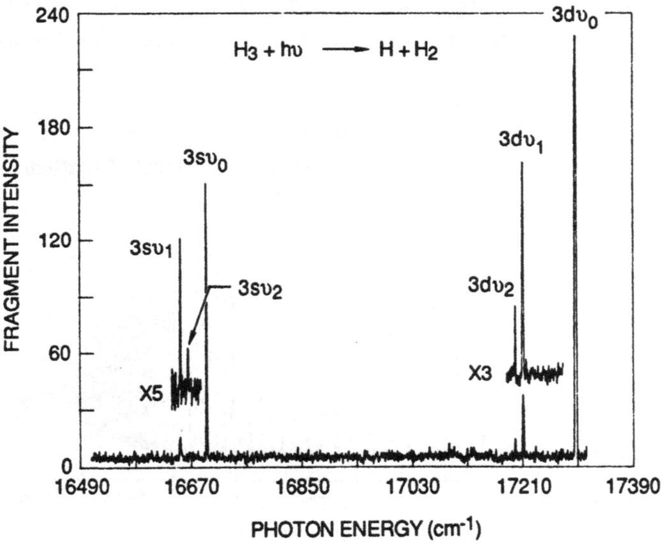

Figure 1: Observed production of photofragments from H_3 as a function of the photon energy of the exciting laser. Only photofragments produced with center-of-mass energies W > 4 eV are recorded for this spectrum. The spectrum at other energy releases is similar, but with a substantial background due to spontaneous dissociation fragments.

sensitive detector allows us to show that predissociation of these two states leads to substantially different distributions of vibrational energy in the H_2 fragment. For the 3s state, H_2 is observed in vibrational levels from v=0 to v>12. For the 3d state, however, a significance preference is observed for the formation of H_2 with v=5-9. In addition to predissociation to H_2 + H products, both the 3s and 3d states are observed to predissociate into three atoms, H + H + H. Also shown in Figure 1 are transitions from vibrationally excited $\tilde{B}$ 2p $^2A_2''$ initial states to 3s and 3d states with the corresponding vibrational excitation (labeled $3s\nu_1$, $3s\nu_2$, $3d\nu_1$, $3d\nu_2$). These indicate that vibrationally excited metastables are present in the beam (their relative abundance depends on the ion source conditions) and provides the first direct information about the vibrational frequencies of the excited states of H_3.

Field Ionization in the Ultraviolet

The vibrational structure described above provides an indication of the role to be played by the H_3^+ core in making more interesting the pseudo-atomic H_3^* absorption spectrum. We can explore this core structure in somewhat more detail by gradually promoting the outer electron to higher principal quantum numbers in the Rydberg series nd ← 2p, and detecting the photoabsorption by field ionization. The resulting spectrum is shown in Figure 2, in which nd levels from

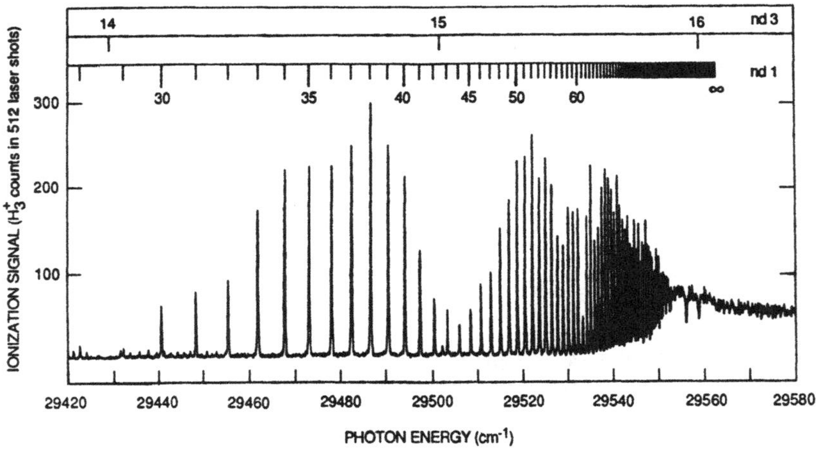

Figure 2: Photoabsorption of the lowest metastable level of H_3, detected by field ionization of the excited d-electron Rydberg series. Series members converging to the N^+=1, K^+=0 level of H_3^+ are marked at the top of the figure as nd1. A series of interlopers, nd3, converging to the rotational level N^+=3, K^+=0 is also indicated. Three additional window resonances appear within 30 cm^{-1} of the threshold.

n = 28 to n = 100 have been assigned [He88]. Field ionization has also been used by Dodhy _et al._ [DKM89] to detect nd states for n as low as 14. All n values from n = 3 to n = 100 have been detected by double-resonance ionization-depleting spectroscopy [LHH89b]. Extrapolation of this series has enabled the accurate determination of the ionization potential of the rotationless $\tilde{B}$ 2p $^2A_2''$ state as 29562.58 ± 0.5 cm^{-1}. The intensity modulations shown in Figure 2 are significant, although weakly dependent on laser power, and show the influence of having multiple Rydberg series converging to different rotational states of the H_3^+ core. A particular example is the broad dip in signal near n = 42, which appears to correspond to an n = 15 interloper from the N^+ = 3 level of H_3^+. Attempts to model this spectrum suggest that the actual situation is not quite so simple.

Vibrational Autoionization

Just to the red of the spectrum shown in Figure 2, additional photoionization peaks appear that reveal photoabsorption into levels that ionize spontaneously without an external field. These are shown in Figure 3, and as indicated by the labels, are the higher members of the ns ← 2p and nd ← 2p Rydberg series from the vibrationally excited metastables. These peaks appear in the ionization spectrum because the upper levels lie energetically above that of the vibrationless H_3^+

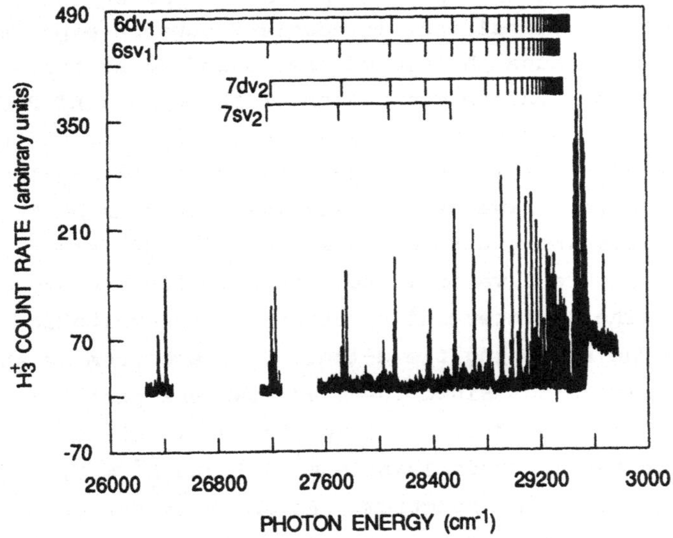

Figure 3. Vibrational autoionization of s- and d-electron Rydberg series converging to the bending excited core level (ν_2) and symmetric stretch excited core level (ν_1) of H_3^+.

core plus a free electron. The autoionization mechanism is depicted in Figure 4. Note that n = 6 in the symmetric-stretch (ν_1) series and n = 7 in the bending vibration (ν_2) series are the lowest members that are subject to vibrational autoionization.

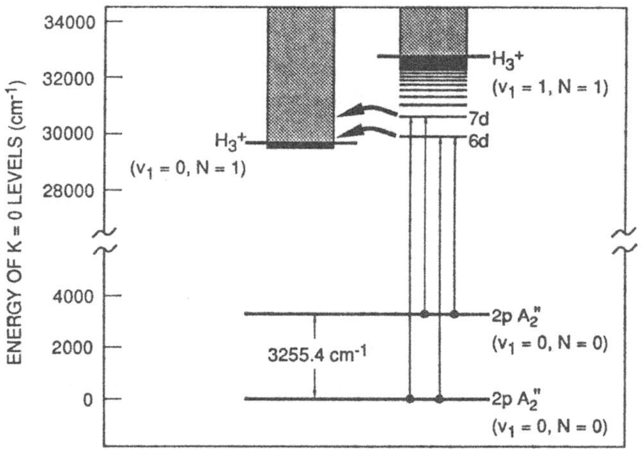

Figure 4. Energy level diagram of H_3 showing diagonal and off-diagonal excitation to the vibrationally autoionizing 6d and 7d Rydbergs with symmetric stretch excitation.

Two-Photon Excitation of np and nf Rydberg Series

The great strength of the 3s ← 2p and 3d ← 2p transitions suggests using these transitions in double-resonance experiments. In this paragraph we discuss the use of the 3s and 3d states as intermediates for two-photon excitation of the np and nf Rydberg series that are of the same symmetry as the 2p initial state. Two-step excitation also allows us to reach higher total angular momentum, i.e. N=0,1,2, and to influence the relative strengths of the transitions by choosing intermediate states with different core properties. In Figure 5 we indicate schematically the excitation pathways used. The 3s intermediate state favors excitation of np final states. The resulting field-ionization spectrum is shown in the lower trace in Figure 6. Since there are two one-photon accessible 3d states, called 3dE" and $3dA_1'$, and since these two states have core rotational character of predominantly $N^+ = 1$ and $N^+ = 3$, respectively, we can bias the absorption spectrum. As shown in the upper two traces in Figure 6, choosing the 3dE" intermediate leads to a field-ionization spectrum labeled nf1, by which we mean an nf Rydberg orbital outside an $N^+ = 1$ H_3^+ core. Similarly, choosing the $3dA_1'$ intermediate leads to a substantially different spectrum, called nf3.

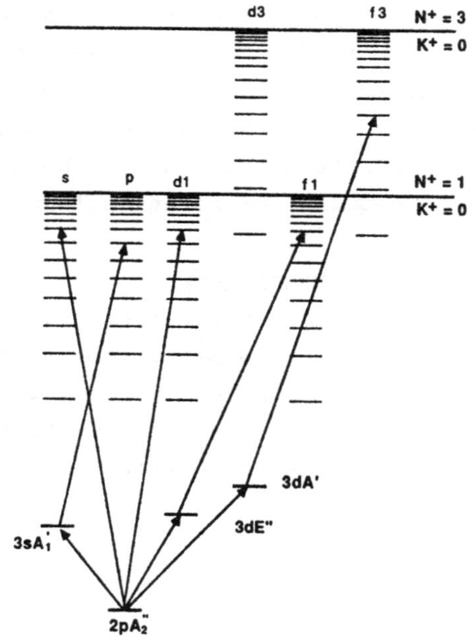

Figure 5. Energy level diagram of N=1 Rydbergs built on the lowest rotational cores of ortho-H_3. Direct and stepwise excitation paths to these Rydbergs from the metastable $\tilde{B}$ 2p $^2A_2$" state are indicated.

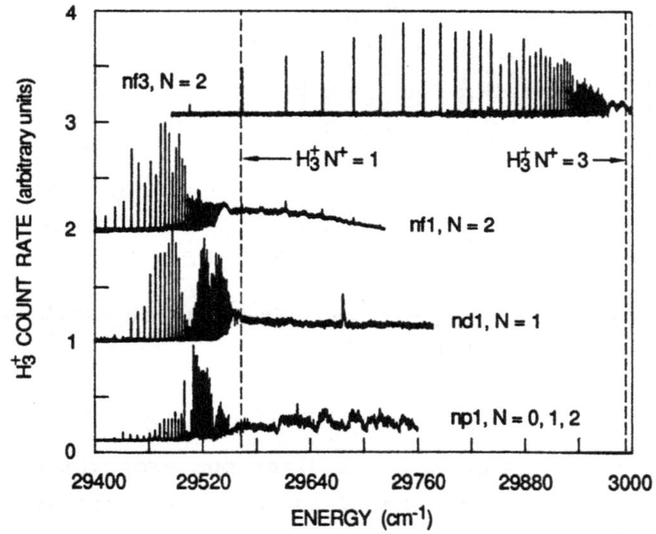

Figure 6: Excitation spectra into the field-ionized and rotationally-autoionized regions of the lowest p-, d-, and f- Rydberg series.

Interestingly, some of the peaks appear in both spectra. The narrow peaks in the right hand part of the trace labeled nf1, above the $N^+ = 1$ ionization limit (dashed line), agree in position with the peaks in the nf3 trace and manifest the weak admixture of the $N^+ = 3$ core in the $3dA_1'$ intermediate state. Above the $N^+ = 1$ threshold the spectra result as a consequence of rotational autoionization.

Vibrationally Off-Diagonal Transitions

Finally, we have examined the spectral region between the n = 3 and n = 4 members of the Rydberg series. Absorptions are detected by one- or two-color photoionization, or by ionization depletion of another transition. Some of the transitions observed are shown schematically in Figure 7 and as a spectrum in Figure 8. The peaks

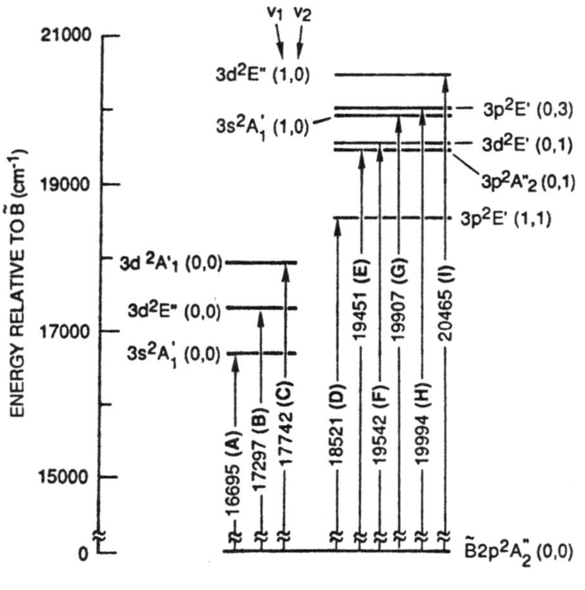

Figure 7: A selection of diagonal and off-diagonal excitations observed into the n=3 manifold of H_3

labeled "G" and "I" correspond to simultaneous excitation of the electron, 3s ← 2p or 3d ← 2p, and of the symmetric vibration mode, ν_1. Peaks "D", "E", "F", and "H" result from excitation of the bending vibrational mode, ν_2. But here the simultaneous electronic excitation is unusual. Only electronic transitions that previously were forbidden by parity, i.e. 3p ← 2p, are now possible. This is because the bending vibrational wavefunction changes the overall parity of the

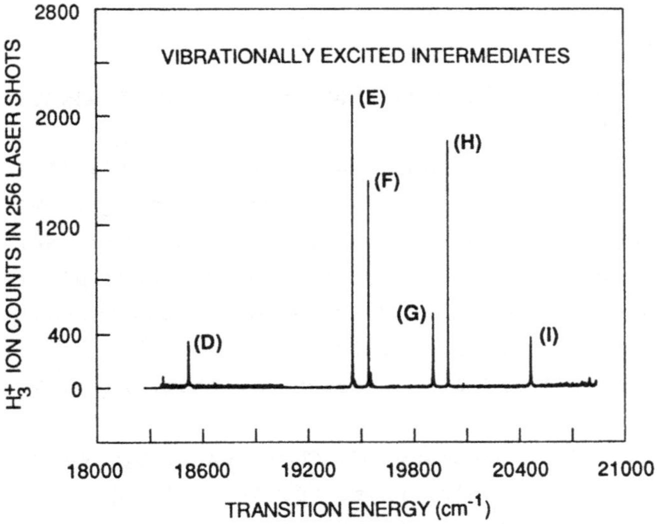

Figure 8: One-photon resonant two-photon ionization spectra involving rovibronic transitions marked in Figure 7.

state. Such vibronically allowed transitions have also been observed in other polyatomic molecules and provide a detailed and specific probe of the electron-core interactions.

Summary

We have seen that the $\tilde{B}$ 2p $^2A_2''$ state of H_3 is metastable only for N = 0, but for any v. It is easily produced by charge transfer and is straightforward to investigate using photodissociation and photoionization. It shows a very simple, atomic-like absorption spectrum, with relatively weak, but revealing, interactions between the outer Rydberg electron and the vibrational and rotational motions of the underlying H_3^+ core.

Acknowledgments

This research was supported by the Air Force Office of Scientific Research and the National Science Foundation.

References

Bu79 P. R. Bunker, Molecular Symmetry and Spectroscopy, (Academic Press, NY, 1979).

CF72 E. S. Chang and U. Fano, Phys. Rev. A 6, 173 (1972).

CH88 P. C. Cosby and H. Helm, Phys. Rev. Lett. 61, 298 (1988).

CP77 C. D. Carney and R. N. Porter, J. Chem. Phys. 66, 2758 (1977).

DH80 I. Dabrowski and G. Herzberg, Can. J. Phys. 58, 1238 (1980).

DKM89 A. Dodhy, W. Ketterle, H.-P. Messmer, and H. Walther, Chem. Phys. Lett. 151, 133, (1988).

GP83 G. I. Gellene and R. F. Porter, J. Chem. Phys. 79, 5975 (1983).

HC87 H. Helm and P. C. Cosby, J. Chem. Phys. 86, 6813 (1987).

He79 G. Herzberg, J. Chem. Phys. 70, 4806 (1979).

He86 H. Helm, Phys. Rev. Lett. 56, 42 1986).

He88 H. Helm, Phys. Rev. A 38, 3425 (1988).

HHW84 G. Herzberg, J. T. Hougen, and J.K.G. Watson, Can. J. Phys. 60, 1238 (1984).

Ho62 J. T. Hougen, J. Chem. Phys. 37, 1433 (1962).

JB84 J. R. Peterson and Y. K. Bae, Phys. Rev. A 30, 2807 (1984).

LHH89a L. J. Lembo, H. Helm, and D. L. Huestis, J. Chem. Phys. (in press).

LHH89b L. J. Lembo, D. L. Huestis, and H. Helm, to be published.

LPH89 L. J. Lembo, A. Petit, and H. Helm, Phys. Rev. A (in press).

TS55 C. H. Townes and A. L. Schawlow, Microwave Spectroscopy (McGraw Hill, NY, 1955; Dover, NY, 1975).

CONTRIBUTED PAPERS

(Abstracts)

MULTIPHOTON IONISATION OF XENON WITH TUNABLE SUBPICOSECOND LASER PULSES

Guillaume Petite, Pierre Agostini, Harm G. Muller (*) and Alain Migus (**)

Service de Physique des Atomes et des Surfaces

CEN Saclay, 91191, Gif sur Yvette France

(*) Permanent Adress: FOM Institut voor Atoom en Molecuulfysica

1098 SJ Amsterdam, The Netherlands

(**) Laboratoire d'Optique Appliquee, Ecole Polytechnique

ENSTA, 91120, Palaiseau France

As recently shown (see chapter by R.R. Freeman in this book), Electron Energy Spectra obtained in experiments with subpicosecond pulses show a series of ATI peaks with pronounced substructures which were assigned to intensity induced resonances occuring during the ionisation process: because of the very high intensity used in such experiments (up to 10^{14} W.cm^{-2}) atomic states can be a.c. stark-shifted into resonance. The electrons released in such processes have energies which, because of the ponderomotive shift of the ionisation threshold, are characteristic of the intensity needed to tune the resonance in.

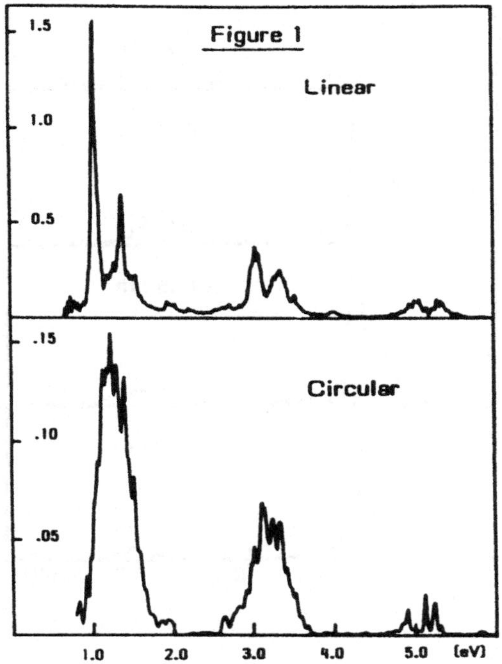

This mechanism is well demonstrated on figure 1. It shows two electron spectra obtained in 7-photon ionisation of Xenon at about 10^{13} W.cm^{-2}, by 120 fs pulse at 615 nm. The top spectrum is obtained with a linear laser polarisation and shows the resonant structure, which is absent of the bottom spectrum, obtained in the same conditions, but with a circular laser polarisation, due to the angular momentum selection rules governing multiphoton resonances. There is a noticeable decrease in the count rate in circular polarisation, which is due to the closure of many resonant channels.

An assignment of the many resonances in a spectrum to atomic states requires the knowledge of each state's stark coefficients. Their computation in the high intensity regime can only be approximate. It is not possible to make an unambiguous choice between the different approximations on the basis of measurements at a single wavelength only. Much more informations can be obtained from an experiment using tunable subpicosecond pulses. A set of spectra obtained for wavelength ranging from 600 to 660 nm are shown on fig. 2. One observes:

(i) very small shifts of the different resonant peaks, showing that the resonant states shift upward, approximately like the continuum limit.

(ii) the disappearance of subpeaks of higher energy when the photon energy is decreased: states shifted upward can only cause a resonance whith a six photon transition of energy larger than their unperturbed energy. This can be used for assigning these peaks.

At 600 nm, six photons are enough to ionise the atom, yielding electrons of 0.28 eV energy at most. Because of the high intensity, the six photon peak is suppressed and only the seven photon peak is observed. It shows, as in the case of a (lowest order) seven photon process at longer wavelengths, resonances at the six photon level. This shows that bound states have been shifted up to an energy higher than the unperturbed ionisation threshold, while keeping a clear 'discrete

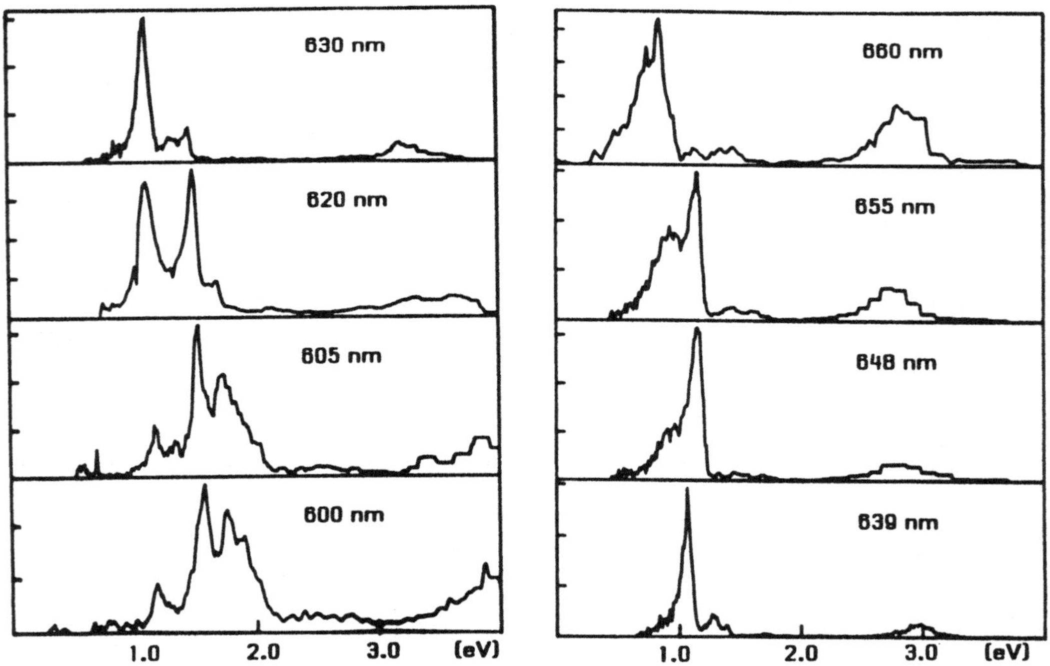

Figure 2

ROLE OF COUNTER-ROTATING WAVE DYNAMICS IN NEAR-THRESHOLD PHOTOIONISATION OF ATOMIC HYDROGEN

B. Piraux*
Laboratoire de Chimie Physique, Université Pierre et Marie Curie
11, rue Pierre et Marie Curie, F75231 Paris Cedex 05 (France)

P.L. Knight
Optics section, Blackett laboratory, Imperial College
London SW7 2BZ, (England)

We consider the excitation of atomic hydrogen in its ground state by an intense radiation field tuned close to the ionisation threshold. We calculate the ac-Stark shift of the 1s and 2s-state and the photoelectron energy spectrum.

Our non perturbative analysis includes the s, p-series of excited states and the Coulomb p-continuum. We omit coupling of excited p-states with opposite parity continua as well as further continuum-continuum transitions; for the frequencies and field intensities considered here ($\hbar\omega \geq$ 10 eV and $I \approx 10^{13}$ W/cm2), we show that these transitions are not dominant. A Laplace-transform solution of the equations of motion for the state probability amplitudes is developed and the non-rotating wave interaction is included by resumming the Floquet type series expansion using continued fraction techniques. Off resonance and in the range of frequencies considered here, the 1s and 2s-state ac-Stark shifts are dominated by the quadratic ac-Stark shift whose evaluation is facilitated by using an integral representation of the Coulomb Green's function.

We find that the dynamics and electron spectra are strongly dependent on the shift and width of the initial state; this shift may be positive or negative depending upon the photon frequency. In particular, we show that for above threshold frequencies, the ground state ac-Stark shift is positive and leads to a slight decrease of the ionisation potential. We also demonstrate that the photoelectron energy spectrum may be strongly affected by the counter-rotating wave dynamics. When the laser is tuned below the ionisation threshold, the p-series of Rydberg states generate a rich structure in the spectrum mainly through the action of the counter-rotating wave terms. Although the coupling between the excited p-states and the s-excited states is very strong, we show that the presence of the

---

* On leave of absence of Imperial College, Optics section, Blackett Laboratory, London SW7 2BZ, England and Clarendon Laboratory, Oxford university, Oxford OXI 3PU, England and Wolfson College, Oxford university, Oxford OX2 6UD, England.

other series of Rydberg states only slightly affects the width of the resonances and does not smear out the structure. In the figure we show the photoelectron energy spectrum produced by a laser tuned 0.54 eV below threshold ($\Delta = -0.54$ eV) resonant with the 5p-state; the laser electric field strength E_0 is equal to 10^8 Volt cm-1. The electron energy E varies between E = 22.7 eV and E = 25.8 eV; in this energy interval, we observe five peaks : the first three peaks are generated by the 2p, 3p and 4p-state respectively while the last two peaks represent an Autler-Townes doublet produced by the 5p state. When the laser is tuned above threshold, the spectrum is dominated by a single-photon rotating wave term which greatly exceeds the multiphoton terms responsible for the Rydberg structure.

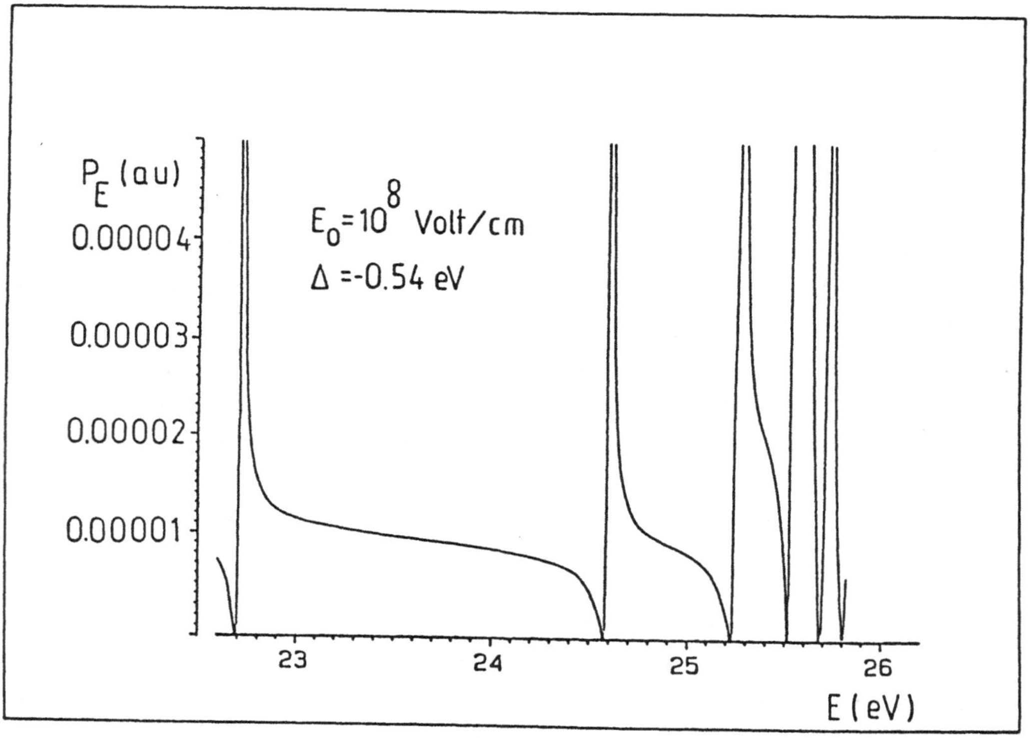

Photoelectron energy spectrum in atomic unit produced by a laser tuned 0.54 eV below threshold ($\Delta = -0.54$ eV) resonant with the 5p-state (the photon energy is equal to 13.06 eV). The laser electric field strength Eo is equal to 10^8 Volt cm-1. The electron energy E varies between E = 22.7 eV and E = 25.8 eV. For clarity, the peak heights have been truncated; the actual values are several orders of magnitude higher.

LASER-ASSISTED ELECTRON-ION RECOMBINATION
AND HIGH-FREQUENCY GENERATION

S. Basile[+], E. Fiordilino[*] and F. Trombetta[*]

[+]Istituto di Fisica Teorica dell'Università,
C.P. 50, 98166 Sant'Agata di Messina, Messina, Italy
[*]Istituto di Fisica dell'Università,
Via Archirafi 36, 90123 Palermo, Italy

Considerable theoretical and experimental efforts have been devoted in these years to the study of highly non linear interaction processes of radiation with matter. Among these the "above threshold ionization" has focused the most interest; moreover, the possibility to obtain, via this latter process, very energetic electrons, has stimulated researches in the direction of studying the recombination processes and the generation of high frequency electromagnetic radiation[1]. The laser-assisted electron-ion recombination has been already considered with reference to photon gain[2] and laser diagnostics in plasmas[3] and more recently experiments and theoretical treatments on multiple harmonic conversion in rare gases have appeared[4].

We have considered two models of the process: the free-bound transition of an electron embedded in a high intensity radiation field with emission of one high frequency photon and the two-step bound-free-bound transition; a Coulomb-Volkov wave is used for the electron state in the continuum (this wavefunction has quite successfully been used in multiphoton ionization processes[5]).

In any case we deal with a structured and dressed continuum; so that transitions from virtual multiphoton continuum states will contribute the process; the energy conservation relation will then allow for an infinite set of discrete frequencies of the emitted photon:

$$\hbar\Omega = E_k + J_o + \Delta + n\hbar\omega \quad , \quad n = -|n_{min}|, \ldots, \infty$$

where E_k, J_o and Δ are the kinetic, ionization and quiver energies; assuming the multiphoton ionization "origin" of the kinetic term, $E_k = N\hbar\omega - J_o - \Delta$, we may describe also the "harmonic generation" of radiation from continuum states.

We have calculated the cross sections for the allowed multiphoton channels of the process for e^--H^+ and e^--He^+ recombination as function of the electron energy and laser parameters (intensity and polarization); preliminary results show that while at low intensities the n = 0 channel is obviously the dominant one, the two sidebands (negative and positive values of n) of the spectrum become dominant with the cross sections for negative n's larger than the corresponding positive counterpart. We are considering in more detail the two-step process that could be in competition with the ionization process itself; parenthetically a high efficiency of the laser-assisted recombination process could provide an additional source of high frequency electromagnetic radiation suitable for ionization; moreover realistic energy distributions (e.g. a distribution emerging from ATI processes or a thermal one for a plasma[3]) will be considered to obtain information of experimental interest.

REFERENCES

1. B.W. Shore and P.L. Knight, J. Phys. B 20, 413 (1987).
2. F.H.M. Faisal, A. Lami and N.K. Rahman, J. Phys. B 14, L569 (1981); A. Lami, N.K. Rahman and F.H.M. Faisal, Phys. Rev. A 30, 2433 (1984).
3. L. Schlessinger and J. Wright, Phys. Rev. A 24, 2541 (1981).
4. A. McPherson, G. Gibson, H. Jara, U. Johann, T.S. Luk, I.A. McIntyre, K. Boyer and C.K. Rhodes, J. Opt. Soc. Am. B 4, 595 (1987); M. Ferray, A. L'Huillier, X.F. Li, L.A. Lompré, G. Mainfray and C. Manus, J. Phys. B 21, L31 (1988); K.C. Kulander and B.W. Shore, Phys. Rev. Lett. 62, 524 (1989).
5. S. Basile, F. Trombetta and G. Ferrante, Phys. Rev. Lett. 61, 2435 (1988).

MULTIPHOTON IONIZATION OF HYDROGEN INDUCED BY TWO STRONG LASER FIELDS

S. Bivona, R. Burlon, G. Ferrante and C. Leone
Dipartimento di Energetica ed Applicazioni di Fisica
Viale delle Scienze 90128 Palermo Italy

In this Communication we present a theoretical approach to the problem of multiphoton ionization of hydrogen atom by two strong laser fields. A restriction is that the frequencies of the two fields are thought to be different (but not much), and the field of the higher frequency to be somewhat weaker than the other. More extreme situations have been considered also previously in [1,2] .

The physical picture arising from our treatment is that of a field (the high frequency one) acting as a perturbation on the atomic states dressed by the other field (below referred to as low-frequency field). From this viewpoint, the theory resembles that of collisions in the presence of lasers, and is an extension to the two field ionization problem of a treatment originally devised by Kowarskii and Perel'man for the single-frequency ionization [3] .

The hamiltonian of the system may be written as

$$H= H_o+W_L(t)+W_H(t) \tag{1}$$

where H_o is the unperturbed atomic hamiltonian and $W_i(t)=e\underline{E}_i \cdot \underline{r} \cos(\omega_i t)$ are the interactions of the e.m. fields with the atom. Both the radiation fields are taken in dipole approximation (i=L, H).

By expanding the full retarded Green function G^+ in terms of the retarded Green function g^+ belonging to the hamiltonian of the atom in the presence of the low-frequency field only, an expansion of the S-matrix of the process is obtained whose two first terms are

$$S^{(1)}_{qi} = (-i/h)\int_{-\infty}^{\infty} dt < \Phi_q(t) \mid W_H(t) \mid \Phi_i(t) > \tag{2}$$

$$S^{(2)}_{qi}=(-i/\hbar)^2 \int_{-\infty}^{\infty} dt \int_{-\infty}^{t} dt' < \Phi_q(t) \mid W_H(t)\, g^+(t,t')\, W_H(t') \mid \Phi_i(t') > \tag{3}$$

with $\Phi_i(t)$ the exact wavefunction of the electron in the initial state and $\Phi_q(t)$ the wavefunction of the ionized electron having asymptotic average momentum $\hbar q$. Both the wavefunctions are dressed by the low-frequency field. By assuming the low-freqency field to be linearly polarized, the hamiltonian $H_L=H_o + W_L(t)$ is diagonalized in each manifold characterized by the principal quantum number n by using wavefunctions written in terms of parabolic states $\mid n_1,n_2,m >$

$$\Psi_{n_1,n_2,m} = \mid n_1,n_2,m > \exp[(-i/\hbar)(E_n\, t + D \sin(\omega_L\, t)] \tag{4}$$

with $n= n_1+n_2+ |m| +1$ and $D = 1.5\, eE_L\, n\, (n_1-n_2)/\omega_L$

By expanding $\exp(-i\, D \sin(\omega_L\, t)/\hbar)$ in terms of Bessel functions, it appears that these states are dressed with all the integer numbers of photons k of frequency ω_L, the dressing strength being measured by the Bessel function of order k .

Expanding g^+ in terms of the zero order Green function g_0^+ constructed by means of the dressed states $\Psi_{n1,n2,m}$, one can give a diagrammatic representation of the S-matrix in which both the interaction terms W_H and W_L causes transitions between dressed states. For instance ,we have for $S^{(2)}_{qi}$

$$S^{(2)}_{qi} = (-i/\hbar)^2 \int_{-\infty}^{\infty} dt \int_{-\infty}^{t} dt' \, \langle \Phi_q(t) \mid W_H(t)\, g_0^+(t,t')\, W_H(t') \mid \Phi_i(t') \rangle \quad +$$

$$(-i/\hbar)^3 \int_{-\infty}^{\infty} dt \int_{-\infty}^{t} dt' \int_{-\infty}^{t'} dt'' \, \langle \Phi_q(t) \mid W_H(t)\, g_0^+(t,t')\, W_L(t')\, g_0^+(t',t'')\, W_H(t'') \mid \Phi_i(t'') \rangle + \ldots \qquad (5)$$

Below we show the angular distributions (AD) of the photoelectrons in the plane of vibration of the low-frequency field calculated by using eq.(2) for the extreme case in which the high-frequency field is comparatively very weak and its frequency so high that a single photon may ionize the atom.The high-frequency field has been assumed to be linearly polarized, and the low-frequency laser circularly polarized, with both the fields vibrating on the same plane. The AD shape exhibits asymmetries which are similar to those observed in multiphoton ionization of noble gases by a single-frequency elliptically polarized light. The explanation of this asymmetries may be given in terms of lowering of symmetry caused by the simultaneous action of the two fields during the ionization of the atom. In the figure, $|n|$ denotes the number of the low-frequency photons exchanged during the ionization, and the angle is measured from the direction of the polarization of the high-frequency field. Dashed lines correspond to emission of the low-frequency photons.

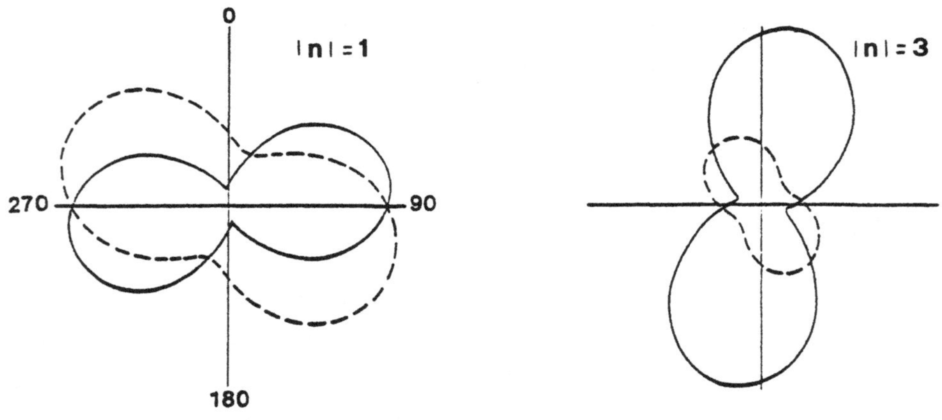

REFERENCES

[1] M.Door and R.Shakeshaft, Phys.Rev.A 36, 421 (1987).
[2] C.Leone, S.Bivona, R.Burlon and G.Ferrante, Phys.Rev.A 37 , 5642 (1988).
[3] V.A.Kovarskii and N.F.Perel'man, Soviet Pysics JEPT 34, 734 (1972).

ENERGY SPECTRUM OF PHOTOELECTRONS PRODUCED BY PICOSECOND LASER INDUCED SURFACE MULTIPHOTON PHOTOEFFECT

Gy. Farkas and Cs. Tóth
Central Research Institute for Physics
Budapest, P.O. Box 49, H-1525, HUNGARY

1. Introduction

The fundamentals of multiphoton theory are the same for atoms and for metals [e.g. 1]. Consequently, the question arises, how the ATI, the ponderomotive effects, etc. [e.g. 2] manifest themselves at high laser intensities for metals?

Existing experimental evidences, connected with these effects:

a, at low laser intensities no electrons with energy higher than $Nh\nu - A$ was found [3].

b, at somewhat higher light intensities electrons having about 10 eV energy were observed [4].

c, at high laser intensity electrons with energy around 20 eV were produced from semiconductor targets [5].

In all the above three cases the inherent space charge effects were not eliminated, which may wash out the high intensity effects of metals, even if they exist.

Our present task is to search for energetic multiphoton photoelectrons emitted from metals at high laser intensities in the absence of space charge. The basic features of our experiment are the following:

1. Application of high intensity, psec, Nd-laser pulses
2. Application of high extracting electrostatic fields
3. Use of an appropriate energy analyser (retarding field method with grids, small solid angle for detecting and electronmultiplier [e.g. 6,7])
4. Use of a good vacuum system ($5*10^{-8}$ mbar, turbomolecular pump)
5. For data evaluation we used the practically simple and reliable numerical metods as in Refs. 6 and 7.

2. Experiment

Single pulses (8 ps, 1 mJ, 1 Hz repetition rate) of a passive mode-locked Nd:glass oscillator-amplifier laser system were directed onto a well polished and cleaned thick gold plate (reflectivity is ~99%) at a grazing incidence (~85°). The electric field of the laser was p-polarized to be perpendicular to the gold surface. In front of the gold plate a mesh, kept on 0- +15 kVolts extracting potential, was placed to eliminate the space charge. Behind this mesh a usual retarding field energy analyser system [6,7] with 0.01 steradian observing solid angle was used. The entrance diaphragm of the analyser was on the same potential (i.e. earth) as the gold target for reobtaining the original, space charge free electron energy spectrum. In this way the accelerating high voltage electrode only eliminates the space charge effect, however it has no effect on the electron energy distribution.

The total number of the electrons emitted from the gold plate was measured directly on a 50 ohm load resistance of an oscilloscope (S_{DIR}), while the analyzed electrons were detected through the analyzer by an electron multiplier (S_{EMP}). The overall interaction system was situated in a vacuum of $5*10^{-8}$ mbar, produced by a turbomolecular pump.

In the course of the experiment we measured the multiplier signal (S_{EMP}) and the direct signal (S_{DIR}) of the gold plate as a function of the extracting field (V_{ext}: 0- +15 kV), as a function of the retarding potential (U_{ret}: 0- -1000 V) and as a function of the incident laser intensity I, which was deduced from the monitoring photocell signal S_I. After the slight focussing the maximum light intensity at the metal surface was of I_0=25 GW/cm². Due to the shot-to-shot fluctuation of the laser pulses, to ensure the use of the actual yields, the obtained S_{EMP} signals were normalized to S_{DIR} which latter was proportional to the total number of the emitted electrons (more or less similarly as in Ref.6).

Figure 1

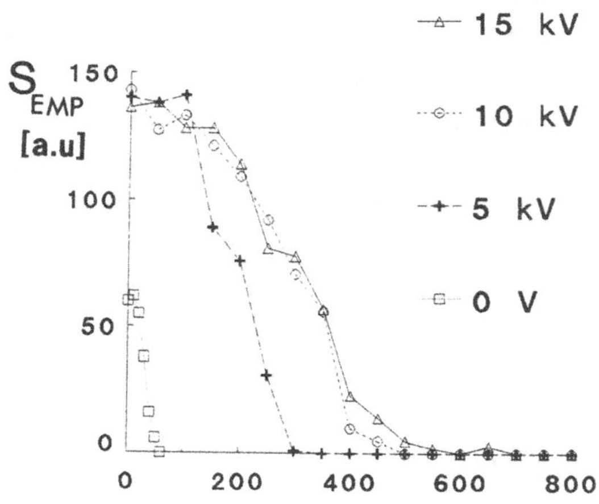

3. Results

a, Before performing each experimental set the standard check of the theoretical multiphoton order was done by measuring the photocurrent (S_{EMP}) - laser intensity (S_I) dependence. The multiphoton order at V_{ext}=+15 kVolts and U_{ret}=0 V was found to be 4, corresponding to [A/hν+1]=4, where A is the work function of gold an hν is the photon energy of the Nd laser light.

b, According to Fig. 1, our experiment really shows the existence of high energy electrons. When we continuously increase the extracting voltage, the high energy electron distribution appears. At V_{ext}>10 kV all curves coincide: the space charge effect is really eliminated!

Figure 2

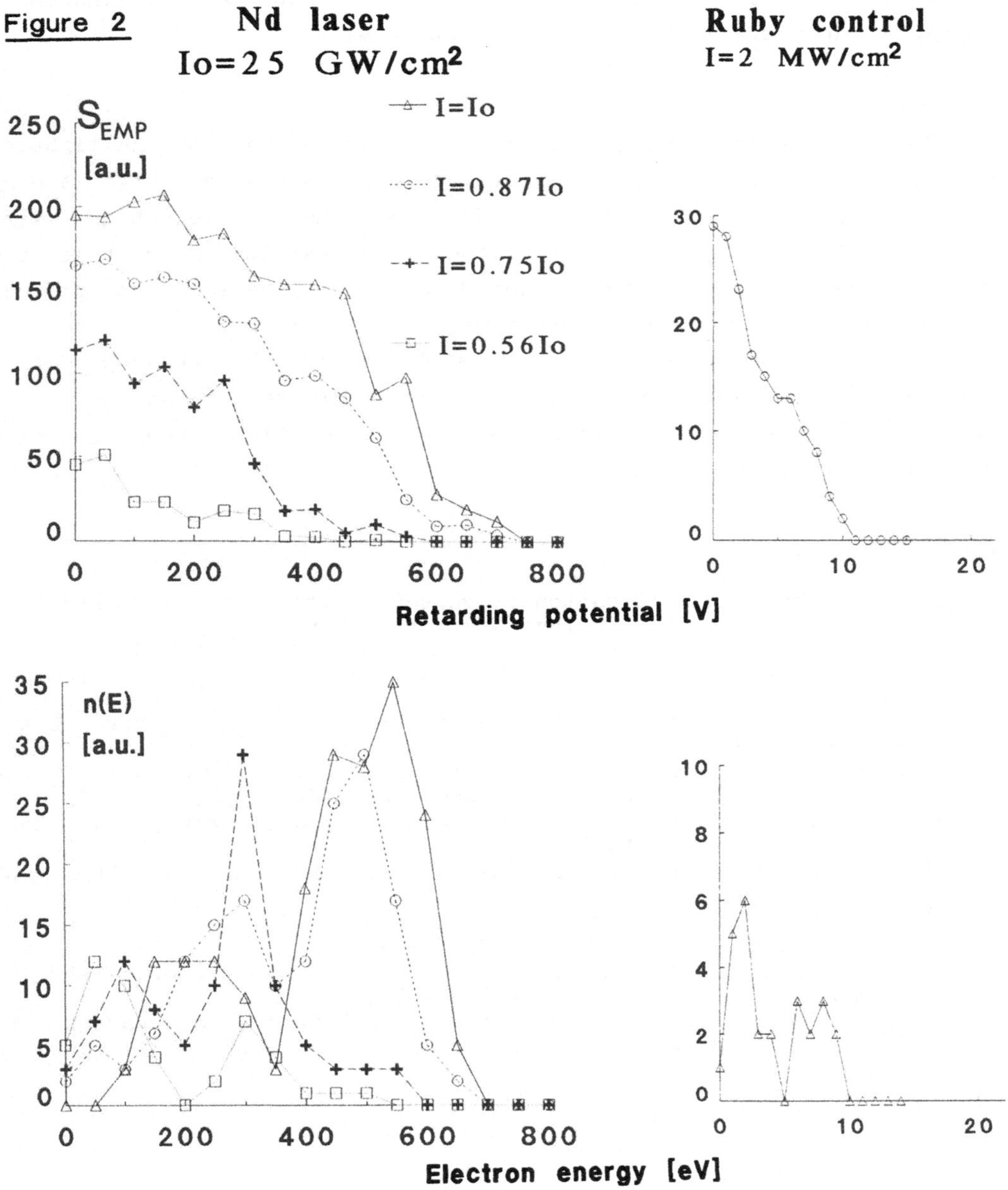

c, The obtained integral electron energy distribution taken at V_{ext}=15 kV and the corresponding derivated differential electron energy distribution curves at different laser intensities can be seen on Fig. 2. Horizontal axis: U_{ret} corresponds to the electron energy E; vertical axis: S_{EMP}, which corresponds to the number of electrons with energies above a given E value ($\sim \int_E^\infty n(E')dE'$) for the upper curves, and the n(E) electron numbers having the energy of E for the bottom curves, respectively. The presented ruby laser curves on the right side of Fig. 2 were taken for the sake of control: low intensity (MW/cm^2) and long (nsec) laser pulses do not induce energetic electrons as was found earlier [3,4].

4. Discussions

The photoelectron energy distribution induced by the multiphoton photoeffect of gold was investigated experimentally. High energy electrons appeared even at unexpectedly low laser intensities (with respect to those used recently for atoms). The results show that the usual theoretical interpretation based on ATI, ponderomotive effect, inverse bremsstrahlung, etc., which valid for atoms [2,8-11], can not be used here directly. Due to the image force and the presence of a macroscopic third body these effects may manifest themselves in a modified form. The further experimental and theoretical study of their role is in progress.

References

[1] Gy. Farkas, in Multiphoton Processes, edited by J.H. Eberly and P. Lambropoulos (Wiley, New York, 1978) p. 81

[2] P. Agostini and G. Petite, Contemp. Phys. 29, 57 (1988)

[3] E.M. Logothetis and P.L. Hartman, Phys. Rev. 187, 460 (1969)

[4] Gy. Farkas, I. Kertész, Zs. Náray and P. Varga, Phys. Lett. 24A, 475 (1967)

[5] G. Petite, P. Agostini, G. Boiziau, J.P. Vigouroux, C. Le Gressus and J.P. Duraud, Opt. Comm. 53, 189 (1985)

[6] L.A. Lompré, A. L Huillier, G. Mainfray and C. Manus, J. Opt. Soc. Am. B2, 1906 (1985)

[7] W. Xiong, F. Yergeau, S.L. Chin and P. Lavigne, J. Phys. B. 21, L159 (1988)

[8] F. Ehlotzky, Can. J. Phys. 63, 907 (1985)

[9] I.J. Berson, J. Phys. B, 8 2078 (1975)

[10] P. Agostini, J. Kupersztych, L.A. Lompré, G. Petite and F. Yergeau, Phys. Rev. A. 36, 4111 (1987)

[11] R.R. Freeman, P.H. Bucksbaum, H. Milchberg, S. Darack, D. Schumacher and M.E. Geusic, Phys. Rev. Lett. 59, 1092 (1987)

HYDROGEN ATOM IN A STRONG RADIATION FIELD: NONPERTURBATIVE ANALYTIC SOLUTIONS

Sândor Varró*
Max-Planck-Institut für Quantenoptik
D-8046 Garching,POBox 1513,West Germany

In the present paper we give nonperturbative analytic solutions to the Schrödinger equation of the system "electron in a Coulomb field plus a strong quantized radiation mode".The importance of this problem needs no particular explanation since any nontrivial solution to it can give a deeper understanding of fundamental strong field phenomena.

The equation we wish to solve reads in atomic units as

$$\left[\frac{1}{2}\Delta - \frac{1}{r} - i\frac{\partial}{\partial x_1}\left(\frac{e_0}{\nu}\right)(a + a^+) + \frac{1}{2}\left(\frac{e_0}{\nu}\right)^2(a+a^+)^2 + \nu\left(a^+a + \frac{1}{2}\right)\right]|\Psi\rangle = \\ = E|\Psi\rangle, \qquad (1)$$

where

$$e_0 \equiv |e|\ell_0(2\pi\hbar\omega/V)^{1/2}(\hbar\omega_0)^{-1}, \quad \nu \equiv \omega/\omega_0\,, \quad E \equiv E_t/\hbar\omega_0. \qquad (1a)$$

$\ell_0 = Z^{-1}(\hbar^2/Me^2) = Z^{-1}0.53\text{Å}$ denotes the atomic unit of length and $\hbar\omega_0 = Z^2(Me^4/\hbar^2) = Z^2 27.2\,\text{eV}$ is the atomic unit of energy. ν measures the frequency ω of the radiation field and E measures the total energy E_t of the electron plus field system.The radiation field is represented by a linearly polarized plane wave in dipole approximation. $x_1 \equiv z\,\ell_0^{-1}$ denotes the scaled Cartesian component of the electron's position along the polarization direction of the radiation field.All the other symbols have their usual meaning.

As a first step towards the solution of eq(1)we introduce the Kustaanheimo and Stiefel (KS) coordinates (u_1, u_2, u_3, u_4) to parametrize the electron's position.

$$x_1 = u_1^2 - u_2^2 - u_3^2 + u_4^2\,, \quad x_2 = 2(u_1u_3 - u_2u_4)\,, \quad x_3 = 2(u_1u_2 - u_3u_4). \qquad (2)$$

This transformation has long been known and used in celestial mechanics to regularize the Kepler problem[1]. In the KS space eq(1) becomes

$$\left[\frac{1}{4}(p_1^2 + p_2^2 + p_3^2 + p_4^2) + (e_0/\nu)(a + a^+)(u_1p_1 - u_2p_2 - u_3p_3 + u_4p_4) + \\ + (u_1^2 + u_2^2 + u_3^2 + u_4^2)(2\hat{E}_B)\right]|\Psi\rangle = 2|\Psi\rangle, \qquad (3)$$

where $p_k \equiv -i(\partial/\partial u_k)$ $(k = 1, 2, 3, 4)$ and

$$\hat{E}_B \equiv \left[\nu(a^+a + \tfrac{1}{2}) + \tfrac{1}{2}(e_0/\nu)^2(a + a^+)^2 - E\right]. \qquad (3a)$$

In addition to eq(3), $|\Psi\rangle$ is subjected to the constraint equation [2] in the KS space

$$[(u_1 p_4 - u_4 p_1) - (u_2 p_3 - u_3 p_2)]|\Psi\rangle = 0. \tag{3b}$$

Without terms representing the radiaton mode, eq(3) is just the eigenvalue equation of a four-dimensional isotropic oscillator.

Via the operator scaling

$$\tilde{u}_k = 2^{1/2}(2\hat{E}_B)^{1/4} u_k \quad , \quad \tilde{p}_k = 2^{-1/2}(2\hat{E}_B)^{-1/4} p_k \; , \tag{4}$$

and the definitions

$$b_k \equiv 2^{-1/2}(\tilde{u}_k + i\tilde{p}_k) \; , \; b_k^+ \equiv 2^{-1/2}(\tilde{u}_k - i\tilde{p}_k) \tag{5}$$

we introduce the collective boson operators b_k and b_k^+ ($[b_k, b_{k'}^+] = \delta_{kk'}$) in terms of which eq(3) reads

$$\Big\{\tfrac{1}{2}\big[(b_1^+b_1 + 1/2) + (b_2^+b_2 + 1/2) + (b_3^+b_3 + 1/2) + (b_4^+b_4 + 1/2)\big] + \\ + \tfrac{1}{2}\hat{A}\tfrac{1}{2i}\big[(b_1^2 - b_1^{+2}) - (b_2^2 - b_2^{+2}) - (b_3^2 - b_3^{+2}) + (b_4^2 - b_4^{+2})\big]\Big\}|\Psi\rangle = \\ = (2\hat{E}_B)^{-1/2}|\Psi\rangle \; , \tag{6}$$

where $\hat{A} \equiv (2\hat{E}_B)^{-1/2}(e_0/\nu)(a + a^+)$. (6a)

It can be shown [3] that the squeezing transformation

$$|\Psi\rangle = \Sigma(\lambda)|\psi\rangle \; , \; \Sigma(\lambda) = \Sigma_1(\lambda)\Sigma_2(-\lambda)\Sigma_3(-\lambda)\Sigma_4(\lambda) \; , \tag{7}$$

with $\Sigma_k(\lambda) \equiv \exp[-i\tfrac{\lambda}{2}(b_k^2 + b_k^{+2})] \; , \; \lambda = \log[(1+\hat{A})/(1-\hat{A})]^{1/4}$ (7a)

diagonalizes eq(6). The transformed state vector $|\psi\rangle$ is given as

$$|\psi\rangle = \hat{\psi}_C|N\rangle \; , \tag{8}$$

where $|N\rangle$ is a photon number state and the operator $\hat{\psi}_C$ satisfies the unperturbed Coulomb eigenvalue equation. The approximation we made in obtaining eq(8) was that we considered $\hat{E}_B^{-1/2}$ and $(e_0/\nu)(a + a^+)$ as commuting quantities on the solution space.

For bound states the explicit form of $\hat{\psi}_C$ (apart from a normalization factor) reads as $\hat{\psi}_C(n) = e^{im\varphi}\tilde{\rho}_1^{|m|}e^{-\tilde{\rho}_1^2/2}L_{n_1}^{|m|}(\tilde{\rho}_1^2)\tilde{\rho}_2^{|m|}e^{-\tilde{\rho}_2^2/2}L_{n_2}^{|m|}(\tilde{\rho}_2^2)$, (8a)

where $\tilde{\rho}_1^2 = \tilde{u}_1^2 + \tilde{u}_4^2$ and $\tilde{\rho}_2 = \tilde{u}_2^2 + \tilde{u}_3^2$. In eq(8a) m is the angular momentum quantum number and n_1, n_2 are parabolic quantum numbers. The principal quantum number n is expressed as $n = n_1 + n_2 + m + 1$. According to eqs (8) and (7) the complete solutions to eq(1) are given by

$$|\Psi\rangle = \Sigma(\lambda)\,\hat{\psi}_C(n)|N\rangle \; , \tag{9}$$

with

$$E = -(1/2n^2) + \nu(N + 1/2). \tag{9a}$$

That the energy spectrum coincides with the unperturbed spectrum might seem to be strange. In fact, the present example is far not unique in quantum electrodynamics [4]. As a simple example to this situation we can think of a shifted harmonic oscillator which can have an enormous interaction energy, but in the same time its energy spectrum is exactly the same as the unperturbed spectrum.

The physical interpretation and some applications of the solutions (9) is in progress and will be published elsewhere [3].

*Alexander von Humboldt Fellow on leave from Central Research Institute for Physics H-1525 Budapest, POBox 49, Hungary

[1] P.Kustaanheimo and E.Stiefel: J.Reine Angew.Math. 218, 2o4(1965)
[2] A.C.Chen and M.Kibler: Phys.Rev. A31, 3960(1985)
[3] S.Varró: manuscript in preparation
[4] F.Bloch and A.Nordsieck: Phys.Rev.52, 54(1937)

Laser excitation of Stark-induced resonances

G. Alber
Fakultät für Physik, Albert-Ludwigs-Universität,
Hermann-Herder-Str.3, D-7800 Freiburg, W.Germany

Recently a systematic theory of laser excitation processes close to a photoionization threshold has been developed /1/. It is based on the observation that in the optical frequency regime and for moderate laser intensities $I << I_0 = 1.4 \cdot 10^{17} W/cm^2$ the atom-laser interaction takes place in a finite reaction zone, which typically extends only a few Bohr radii around the atomic nucleus. However, so far these studies have concentrated on cases, where outside the reaction zone the dynamics of the excited electron is completely determined by the Coulomb potential of a positively charged ionic core.

A Hydrogen atom placed in a homogeneous static electric field of typical laboratory field strength $F << F_0 = 5.142 \cdot 10^9 V/cm$ is one of the simplest physical systems, where the motion of the excited electron far from the atomic nucleus is strongly modified by the presence of the static electric field. Above the zero-field photoionization threshold classically this leads to a periodic motion of the excited electron in the electric field direction. It has recently been shown /2/ that the experimentally discovered polarization- dependent modulations of photoabsorption cross sections /3/, which characterize one photon absorption by a long and weak laser pulse, can completely be described in terms of properties of this isolated classical orbit. Here we want to adress the question how the presence of this isolated periodic orbit manifests itself in the dynamics of laser excitation by an *intense* pulse.

For this purpose we have studied one photon excitation of Hydrogen in the energy region slightly above the zero-field photoionization threshold /4/. As typical laboratory electric fields become significant only at distances $r > F^{-1/2} >> 1(a.u.)$, inside the reaction zone the static electric field is negligible and the photonabsorption process can be characterized by an approximately energy- and field-independent ionization rate Γ. Exploiting the separability of the Hydrogenic Stark problem in parabolic coordinates, we can derive analytical expressions for atomic transition amplitudes, which can be represented as a sum over contributions due to all returns of the excited electron to the reaction zone, where two alternative processes may occur, namely laser assisted electron-proton scattering and stimulated recombination back to the initial state. However, in the asymptotic limit of small electric fields, i.e. $\lambda = F^{-1/4} >> 1$ (λ measures the magnitude of the classical action of the isolated periodic orbit), laser assisted electron-proton scattering may be neglected, because any scattering event leads to electron emission into all possible directions and only an electron emitted in the direction of the applied electric field is able to return again to the reaction zone. In this limit thus we may derive simple asymptotic expressions, where the atom-laser interaction inside the reaction zone is characterized by the total ionization rate Γ and the dipole matrix element between the initial state and an energy normalized continuum state, which describes asymptotically a (zero-energy) electron moving in the direction of the external field. Besides these energy- and field-independent quantities, the atomic transition amplitudes depend on characteristics of the isolated classical periodic orbit, namely its action, its orbit time and its Ljapunov exponent.

If the laser pulse is so intense that the initial state is depleted on a time scale smaller than the mean classical orbit time $T_{\bar{\epsilon}}$ of the excited electron, the excitation process is not only localized in space but also in time so that an electronic wave packet is generated (Fig.1). It leaves the reaction zone in a time of the order of $1/\Gamma$. After a time $t \approx T_{\bar{\epsilon}}$ only the small fraction of this wave packet, which has been emitted in the direction of the applied field, returns again to the reaction zone and the initial state probability is increased by stimulated recombination. The heights of the recombination peaks at $t \approx MT_{\bar{\epsilon}}$ reflect the instability of the isolated classical periodic orbit, which is characterized by the Ljapunov exponent $\lambda(\epsilon)$(Fig.2).

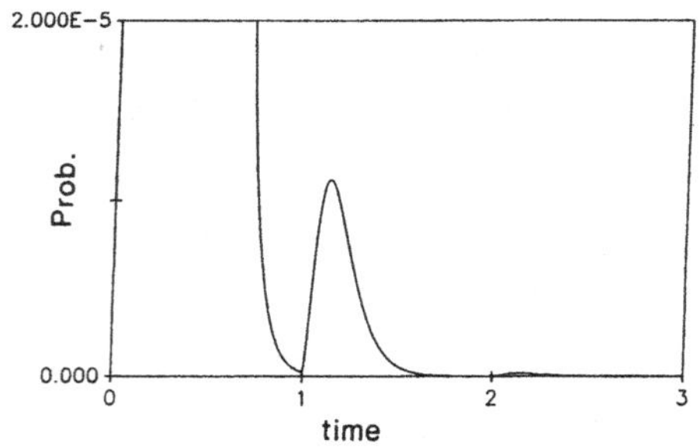

Fig.1: Initial state probability as a function of time (in units of the mean classical orbit time) for $F = 10^{-8}$(a.u.), $\Gamma = 7 \cdot 10^{-6}$(a.u.), $\bar{\epsilon} = 5 \cdot 10^{-5}$(a.u.)(mean excited energy).

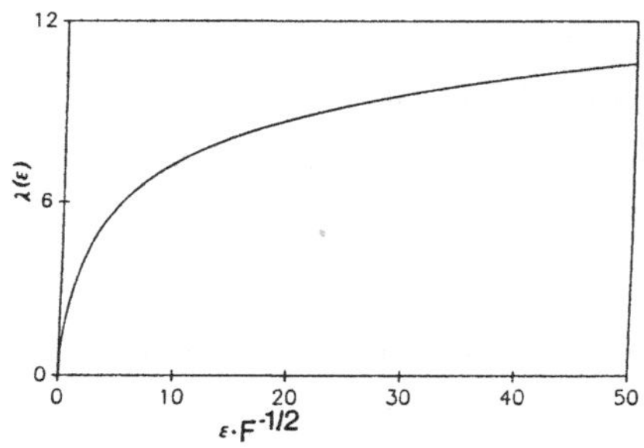

Fig.2: Ljapunov exponent $\lambda(\epsilon)$ as a function of energy ϵ.

References:

(1) G.Alber and P.Zoller, Phys.Rev.A37,377(1988)

(2) E.G.Bogomol'nyi,Pis'ma Zh.Eksp.Teor.Fiz.47,445 (1988) (JETP Lett. 47,526 (1988))

(3) R.R.Freeman and N.P.Economu, Phys.Rev.A20,2356(1979)

(4) G.Alber, Phys.Rev.A (submitted)

Two–Photon Ionisation of Rare–Gas Atoms: An Application of the Relativistic Time–Dependent Dirac–Fock Theory.

Michael GJ Fink, Walter R Johnson
Department of Physics, University of Notre Dame
Notre Dame, IN 46556, USA

Multiphoton ionisation (MPI) of atoms is an example of a development in atomic physics where the nonlinear interaction of an atomic manyelectron system with a strong radiation field is studied to gain deeper insight into atomic structure and its role in radiative processes. Theoretical understanding of MPI processes has now definitively evolved beyond mere phenomenology, but there remain still many open questions in the quantitative prediction of MPI phenomena. It is the purpose of the study summarised in this paper to attempt an *ab initio* approach to two–photon ionisation (2PI) of closed–shell atoms. The choice of the theoretical approach taken here (Relativistic Time-Dependent Dirac-Fock (TDDF) model) is based on fundamental and practical reasons: spin-orbit coupling is included in the relativistic Hamiltonian, so the fine-structure splitting is properly accounted for. The relativistic interaction Hamiltonian is linear in the strength of the external field. The TDDF contains Many-Body Perturbation Theory (MBPT) matrix elements known to be dominant in 1PI of rare gases, to which this model has been applied successfully[1].

Formally, the TDDF equations[2] can be obtained from a straightforward extension of the usual, time-independent Hartree-Fock model: The time-dependent many-body Dirac-Coulomb equation for an atom subject to an external perturbation $V(t)$

$$i\hbar\frac{\partial}{\partial t}\Psi(t) = \left[\sum_i \left[h_0(i) - \frac{Ze^2}{r_i} + \sum_{j<i}\frac{e^2}{r_{ij}}\right] + \sum_i V(x_i,t)\right]\Psi(t) \tag{1}$$

is to be solved approximately in a model space defined by all many-particle wave functions that can be represented by a single Slater determinant of time-dependent orbitals $u_a(x,t)$

$$\Psi(t) = \det\left(u_a(x_i, t)\right) \ . \tag{2}$$

This leads to the time-dependent Hartree-Fock equation (the Lagrange multipliers ensure normalisation)

$$i\hbar\frac{\partial}{\partial t}u_a(t) = \left[h_0 - \frac{Ze^2}{r_i}\right]u_a(t) + V_{HF}[u_b(t),u_b(t)]u_a(t) + V(t)u_a(t) - \sum_b\lambda_{ab}u_b(t) \ , \tag{3}$$

where

$$V_{HF}[u_b,v_b]w := \sum_b\left(\langle u_b|\frac{1}{r_{12}}|v_b\rangle w - \langle u_b|\frac{1}{r_{12}}|w\rangle v_b \right) \tag{4}$$

is the Hartree-Fock potential. Since we lay our emphasis on the electron–electron interaction and not on strong-field effects, we expand the equations up to second order perturbation theory in V. Lowest order perturbation theory, characterised by an I^n intensity-dependence of the n–photon ionisation signal, describes MPI up to quite high intensities. We assume V to be a purely harmonic dipole potential

$$V(x,t) = V^+(x)\, e^{-i\omega t} + V^-(x)\, e^{+i\omega t} \tag{5}$$

and expand $u_a(x,t)$ consistent with the time-dependence of V and in powers of its amplitude:

$$u_a(x,t) = e^{-i\varepsilon_a t} \sum_{k=0}^{\infty} \sum_{f=-k}^{k} u_a^{(k)f}(x)\, e^{-if\omega t} \; , \tag{6}$$

where k is the order of perturbation theory in V and f the harmonic coefficient. Inserting (6) into (3), we obtain a hierarchy of equations. The k=0, f=0 equation (we denote $u_a^{(0)0} =: u_a$) is the Dirac–Fock equation for the ground state (we omit the Lagrange multipliers in the following)

$$(h_0 - \varepsilon_a + V_{\mathrm{HF}}[u_b,u_b])\, u_a = 0 \; . \tag{7}$$

The first– and second order equations have a common structure (we write $u_a^{(k)f} =: v_a^{kf}$)

$$(h_0 - (\varepsilon_a \pm \omega) + V_{\mathrm{HF}}[u_b,u_b])\, v_a^{k\pm} + V_{\mathrm{RPA}}^{\pm}[v^k]\, u_a + \mathcal{R}_a^{\pm} = 0 \; , \tag{8}$$

where

$$V_{\mathrm{RPA}}^{\pm}[v^k] := \mathrm{V}_{\mathrm{HF}}[v_b^{-},u_b] + \mathrm{V}_{\mathrm{HF}}[u_b,v_b^{+}] \tag{9}$$

is the RPA-potential (Random Phase Approximation), which yields together with the ground state Hartree-Fock potential a V(N–1) potential. The driving term $\mathcal{R}_a^{\pm}$ has, for the first-order equation the form

$$\mathcal{R}_a^{\pm} = V^{\pm} u_a \; , \tag{10}$$

and for second order in the external perturbation V

$$\mathcal{R}_a^{\pm} = V^{\pm} v_a^{1\pm} + V_{\mathrm{RPA}}^{\pm}[v^1] v_a^{1\pm} + V_{\mathrm{HF}}[v_a^{1-}, v_a^{1+}] u_a \; . \tag{11}$$

The perturbed orbitals $v_a^{k\pm}$ correspond to one–particle one–hole excited/scattering states. The spectroscopy of such states is correctly predicted with a typical error of at most 10% of the level spacing in Rydberg series. Note that the TDDF wave function goes, despite the appearance of Eq. (2), beyond the single-configuration picture, since it contains MBPT corrections for both the ground and excited states.

The method, as shown here, can be applied directly to 2PI of rare gases with final states in the flat continuum. An Quantum Defect analysis can be applied to extend the energy range for final states into the autoionising range between the two fine strucutre components of the core.

As an example for typical results we include the two–photon ionisation cross section and the corresponding Lu–Fano MQDT plot for the J=2 final state of argon.

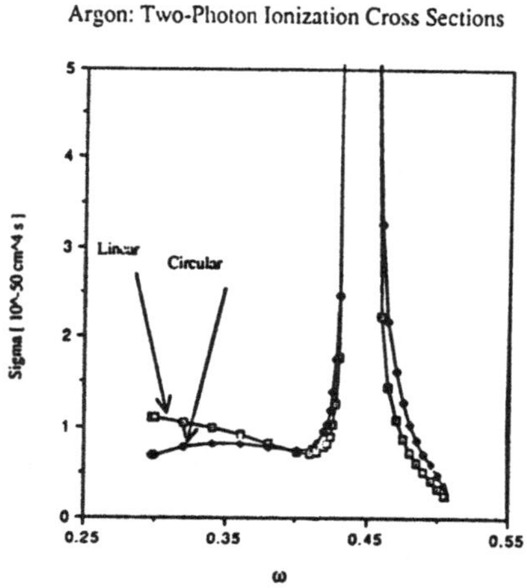

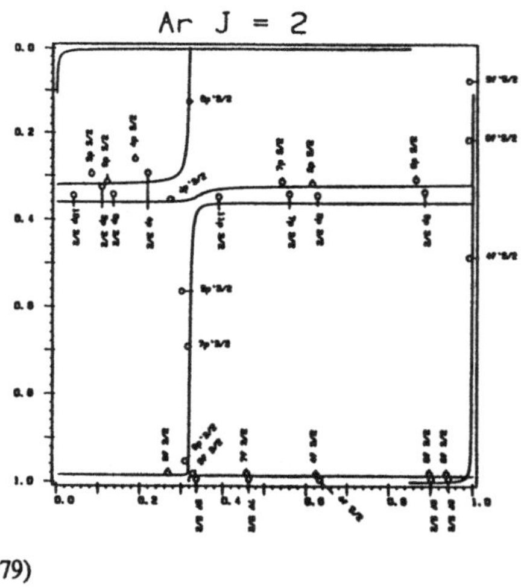

References: [1]W.R. Johnson, C.D. Lin, Phys. Rev. A 20, 964(1979)

[2]F. Smet, J. Tillieu, A. van Groenendael, Int. J. Quantum Chem., 17, 531 (1980)

DIAGRAMMATIC PERTURBATION THEORY FOR A TWO-LEVEL SYSTEM IN AN INTENSE LIGHT PULSE

Lars Jönsson* and Göran Wendin*
Laboratory for Atomic, Molecular, and Radiation Physics
Department of Physics, University of Illinois at Chicago
P. O. Box 4348, Chicago, Illinois 60680

ABSTRACT

We have used diagrammatic many-body perturbation theory (DMBPT) to derive results, valid at high intensities, for a one-electron two-level atom in an exponentially-turned-on laser field. For such a system, the solution is known[1] if the laser is close to resonance (rotating wave approximation (RWA)) and the turn-on rate is slow enough. The system passes through a succession of Floquet states[2] as the field is turned on. We show how this can be derived within DMBPT using both the linked and the unlinked expansions. We also derive some results outside RWA including the Bloch-Siegert shift and harmonic generation. The purpose of the work is not to learn anything new about the two-level system, which is better treated by other methods, but to learn how to sum infinite sets of diagrams to get results valid at all intensities. The focus of future work will be to generalize the treatment to include many-body effects in more realistic systems (see also Ref. 3).

The interaction Hamiltonian is $\hat{H}_I = 2\hat{D}e^{\Gamma t}\cos\omega t$, where $\hat{D}$ is the electric-dipole operator times the field strength ($D_{00}=D_{11}=0$), Γ is the turn-on rate, ω is the laser frequency, and t is the time. Using the definition $\Delta=\omega_{10}-\omega$, where ω_{10} is the resonance frequency of the two-level system, we define $b=a_1/a_0=-(D_{10}/\Delta)e^{i\Delta t+\Gamma t}\,b'$, where a_1 and a_0 are the upper- and lower-state amplitudes, respectively, and D_{10} is the matrix element of $\hat{D}$. We will in general assume that $\Gamma<<\Delta$.

Figure 1:

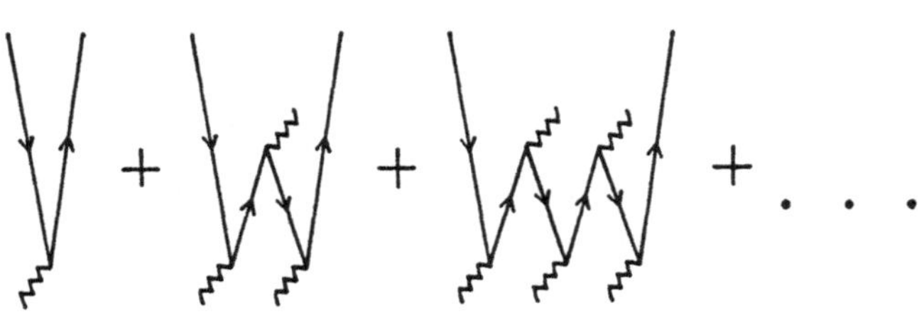

The linked Goldstone diagrams[4] in RWA are shown in Fig. 1. They represent the perturbation expansion of b. For our system the ground state corresponds to an electron in the lower state, and a particle–hole pair corresponds to an electron in the upper state. For vertices with the same time (same horizontal position) all time orders should be taken. If we define $x=(D^2/\Delta^2)e^{2\Gamma t}$, where $D^2=D_{01}D_{10}$, the evaluation of the diagrams (with $\Gamma=0$ in all energy denominators) give[5]

$$b' = 1-x+2x^2-5x^3+14x^4-42x^5+132x^6-429x^7+\ldots = \frac{\sqrt{1+4x}-1}{2x} \qquad (1)$$

If we define ΔE through $a_0=\exp(-i\int_{-\infty}^{t}\Delta E\, dt')$ we get[6] $\Delta E_{RWA}=-\Delta x b'=\Delta(\sqrt{1+4x}-1)/2$, which together with the norm $N^2=1+|b|^2$ give

$$a_0 = \frac{|\delta|+\sqrt{\delta^2+D^2e^{2\Gamma t}}}{\sqrt{D^2e^{2\Gamma t}+\{|\delta|+\sqrt{\delta^2+D^2e^{2\Gamma t}}\}^2}} \exp\left(-i\int_{-\infty}^{t}\{\delta-s\sqrt{\delta^2+D^2e^{2\Gamma t'}}\}dt'\right) \qquad (2)$$

where $\delta=\Delta/2$, and $s=\Delta/|\Delta|$. This is the correct result for small Γ.[1] A simmilar result can be derived for a_1. Eq. (2) shows how the system goes through a succession of Floquet states as the field is turned on.

This work was supported by the U.S. Air Force Office of Scientific Research, the U.S. Office of Naval Research, the Lawrence Livermore National Laboratory, and the Strategic Defense Initiative Organization.

---

* Permanent address: Institute of Theoretical Physics, Chalmers University of Technology, S–412 96 Göteborg, Sweden.

1 N. B. Delone and V. P. Krainov, Atoms in Strong Light Fields, Springer Ser. Chem. Phys., Vol. 28 (Springer, Berlin, 1985).
2 J. H. Shirley, Phys. Rev. B 138, 979 (1965).
3 G. Wendin, L. Jönsson, and A. L'Huiller, J. Opt. Soc. Am. B 4, 833
4 J. Goldstone, Proc. Roy. Soc. (London) A239, 267 (1957).
5 L. Jönsson and G. Wendin, to be submitted to Phys. Rev. A.
6 P. W. Langhoff, S. T. Epstein, and M. Karplus, Rev. Mod. Phys. 44, 602 (1972).

MICROWAVE IONIZATION OF ALKALI RYDBERG ATOMS

Birte L. Christensen-Dalsgaard
Institute of Physics, University of Aarhus
DK-8000 Aarhus C, Denmark

In the present work I want to describe the interaction between alkali atoms and a microwave field. The microwave ionization of high lying Rydberg states in H has been studied extensively in resent years. The quantum mechanical studies in this area has established the usefulness of Floquet states to describe the dynamical properties[1]. Also they have pointed to the exsistense of window-states, which are very active as a bridge between bound and free states. Within static field studies of alkali atoms, the ionization route is well studied. To this end, the Stark-map play a crucial role in the understanding of the critical field, at which the atoms ionizes. Avoided crossings are suggested to be important in microwave ionization of alkali Rydberg atoms[2]. The purpose of the present work is to establish a connection between the Stark-picture and the dynamical Floquet-picture. In doing so, we can identify the physical important mechanisms for ionization. Analytic results can only be obtained for simplified situations; however conclusions based on these seem to be of general nature. Here I will present the two-state case.

The basic idea behind the approximation is, that the time propagation matrix U(T) for one period T can be written as a product of simple matrices. We will operate with two types of matrices:

$\mathbf{P}_l$, which is a diagonal matrix whose l'th element describe the phaseevolution from time t_i to time t_{i+1}: $\exp(-i \cdot \epsilon_l(t)\, dt)$

$\mathbf{LZ}_{k,l}$, which is a unit matrix except for the (k,l) row and column, where it contains the Landau-Zener matrix associated with an avoided crossing of state k and state l. If k denote the upper level, we have: $\mathbf{LZ}_{k,l}(k,k)=\mathbf{LZ}_{k,l}(l,l)=P$, and $\mathbf{LZ}_{k,l}(k,l)=-\mathbf{LZ}_{k,l}(l,k)^{*}=(1-P) \cdot \exp(\phi)$, where P is the Landau Zener transition probability and ϕ the Landau-Zener phase (here set to $-\pi/4$).
Using the above U(T) can for the two-level system be expressed as:

$$U(T)=\mathbf{P}_1 \cdot \mathbf{LZ}_{1,2} \cdot \mathbf{P}_A \cdot \mathbf{LZ}_{2,1} \cdot \mathbf{P}_2 \qquad (1)$$

The eigenvalues for U can be written as $\exp(-i\cdot\epsilon_a\cdot T)$, where the Floquet energy ϵ_a is given by:

$$\epsilon_a\cdot T = (E_u + E_l)\cdot T/2 \pm \operatorname{acos}[\ (1-P)\cdot\cos(\Omega+2\cdot\phi-p_a) + P\cdot\cos(\Omega)\ ] \quad (2)$$

where $\Omega = (E_u - E_l)\cdot T/2$ and p_a is the phase accumulated between the passage of the avoided crossing on the way "out" and on the way "in". The pure diabatic situation arise for P=1, in which case the Floquet eigenvalues coincide with the normal eigenvalues for the stationary Hamiltonian. In the adiabatic case, the Floquet eigenvalues are shifted compared to the stationary situation by $(p_a-2\phi)/T$. As p_a depends on the maximum field amplitude F_0, the floquet states will oscillate as a function of F_0. The same is true for the generale case. In the calculations discussed here, P lies between 0.9 and 1., ie. close to the diabatic situation. The typical behaviour is shown in figure 2, where also the envelope states (defined by replacing $\cos(\Omega+2\phi-p_a)$ by ±1 in (2)) are shown. The oscillations of the Floquet eigenvalue are caused by the phase accumulation p_a and the wavelength is $\sqrt{(F^2-F_c^2)}\cdot\omega/\Delta$, where Δ is the difference in the slope of the two states. We notice that for values of F_0 around the avoided crossing, we will have strong mixing. For larger field strength we have a small, but non-zero mixing. This work therefore point to two mechanisms for ionization, one is due to the small admixture of the high lying |n>-states in low |a>-states; the other, which also has been discussed for the H-case[3], is caused by variations of F_0, either as the field is turned on (laboratory) or by fluctuations in F_0 (plasma).

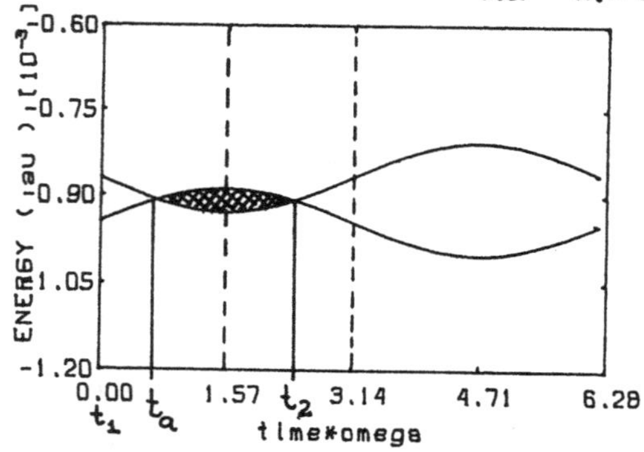

Figure 1: Energies as function of time. p_a is shown as crossed area

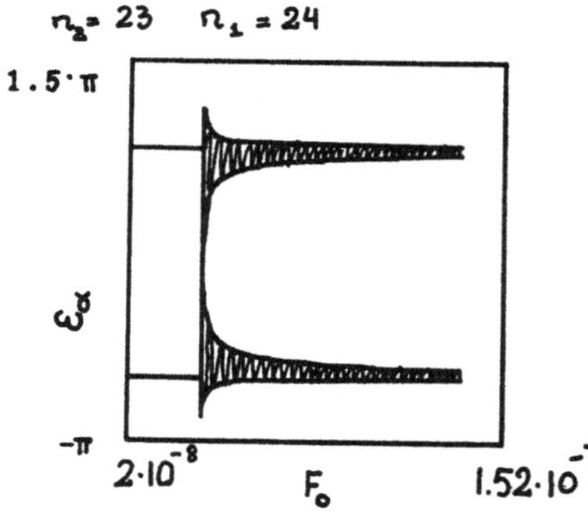

Figure 2: Floquet energies as function of maximal field-strength

[1] R.Blumel, J.Goldberg and U.Smilansky, Z.Phys.D.9, 95 (1988)
[2] H.B.van den Leuvel and T.F.Gallagher, Phys.Rev A32,280 (1985)
[3] H.B.Breuer and M.Holthaus, Z.Phys.D.11, 1 (1989)

AUTOIONIZATION OF LARGE-*l* RYDBERG STATES

Michel POIRIER

Service de Physique des Atomes et des Surfaces, C.E.N.Saclay

F91191 Gif-sur Yvette Cedex, FRANCE

Analysis of doubly excited states is at present one of the most active domains in atomic spectroscopy[1]. Since the work of Wannier[2], several theoretical approaches have developed, using mainly group theoretical methods or hyperspherical coordinates. In particular, the search for doubly excited stable states is one of the most currently exciting problems. An example is provided by interference-stabilized states which may be found when several Rydberg series interact[3]. However, the question of stability and of production of highly-correlated states such as the Wannier states is yet unsolved.

A first analysis of correlated excited states is made using classical mechanics. It has been known for long that one can find rigid rotating solutions to the *N*-body problem. The search for classical stable periodic motions is of interest since it can lead to quantization using the Gutzwiller formalism [4]. In a two-electron atom, as pointed out by Klar[5], two types of rigid configuration exist. If one releases the condition that the nucleus charge Z must be integer, for Z smaller than unity a third solution is found which is unsymmetrical versus the electron exchange[6]. All this rotating configurations are unstable, but the instability character is less pronounced in the unsymmetrical configuration.

This indicates that special stability properties may be achieved in two-electron atoms when the radial wavefunctions are spatially separated. Using this physical idea and in connection with recent experimental works[7], the formalism of autoionization of large-*l* Rydberg states is developed[8]. Because the inner and the outer wavefunctions do not overlap, the exchange amplitude can be neglected and the direct autoionization amplitude may be factored in two parts. The first one involves the wavefunction of the Rydberg electron. Since this electron is insensitive to the detail of the atomic core potential, an hydrogenic description is well suited. Exact analytic expressions are available for these matrix elements, as well as asymptotical expressions using simple transcendental functions or derived through the correspondence principle. The inner electron matrix elements are computed using a numerical integration of the Schrödinger equation including core-polarization terms. Interaction of the outer electron with a closed shell polarized by the inner electron is taken into account. Comparison with available experimental data shows improving agreement with increasing l. Possible refinements of this method should include first the exchange amplitude and then other channels interaction and spin-orbit effects.

References

[1] U.Fano, Rep.Progr.Phys. 46, 97 (1983).

[2] G.H.Wannier, Phys.Rev. 90, 817 (1953).

[3] L.D.Van Woerkom, J.G.Story and W.E.Cooke, Phys.Rev.A 34, 3457 (1985); M.Aymar, J.Phys.B 18, L763 (1985).

[4] D.Wintgen, Phys.Rev.Lett. 61, 1803 (1988).

[5] H.Klar, Phys.Rev.Lett. 57, 66 (1986); Z.Phys.D 3, 353 (1986); J.Opt.Soc.Am.B 4, 788 (1987).

[6] M.Poirier, "On the possible rigid rotations in two-electron atoms" (Phys.Rev.A, submitted).

[7] S.M.Jaffe, R.Kachru, H.B.Van Linden Van Den Heuvell and T.F.Gallagher, Phys.Rev.A 32, 1480 (1985); E.A.J.M.Bente and W.Hogervorst Phys.Rev.A 36, 4081 (1987); R.R.Jones and T.F.Gallagher, Phys.Rev.A 38, 2846 (1988).

[8] M.Poirier, Phys.Rev.A 38, 3484 (1988); J.Phys.B (in press).

Fractallike structures and the strongly kicked H-atom

C. Hillermeier, R.Blümel
Max-Planck-Institut für Quantenoptik
D-8046 Garching

We investigate the classical 1-dim. model of H-atoms (1) prepared in extremal Stark-Rydberg states which are exposed to strong periodic electromagnetic kicks (cf.2). An ensemble of initial phase-space points with defined energy decays in time under the effect of the kicks according to a power law. The underlying mechanism also allows an understanding of fractallike patterns made up by the regions of equal stability under the kick-pulses.

Using convenient units the model-Hamiltonian is written:

$$H = p^2/2 - 1/q - \beta \cdot q \sum_k \delta\,(t - k \cdot \tau)$$

The (p,q)-coordinates immediately give the kickmap: $p' = p + \beta$, $q' = q$, whereas the unperturbed Kepler-motion between two kicks, in the following called 'twist' (cf.3), is described canonically in action-angle variables (n, θ).

For every starting point in phase-space the system remains bound under a number of kick-pulses characteristic of that point. We have assigned the same colour to all points which ionize after an equal time, thus making visible the 'equiionizing sets'. The speed of ionization turns out to fluctuate strongly. The pattern of the 'equiionizing sets' shows structure on all scales. One might be led to the conclusion that the ionization mechanism is analogous to the 'recursive removal of the middle-third' which brings about the famous Cantor set. But our complex structure of phase space arises already after very few kicks - which act as removing steps -, whereas the middle-third process only generates structure down to scales that are limited by the number of recurrences already carried out. The mentioned power-law decay of a phase-space ensemble is also in contrast with the behaviour of Cantor-processes: Removing a constant fraction of the surviving set at every stage results in an exponential decay.

To understand these peculiar features we have to look closer at the kick- and twist-maps:

Fig.1:*Result of one kick exerted on the line at* $n = 5$.

Fig.2:*Result of one twist after the kick*

Fig.3:*Ionization pattern, impressed on the original interval at* $n = 5$ *, by the sequence kick – twist – kick.*

fig. 3

fig. 1

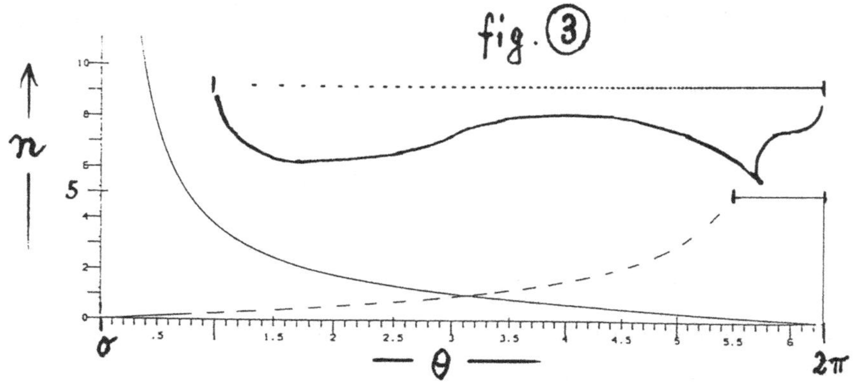

fig. 2

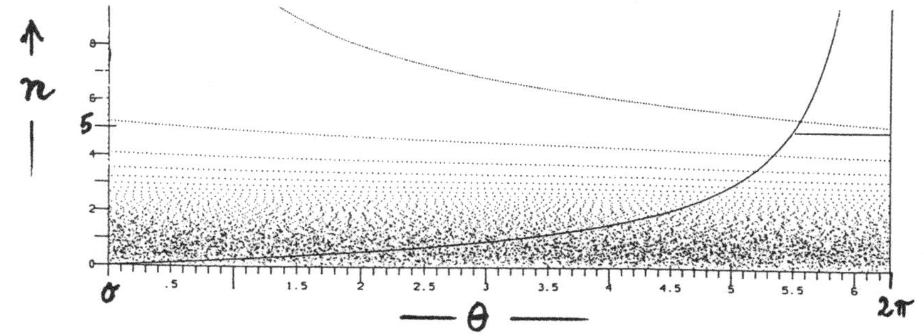

Figures 1 and 2 contain in addition a *tan* $(\theta/4)$- curve. Points to its left are ionized by the next kick.

The singular behaviour of the Kepler-twist at $n = 0$ accounts for the immediate arising of structure at all scales (see figure 3). To give a decay as $1/t^{\gamma}$ the fraction of ionizing points has to decrease as γ/t , t denoting discrete time steps. This is accomplished by a continuous wandering of the (initially δ - shaped) n-distribution towards lower values, where the effect of the twist can be modelled by homogenizing the θ - distribution of phase-space points. Thus a simple comparison with the lowest non-ionizing θ-angle for given n, $\theta = 4 \cdot \arctan(n)$, gives the right behaviour. The stochastic effect of the twist also explains the numerically observed insensitivity of the decay-power against variations of the kick-frequency.

Summarising, we have found that a recursive mechanism which is only structurally, but not quantitatively constant from step to step is responsible for a power-law decay.

1) R. V. Jensen, Phys. Rev. Lett. 49, 1365 (1982)
2) R. Blümel, U. Smilansky, Phys. Rev. A 30, 1040 (1984)
3) G. Troll, U. Smilansky, Physica D, in press

(e,2e) COLLISIONS IN THE PRESENCE OF A LASER FIELD

V. Véniard, P. Martin and A. Maquet
Laboratoire de Chimie Physique, Université Pierre et Marie Curie
11, rue Pierre et Marie Curie, F75231 Paris Cedex 05 (France).

P. Francken
Université Libre de Bruxelles, Faculté des Sciences, Physique Théorique, CP 227,
B1050 Bruxelles (Belgique)

C.J. Joachain
Université Libre de Bruxelles, Faculté des Sciences, Physique Théorique, CP 227,
B1050 Bruxelles (Belgique)
and Université Catholique de Louvain, Département de Physique - FYAM,
chemin du cyclotron 2, B1348 Louvain-la-Neuve (Belgique)

We study the influence of a laser field on the dynamics of fast (e, 2e) collisions on atomic hydrogen. In order to keep the discussion simple, we assume that the laser is treated classically as a spatially homogeneous electric field, linearly polarized and single mode. Specifically, we confine our attention to the case of the Ehrhardt asymmetric coplanar geometry[1].
Particular attention has been paid to the construction of dressed, laser-modified initial and final atomic wave functions. Two methods are discussed for the evaluation of the scattering amplitudes : a nonperturbative approach (valid to all orders in the laser-projectile interaction) and a fully perturbative, second order treatment in the combined electron-atom and laser-(electron plus atom) interactions.

Our results clearly demonstrate the strong influence of the laser parameters (frequency, polarization and intensity) on the dynamics of laser-assisted (e, 2e) reactions. We present the angular distributions of the ejected electrons at two different field strengths $\mathcal{E}_0 = 10^8\ \mathrm{V \cdot m^{-1}}$ (fig. a) and $\mathcal{E}_0 = 5\ 10^9\ \mathrm{V\ m^{-1}}$ (fig. b), which correspond respectively to a perturbative and a nonperturbative regime.

1 H. Ehrhardt, G. Knoth, P. Schlemmer and K. Jung, Phys. Lett. **110A**, 92 (1985).

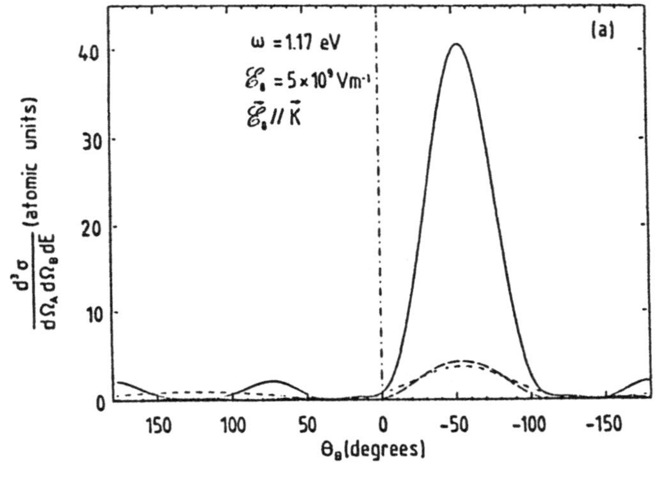

Fig.1(a)

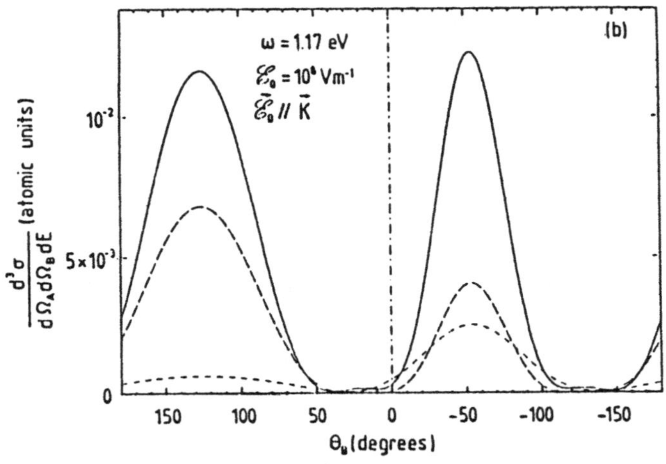

Fig.1(b)

The triple differential cross section (in atomic units) for the ionization of atomic hydrogen from the ground state by electron impact in the presence of a laser field, as a function of the ejected electron angle θ_b. The incident electron energy is 250 eV, the ejected electron energy is 5 eV and the scattering angle is $\theta_a = 3°$. The number of photon absorbed in the reaction is taken to be n = 1. The laser electric field is taken to be parallel to the incident momentum. (a) corresponds to a perturbative regime, the electric field strength being $\mathcal{E}_0 = 10^8$ V • m^{-1}, while (b) corresponds to a nonperturbative regime with $\mathcal{E}_0 = 5\ 10^9$ V • m^{-1}. Solid line : complete calculation. Dashed line : results obtained by using the simplified approach of ref.[2]

---

[2] P. Cavaliere, G. Ferrante and C. Leone, J. Phys. B 13, 4495 (1980).

Atomic deformation and AC-Stark shifts of H under extreme radiation conditions

M. Pont and M. Gavrila
FOM-Institute for Atomic and Molecular Physics,
Kruislaan 407, 1098 SJ Amsterdam, The Netherlands

Since the advent of superintense laser systems attention has been focused on atomic behavior in this new kind of environment. A theoretical description of these phenomena was given some time ago by Gavrila and Kaminski [1], applicable at high frequencies.

It was shown in Ref.[2], that at sufficiently high frequencies the atom becomes stable against decay by multiphoton ionization. Its structure is then governed by a time-independent Schrödinger equation in which the electron-nucleus attraction is replaced by a modified Coulomb potential, the 'dressed potential' V_0. For the case of atomic hydrogen in a circularly polarized laser field the dressed potential is given by [1]:

$$V_0 = -\frac{1}{2\pi}\int_0^{2\pi} \frac{d\phi}{|\mathbf{r} + \alpha_0 \mathbf{e}_x \cos\phi + \alpha_0 \mathbf{e}_y \sin\phi|}.$$

Here we have denoted by $\mathbf{e}_x$ and $\mathbf{e}_y$ the unit vectors in the x and y direction, having chosen the plane of polarization as the xy-plane. It is the electrostatic potential due to a uniformly charged ring (of total charge 1 a.u.) and depends on the intensity I and frequency ω of the field via its radius, α_0, which is given by [2]:

$$\alpha_0 = (I/2)^{1/2}\omega^{-2} \ (a.u.).$$

(The case of linear polarization was discussed earlier in Ref.[3].) The dressed potential is symmetric with respect to rotations about the z-axis and is also invariant under reflections in the origin. Consequently the magnetic quantum number (m) and the parity remain good quantum numbers to describe the states of the atom in the field. Since the symmetry of our problem is in fact the same as encountered in homonuclear diatomic molecules ($D_{\infty h}$), we adopt the standard notation of the electron terms of the latter (e.g. δ_u for states with $|m| = 2$ and negative parity).

We have shown that the Schrödinger equation containing this dressed potential is separable in the toroidal coordinate system for large α_0 (say $\alpha_0 \to \infty$). Explicit expressions are found for its energy eigenvalues and its eigensolutions. They correspond to a

[1] Atomic units will be used throughout.

[2] We define the atomic unit of intensity I_0 by the time-averaged intensity corresponding to a plane wave with an electric field amplitude of one atomic unit. It amounts: $I_0 = 3.51\,10^{16}\ \mathrm{W\,cm^{-2}}$.

drastic reduction of the ionization potential and a dramatic deformation of the atom. For the ground-state energy we find:

$$E_0 = \frac{1}{\pi\alpha_0}\left[-\frac{1}{2}\ln\left(\frac{\alpha_0}{\pi}\right) + \varepsilon_0 - 3\ln 2\right],$$

with ε_0 the lowest energy eigenvalue of a Schrödinger equation with a logarithmic potential ($\varepsilon_0 = 0.179935$). It turns out that the atom in its ground state, placed in the strong field becomes a torus-shaped object.

As a result of the separability new quantum numbers arise analogous to the principal quantum number and the azimuthal quantum number in spherically symmetric problems: the strong-field quantum numbers ν and $|\lambda|$. It turns out that each level associated with a pair of strong-field quantum numbers is in fact (infinitely) degenerate with respect to the magnetic quantum number. At high, but finite α_0 each main level splits up into one or two multiplets, which we refer after as families. Each family consists of states having the same energy position in either of the series $(\sigma_g, \pi_u, \delta_g, \ldots)$ or $(\sigma_u, \pi_g, \delta_u, \ldots)$.

We have established the correlation between the states at low α_0, characterized by the weak-field quantum numbers introduced earlier and the strong-field quantum numbers introduced here. A comparison of our analytical results with a numerical calculation carried out earlier shows excellent agreement up to rather low values of α_0.

Our findings about the atomic deformation have relevant consequences on the (angular dependent) rates at which electrons are ejected from the atom by multi- or excess photon ionization happening under the extreme radiation conditions such as realized by already existing lasers.

References

1) M. Gavrila and J. Z. Kaminski, Phys. Rev. Lett. **52**, 614 (1984).

2) M. Gavrila, in *Atoms in unusual situations*, edited by J.P. Briand, NATO ASI Ser. B, Vol.143 (Plenum, New York, 1987),p.225, and to be published.

3) M. Pont, N. R. Walet, M. Gavrila and C. W. McCurdy, Phys. Rev. Lett. **61**, 939 (1988).

COLLISIONAL TREATMENT OF MULTIPHOTON DISSOCIATION OF SMALL MOLECULES IN STRONG LASER FIELDS

Xin He*, Osman Atabek*, Annick Giusti-Suzor* and Frederic Mies+
*Laboratoire de Photophysique Moléculaire, Bât. 213, Université Paris-Sud,
91405 Orsay France
+NBS, Gaithersburg, USA

Molecular dissociation induced by strong laser fields is viewed as a half-collision multichannel process and studied using the same scattering methods as for laser-assisted collisions in the non-perturbative regime. Any intermediate or final continuum states are strongly coupled by the radiation field and must be described using accurate close-coupled wave functions. In a 'dressed channel' approach the initial bound state appears as a resonance in the scattering calculation and the photodissociation cross-section is related to the resonance width.

Resonances are localized by performing an analytical continuation of the hamiltonian through complex rotation of the radial coordinate. In this technique the resonant state appears as a bound state with complex energy, whose imaginary part gives the level width.

The linear variations predicted by the Fermi golden rule approximation is recovered at low intensities (Figure 1). For the intermediate field regime adiabatic electron-field channels can, in some cases, serve as a guide for the interpretation of non-linearities, while for higher laser intensities very large field-induced mixings between molecular states predominate, requiring the introduction of an increasing number of Floquet blocks fo converged calculations(1). Stimulated transitions between adjacent rotational states in a given electronic state are shown to affect greatly non-linear effects in molecules.

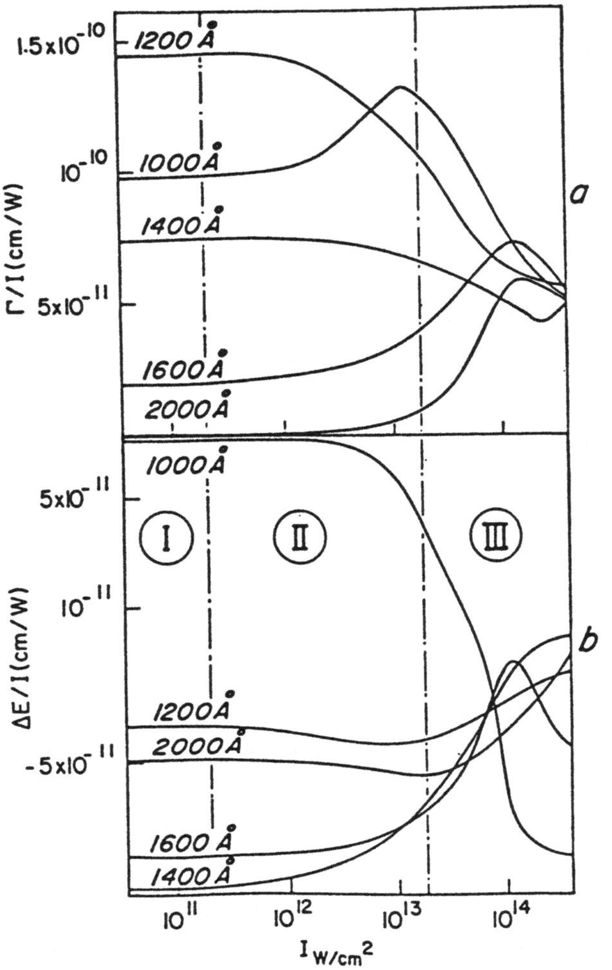

Figure 1 : Reduced ac-Stark width (Γ/I) and shift (ΔE/I) of the laser induced resonance H_2^+ ($1s\sigma_g$, v=0, J=1), as a function of laser intensity.

The formalism is applied to the photodissociation of H_2^+ ($1s\sigma_g$, v=0,J = 1 → $2p\sigma_u$). The corresponding lineshape is altered when the laser intensity increases : the bulk of the absorption is blue shifted and the cross-section decreases more slowly at large wavelengths. Non-linear effects appear for intensities larger than 10^{12} W/cm², for which experiments begin to be available[2].

(1) X. He, O. Atabek and A. Giusti-Suzor, Phys. Rev. A38, 5586 (1988)
(2) C. Cornaggia, D. Normand, J. Morellec, G. Mainfray and C. Manus, Phys. Rev. A 34, 207 (1986).

AUTOIONIZATION OF RYDBERG STATES IN H_2 INDUCED BY A DOUBLY EXCITED STATE

N.Bjerre, S.R. Keiding, L.J. Lembo[+], and H. Helm[+]
Institute of Physics, University of Aarhus, DK-Aarhus C, Denmark
[+]Mol. Phys. Dept.,SRI International, Menlo Park, CA 94025, U.S.A.

Instead of using multiphoton excitation from the ground state one can access the Rydberg states of H_2 with a single photon starting from the metastable $c^3\Pi_u^-$ state. In our experiments, we populate the high vibrational levels of this state by resonant charge exchange in Cs vapor of a fast beam of H_2^+. The resulting neutral H_2 beam is superposed collinearly with the beam from a dye laser operating between 540 and 670 nm. The photoionization spectrum is obtained by detecting the H_2^+ ion signal induced by the laser. The lines in the spectrum in Fig. 1 are labeled according to the lower state v" and N" quantum numbers with P,Q, and R denoting $\Delta N = -1,0,1$.

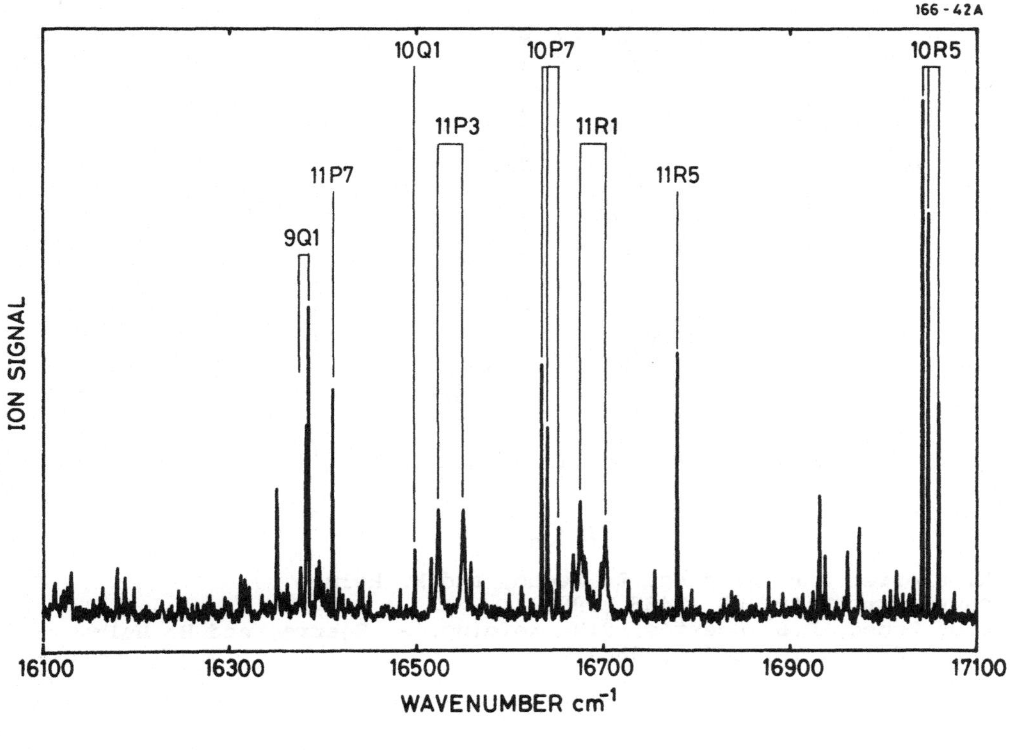

All transitions assigned so far involve a large change in vibrational quantum number. As much as ten vibrational quanta can be converted into electronic energy in the photoionization.[1] This would normally be an extremely unlikely process; here it is made possible by a doubly excited state which diabatically crosses the entire triplet Rydberg series at an internuclear distance R = 4a.u. (Fig. 2). This induces a strong mixing of the Rydberg states of different n and v, thus allowing ionization via higher n and lower v or dissociation via the n = 2 continuum. The latter process has been observed[2] by optical-optical double resonance.

The present technique is complementary to multiphoton ionization. Starting from the electronic and vibrational ground state of H_2 multiphoton excitation probes the molecular properties near R = 1a.u., whereas the interaction here takes place around R = 4a.u.

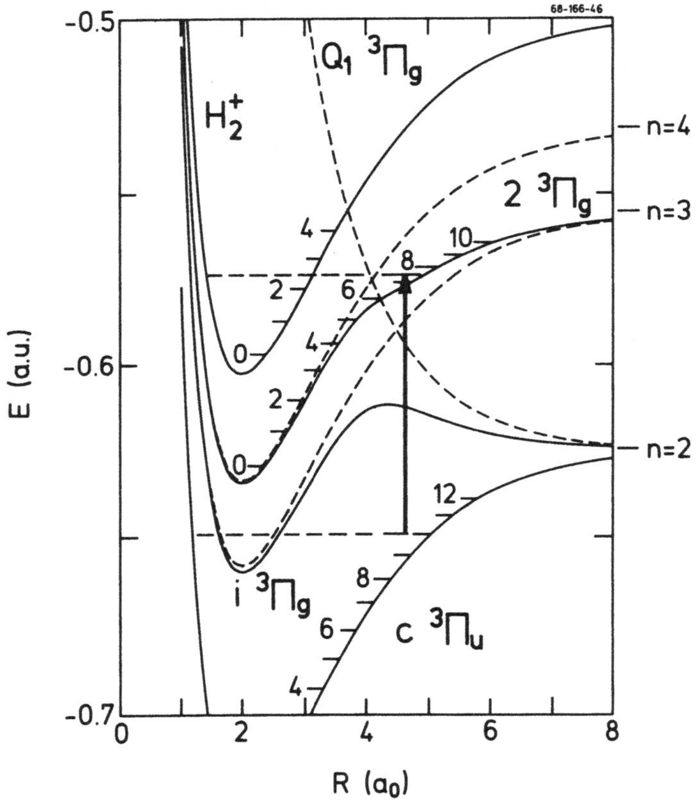

1. N. Bjerre, S.R. Keiding, L. Lembo, and H. Helm, Phys. Rev. Lett. 60, 2465 (1988).
2. L.J. Lembo, D.L. Huestis, S.R. Keiding, N. Bjerre, and H. Helm Phys. Rev. A 38, 3447 (1988).

AUTOIONIZATION AND FRAGMENTATION OF Na_2 STUDIED BY FEMTOSECOND LASER PULSES

T. Baumert, B. Bühler and G. Gerber
Fakultät für Physik, Universität Freiburg, 7800 Freiburg, FRG

We report on first results obtained with femtosecond laser pulses applied to molecular beam studies of the dynamics and the pathways of ionization, autoionization and fragmentation of highly excited molecular states of Na_2. Electronic autoionization of doubly excited molecular states and fragmentation of highly excited neutral and ionic states of Na_2 are hardly investigated and generally not well understood. This is mainly because (1) the final continuum states are usually not analyzed and (2) the dissociative ionization is not distinguished from neutral fragmentation with subsequent photoionization of excited fragments. To study the dynamics of multiphoton processes leading to excitation, autoionization and fragmentation we have applied ultrashort laser pulses to induce the transitions and Time-Of-Flight spectroscopy to determine the mass and initial kinetic energies of the fragments and the energy and angular distributions of ejected electrons. The experiments were carried out with "cold" molecular beams to ensure that only the lowest vibrational level v"=0 in the molecule is predominantly populated.

Femtosecond pulses are generated in a home-built colliding-pulse-mode-locked ring dye laser (CPM) with 4 intracavity prisms to adjust for the group velocity dispersion. The emission peak has been shifted to 616.0 nm by adjusting the DODCI absorber concentration. The output of the CPM laser was amplified at a rate of 100 Hz in a N_2 -laser pumped dye amplifier to produce pulses of 10 nJ energy and t=150 fs time duration. The pulse length was measured using the autocorrelation by second harmonic generation (SHG) in a nonlinear crystal.

The TOF-spectrum in fig.1 clearly shows the observation of Na_3^+, Na_2^+ and "slow" Na^+ as well as "fast" Na^+ fragment ions from sub-ps laser excitation with λ =616.0 nm. "Fast" and "slow" Na^+ ions originate from fragmentation processes occuring at small internuclear distances of Na_2. Predissociation of Na_2^* and photoionization of Na^* as the origin of observed Na^+ ions can be ruled out considering the time duration of the 150 fs laser pulse. Based on this result, which is rather difficult to obtain from other experiments, and with the known molecular potential curves we completely determined for this model case the multiphoton excitation and -fragmentation pathways. With a pulsed tunable dye laser we observe in the wavelength dependent ionization spectra a very prominent peak at 616.08 nm in both the Na_2^+ and in the Na^+ channel. This is due to a resonance enhanced three-photon ionization of Na_2 which is shown in the potential energy diagram in fig.2. Dimer ions a preferentially formed in the v^+=24,25 and v^+=14,15 states of the electronic ground state X ($^2\Sigma_g^+$) leading to electrons with kinetic energies of E=810±10 meV and E=940±10 meV which are actually observed in the electron spectrum. By absorption of one more photon from the fs-laser pulse the Na_2^+-ion undergoes the bound-free transition $Na_2^+(X,v^+)$ + hv --> $Na_2^{+*}(^2\Sigma_u^+)$ --> Na^+ + Na(3s) + W. Taking into account the populated v^+-levels and the known potential curves for the ionic ground and first excited states the corresponding bound-free transitions lead to recoil energies W between 10000cm^{-1} and 11000 cm^{-1}. The energy W=10500 ± 500 cm^{-1} obtained from the analysis of the TOF-spectrum perfectly agrees with that.

The observation of "slow" Na^+ ions however cannot be explained within this framework since photoionization of v^*=20 of the $^1\Pi_g$ Rydberg state directly into the $^2\Sigma_u^+$-continuum is energetically forbidden for the applied laser wavelength. Based on the measured electron energy distribution and the recoil energy W=900±500 cm^{-1}, obtained from the analysis of the "slow" Na^+ fragment ions, a consistent explanation is found by considering the excitation of doubly excited molecular states of Na_2. The doubly excited states Na_2^{**} (nl,n'l') form a Rydberg series converging versus the $^2\Pi_u$ state of Na_2^+, whose potential curve is known theoretically. Assuming that the shape of a $^1\Pi_u$ state potential curve, correlated to Na(3p)+Na(4s), is similar to the ionic $^2\Pi_u$ curve and that it has a potential barrier at large internuclear distances like the $B^1\Pi_u$ state from Na(3s)+Na(3p), we believe that in the resonance enhanced three-photon process vibronic levels close to the dissociation limit of the doubly excited $^1\Pi_u$(3p+4s)state are excited. The wavefunctions of these vibronic levels extend from 3 A to approximately 10 A. These doubly excited levels may autoionize into the $X(^2\Sigma_g^+)$ ground state of Na_2^+ giving rise to electron energies between 260 meV and 500 meV which are actually observed. For internuclear distances greater than 6 A the vibronic levels cross into the continuum of the repulsive $^2\Sigma_u^+$ state of Na_2^+. Therefore for R ≥ 6 A there is a second open autoionization channel which is responsible for the observed electrons having energies in the range from 0 meV to 160 meV. This autoionization process

$$Na_2^{**}\ ^1\Pi_u(3p+4s) \longrightarrow Na_2^{+*}(^2\Sigma_u^+) + e^-$$
$$\longrightarrow Na^+ + Na(3s) + W$$

of course simultaneously produces "slow" Na^+-ions whose kinetic energies depend on the internuclear distance R where the autoionization takes place.

In conclusion, this is the first reported experiment where a fs-laser has been used in combination with ion-and electron spectroscopy to clarify the ionization and fragmentation pathways of a multiphoton process investigated in a molecular beam experiment.

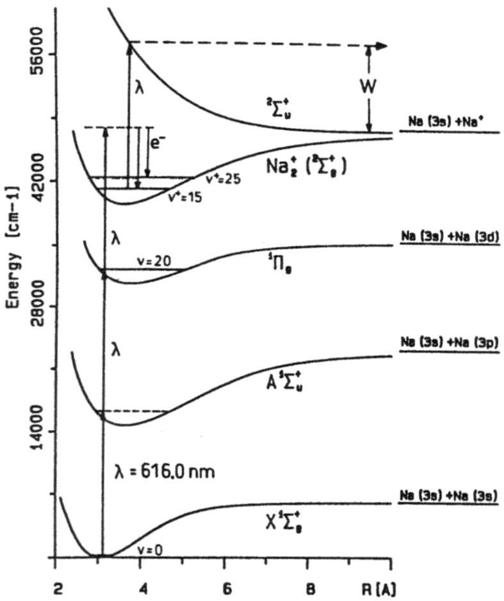

Fig.2: Potential energy diagram explaining the origin of "fast" Na^+ ions

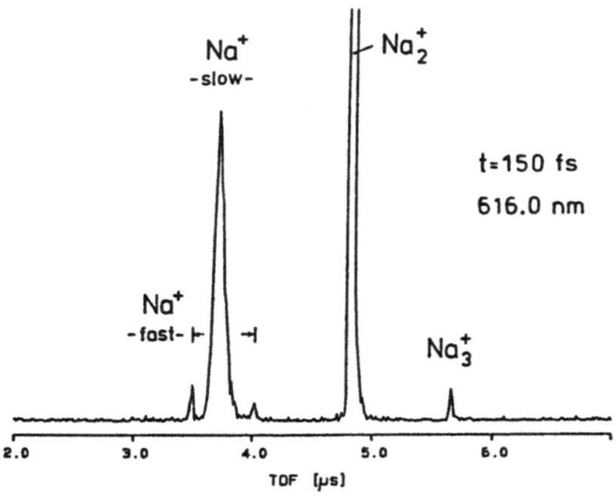

Fig.1: Time-of-Flight spectrum of a sodium molecular beam excited by femtosecond laser pulses at λ = 616.0 nm

LIST OF PARTICIPANTS

G. Alber	Univ. Freiburg, Freiburg, FRG
Mireille Aymar	Lab. Aimé Cotton, Orsay, France
J.E. Bayfield	Univ. of Pittsburgh, Pittsburgh, USA
S. Basile	Univ. Palermo, Palermo, Italy
S. Bivona	Univ. Palermo, Palermo, Italy
N. Bjerre	Univ. Århus, Århus, DK
R. Blümel	MPQ, Garching, FRG
A.Z. Capri	Univ. of Alberta, Edmonton, Canada
S.L. Chin	Laval Univ., Quebec, Canada
Birte Christensen–Dalsgaard	Univ. Århus, Århus, DK
W.E. Cooke	Univ. of Southern California, Los Angeles, USA
Michèle Crance	Lab. Aimé Cotton, Orsay, France
F. Ehlotzky	Univ. Innsbruck, Innsbruck, Austria
K. Ellinger	Univ. Innsbruck, Innsbruck, Austria
G. Farkas	Central Res. Inst. for Physics, Budapest, Hungary
M.V. Fedorov	General Physics Inst., Moscow, USSR
D. Feldmann	Univ. Bielefeld, Bielefeld, FRG
M.G.J. Fink	Univ. Notre Dame, Notre Dame, USA
R.R. Freeman	AT&T Bell Lab., Murray Hill, USA
M. Gavrila	FOM Institute, Amsterdam, The Netherlands
G. Gerber	Univ. Freiburg, Freiburg, FRG
Annick Giusti–Suzor	Univ. Paris–Sud, Orsay, France
H. Gratl	Univ. Innsbruck, Innsbruck, Austria
H. Helm	SRI Internat., Menlo Park, USA
C. Hillermeier	MPQ, Garching, FRG
L. Jönsson	Univ. of Ill. at Chicago, Chicago, USA
H. Klar	Univ. Freiburg, Freiburg, FRG
H.B. van Linden van den Heuvell	FOM Institute, Amsterdam, The Netherlands
I. Littler	Univ. Kaiserslautern, Kaiserslautern, FRG
L.–A. Lompré	CEN de Saclay, Gif–sur–Yvette, France
Monika Marte	Univ. Innsbruck, Innsbruck, Austria
P. Marte	Univ. Innsbruck, Innsbruck, Austria
G. Petite	CEN de Saclay, Gif–sur–Yvette, France
B. Piraux	Univ. P. et M. Curie, Paris, France
M. Poirier	CEN de Saclay, Gif–sur–Yvette, France
M. Pont	FOM Institute, Amsterdam, The Netherlands
C.K. Rhodes	Univ. of Ill. at Chicago, Chicago, USA
H. Ritsch	Univ. Innsbruck, Innsbruck, Austria
H. Rottke	Univ. Bielefeld, Bielefeld, FRG
W. Sandner	Univ. Freiburg, Freiburg, FRG
R. Shakeshaft	Univ. of Southern California, Los Angeles, USA
D. Steinmüller	Univ. Innsbruck, Innsbruck, Austria
Doris Steinmüller–Nethl	Univ. Innsbruck, Innsbruck, Austria
C. Tóth	Central Res. Inst. for Physics, Budapest, Hungary
S. Varró	MPQ, Garching, FRG
Valerie Véniard	Univ. P. et M. Curie, Paris, France
K.H. Welge	Univ. Bielefeld, Bielefeld, FRG
G. Wendin	Chalmers Univ. of Tech. Göteborg, Sweden
Lyn Westling	MPQ, Garching, FRG
G. Wiebusch	Univ. Bielefeld, Bielefeld, FRG
Beate Wolff	Univ. Bielefeld, Bielefeld, FRG
P. Zoller	Univ. Innsbruck, Innsbruck, Austria

Lecture Notes in Mathematics

Vol. 1236: Stochastic Partial Differential Equations and Applications. Proceedings, 1985. Edited by G. Da Prato and L. Tubaro. V, 257 pages. 1987.

Vol. 1237: Rational Approximation and its Applications in Mathematics and Physics. Proceedings, 1985. Edited by J. Gilewicz, M. Pindor and W. Siemaszko. XII, 350 pages. 1987.

Vol. 1250: Stochastic Processes – Mathematics and Physics II. Proceedings 1985. Edited by S. Albeverio, Ph. Blanchard and L. Streit. VI, 359 pages. 1987.

Vol. 1251: Differential Geometric Methods in Mathematical Physics. Proceedings, 1985. Edited by P.L. García and A. Pérez-Rendón. VII, 300 pages. 1987.

Vol. 1255: Differential Geometry and Differential Equations. Proceedings, 1985. Edited by C. Gu, M. Berger and R.L. Bryant. XII, 243 pages. 1987.

Vol. 1256: Pseudo-Differential Operators. Proceedings, 1986. Edited by H.O. Cordes, B. Gramsch and H. Widom. X, 479 pages. 1987.

Vol. 1258: J. Weidmann, Spectral Theory of Ordinary Differential Operators. VI, 303 pages. 1987.

Vol. 1260: N.H. Pavel, Nonlinear Evolution Operators and Semigroups. VI, 285 pages. 1987.

Vol. 1263: V.L. Hansen (Ed.), Differential Geometry. Proceedings, 1985. XI, 288 pages. 1987.

Vol. 1265: W. Van Assche, Asymptotics for Orthogonal Polynomials. VI, 201 pages. 1987.

Vol. 1267: J. Lindenstrauss, V.D. Milman (Eds.), Geometrical Aspects of Functional Analysis. Seminar. VII, 212 pages. 1987.

Vol. 1269: M. Shiota, Nash Manifolds. VI, 223 pages. 1987.

Vol. 1270: C. Carasso, P.-A. Raviart, D. Serre (Eds.), Nonlinear Hyperbolic Problems. Proceedings, 1986. XV, 341 pages. 1987.

Vol. 1272: M.S. Livšic, L.L. Waksman, Commuting Nonselfadjoint Operators in Hilbert Space. III, 115 pages. 1987.

Vol. 1273: G.-M. Greuel, G. Trautmann (Eds.), Singularities, Representation of Algebras, and Vector Bundles. Proceedings, 1985. XIV, 383 pages. 1987.

Vol. 1275: C.A. Berenstein (Ed.), Complex Analysis I. Proceedings, 1985–86. XV, 331 pages. 1987.

Vol. 1276: C.A. Berenstein (Ed.), Complex Analysis II. Proceedings, 1985–86. IX, 320 pages. 1987.

Vol. 1277: C.A. Berenstein (Ed.), Complex Analysis III. Proceedings, 1985–86. X, 350 pages. 1987.

Vol. 1283: S. Mardešić, J. Segal (Eds.), Geometric Topology and Shape Theory. Proceedings, 1986. V, 261 pages. 1987.

Vol. 1285: I.W. Knowles, Y. Saitō (Eds.), Differential Equations and Mathematical Physics. Proceedings, 1986. XVI, 499 pages. 1987.

Vol. 1287: E.B. Saff (Ed.), Approximation Theory, Tampa. Proceedings, 1985–1986. V, 228 pages. 1987.

Vol. 1288: Yu. L. Rodin, Generalized Analytic Functions on Riemann Surfaces. V, 128 pages, 1987.

Vol. 1294: M. Queffélec, Substitution Dynamical Systems – Spectral Analysis. XIII, 240 pages. 1987.

Vol. 1299: S. Watanabe, Yu.V. Prokhorov (Eds.), Probability Theory and Mathematical Statistics. Proceedings, 1986. VIII, 589 pages. 1988.

Vol. 1300: G.B. Seligman, Constructions of Lie Algebras and their Modules. VI, 190 pages. 1988.

Vol. 1302: M. Cwikel, J. Peetre, Y. Sagher, H. Wallin (Eds.), Function Spaces and Applications. Proceedings, 1986. VI, 445 pages. 1988.

Vol. 1303: L. Accardi, W. von Waldenfels (Eds.), Quantum Probability and Applications III. Proceedings, 1987. VI, 373 pages. 1988.

Lecture Notes in Physics

Vol. 316: W. Kundt (Ed.), Supernova Shells and Their Birth Events. Proceedings, 1988. VIII, 253 pages. 1988.

Vol. 317: C. Signorini, S. Skorka, P. Spolaore, A. Vitturi (Eds.), Heavy Ion Interactions Around the Coulomb Barrier. Proceedings, 1988. X, 329 pages. 1988.

Vol. 318: B. Mercier, An Introduction to the Numerical Analysis of Spectral Methods. V, 154 pages. 1989.

Vol. 319: L. Garrido (Ed.), Far from Equilibrium Phase Transitions. Proceedings, 1988. VIII, 340 pages. 1988.

Vol. 320: D. Coles (Ed.), Perspectives in Fluid Mechanics. Proceedings, 1985. VII, 207 pages. 1988.

Vol. 321: J. Pitowsky, Quantum Probability – Quantum Logic. IX, 209 pages. 1989.

Vol. 322: M. Schlichenmaier, An Introduction to Riemann Surfaces, Algebraic Curves and Moduli Spaces. XIII, 148 pages. 1989.

Vol. 323: D.L. Dwoyer, M.Y. Hussaini, R.G. Voigt (Eds.), 11th International Conference on Numerical Methods in Fluid Dynamics. XIII, 622 pages. 1989.

Vol. 324: P. Exner, P. Šeba (Eds.), Applications of Self-Adjoint Extensions in Quantum Physics. Proceedings, 1987. VIII, 273 pages. 1989.

Vol. 325: E. Brändas, N. Elander (Eds.), Resonances, Proceedings, 1987. XVIII, 564 pages. 1989.

Vol. 326: A. Grauel, Feldtheoretische Beschreibung der Thermodynamik für Grenzflächen. IX, 317 Seiten. 1989.

Vol. 327: K. Meisenheimer, H.-J. Röser (Eds.), Hot Spots in Extragalactic Radio Source. Proceedings, 1988, XII, 301 pages. 1989.

Vol. 328: G. Wegner (Ed.), White Dwarfs. Proceedings, 1988. XIV, 524 pages. 1989.

Vol. 329: A Heck, F. Murtagh (Eds.), Knowledge Based Systems in Astronomy. IV, 280 pages. 1989.

Vol. 330: J.M. Moran, J.N. Hewitt, K.Y. Lo (Eds.), Gravitational Lenses. Proceedings, 1988. XIV, 238 pages. 1989.

Vol. 331: G. Winnewisser, J.T. Armstrong (Eds.), The Physics and Chemistry of Interstellar Molecular Clouds mm and Sub-mm Observations in Astrophysics. Proceedings, 1988. XVIII, 463 pages. 1989.

Vol. 332: P. Flin, H.W. Duerbeck (Eds.), Morphological Cosmology. Proceedings, 1988. VII, 438 pages. 1989.

Vol. 333: I. Appenzeller, H.J. Habing, P. Léna (Eds.), Evolution of Galaxies – Astronomical Observations. Proceedings, 1988. X, 391 pages. 1989.

Vol. 334: L. Maraschi, T. Maccacaro, M.-H. Ulrich (Eds.), BL Lac Objects. XIII, 497 pages. 1989.

Vol. 335: A. Lakhtakia, V.K. Varadan, V.V. Varadan, Time-Harmonic Electromagnetic Fields in Chiral Media. VII, 121 pages. 1989.

Vol. 336: M. Müller, Consistent Classical Supergravity Theories. VI, 125 pages. 1989.

Vol. 337: A.P. Maclin, T.L. Gill, W.W. Zachary (Eds.), Magnetic Phenomena. Proceedings, 1988. VI, 142 pages. 1989.

Vol. 338: V. Privman, N.M. Švrakić, Directed Models of Polymers, Interfaces, and Clusters: Scaling and Finite-Sizes Properties. VI, 120 pages. 1989.

Vol. 339: F. Ehlotzky (Ed.), Fundamentals of Laser Interactions II. Proceedings, 1989. XI, 317 pages. 1989.

Lasers in Theory and Practice

R. R. Alfano, City College of the City University of New York, NY, USA (Ed.)

The Supercontinuum Laser Source

1989. XIV, 458 pp. 250 figs.
ISBN 3-540-96946-2

W. Kaiser, Technical University of Munich, FRG (Ed.)

Ultrashort Laser Pulses and Applications

With contributions by D. H. Auston, K. B. Eisenthal, R. M. Hochstrasser, C. K. Johnson, W. Kaiser, A. Laubereau, D. von der Linde, A. Seilmeier, C. V. Shank, W. Zinth

1988. XV, 424 pp. 196 figs. (Topics in Applied Physics, Volume 60).
ISBN 3-540-18605-0

P. Lambropoulos, University of Crete, Iraklion, Crete, Greece and University of Southern California, Los Angeles, CA, USA; S. J. Smith, University of Colorado, Boulder, CO, USA (Eds.)

Multiphoton Processes

Proceedings of the 3rd International Conference, Iraklion, Crete, Greece September 5–12, 1984

1984. VIII, 201 pp. 101 figs. (Springer Series on Atoms and Plasmas, Volume 2).
ISBN 3-540-15068-4

V. S. Letokhov, Moscow, USSR

Nonlinear Laser Chemistry

Multiple-Photon Excitation

1983. XIV, 417 pp. 152 figs. (Springer Series in Chemical Physics, Volume 22).
ISBN 3-540-11705-9

N. B. Delone, V. P. Krainov, Moscow, USSR

Atoms in Strong Light Fields

1985. XII, 339 pp. 49 figs. (Springer Series in Chemical Physics, Volume 28).
ISBN 3-540-12412-8

Springer-Verlag Berlin Heidelberg New York London Paris Tokyo Hong Kong

Springer

GPSR Compliance

The European Union's (EU) General Product Safety Regulation (GPSR) is a set of rules that requires consumer products to be safe and our obligations to ensure this.

If you have any concerns about our products, you can contact us on ProductSafety@springernature.com

In case Publisher is established outside the EU, the EU authorized representative is:

Springer Nature Customer Service Center GmbH
Europaplatz 3
69115 Heidelberg, Germany

Batch number: 10640288

Printed by Printforce, the Netherlands